T0258040

Mechanical Properties and Performance of Engineering Ceramics and Composites VI

Mechanical Properties and Performance of Engineering Ceramics and Composites VI

A Collection of Papers Presented at the 35th International Conference on Advanced Ceramics and Composites January 23–28, 2011 Daytona Beach, Florida

Edited by
Dileep Singh
Jonathan Salem

Volume Editors
Sujanto Widjaja
Dileep Singh

The American Ceramic Society

A John Wiley & Sons, Inc., Publication

Library of Congress Cataloging-in-Publication Data is available.

ISBN 978-1-118-05987-6

oBook ISBN: 978-1-118-09535-5
ePDF ISBN: 978-1-118-17308-4

ISSN: 0196-6219

Printed in the United States of America.

10 9 8 7 6 5 4 3 2 1

Contents

NONDESTRUCTIVE EVALUATION

PROCESSING-MICROSTRUCTURE-PROPERTIES CORRELATIONS

Preface

This volume is a compilation of papers presented in the Mechanical Behavior and Performance of Ceramics & Composites symposium during the 35th International Conference & Exposition on Advanced Ceramics and Composites (ICACC) held January 23–28, 2011, in Daytona Beach, Florida.

The Mechanical Behavior and Performance of Ceramics & Composites symposium was one of the largest symposia in terms of the number (>100) of presentations at the ICACC'11. This symposium covered wide ranging and cutting-edge topics on mechanical properties and reliability of ceramics and composites and their correlations to processing, microstructure, and environmental effects. Symposium topics included:

- Composites: Fibers, Matrices, Interfaces and Applications
- Environmental Effects of Ceramics and Composites
- Fracture Mechanics, Modeling, and Mechanical Testing
- Nondestructive Evaluation
- Processing-Microstructure-Properties Correlations
- Tribological Properties of Ceramics and Composites

Significant time and effort is required to organize a symposium and publish a proceeding volume. We would like to extend our sincere thanks and appreciation to the symposium organizers, invited speakers, session chairs, presenters, manuscript reviewers, and conference attendees for their enthusiastic participation and contributions. Finally, credit also goes to the dedicated, tireless and courteous staff at The American Ceramic Society for making this symposium a huge success.

DILEEP SINGH
Argonne National Laboratory

JONATHAN SALEM
NASA Glenn Research Center

Introduction

This CESP issue represents papers that were submitted and approved for the proceedings of the 35th International Conference on Advanced Ceramics and Composites (ICACC), held January 23–28, 2011 in Daytona Beach, Florida. ICACC is the most prominent international meeting in the area of advanced structural, functional, and nanoscopic ceramics, composites, and other emerging ceramic materials and technologies. This prestigious conference has been organized by The American Ceramic Society's (ACerS) Engineering Ceramics Division (ECD) since 1977.

The conference was organized into the following symposia and focused sessions:

Symposium 1	Mechanical Behavior and Performance of Ceramics and Composites
Symposium 2	Advanced Ceramic Coatings for Structural, Environmental, and Functional Applications
Symposium 3	8th International Symposium on Solid Oxide Fuel Cells (SOFC): Materials, Science, and Technology
Symposium 4	Armor Ceramics
Symposium 5	Next Generation Bioceramics
Symposium 6	International Symposium on Ceramics for Electric Energy Generation, Storage, and Distribution
Symposium 7	5th International Symposium on Nanostructured Materials and Nanocomposites: Development and Applications
Symposium 8	5th International Symposium on Advanced Processing & Manufacturing Technologies (APMT) for Structural & Multifunctional Materials and Systems

Symposium 9	Porous Ceramics: Novel Developments and Applications
Symposium 10	Thermal Management Materials and Technologies
Symposium 11	Advanced Sensor Technology, Developments and Applications
Symposium 12	Materials for Extreme Environments: Ultrahigh Temperature Ceramics (UHTCs) and Nanolaminated Ternary Carbides and Nitrides (MAX Phases)
Symposium 13	Advanced Ceramics and Composites for Nuclear and Fusion Applications
Symposium 14	Advanced Materials and Technologies for Rechargeable Batteries
Focused Session 1	Geopolymers and other Inorganic Polymers
Focused Session 2	Computational Design, Modeling, Simulation and Characterization of Ceramics and Composites
Special Session	Pacific Rim Engineering Ceramics Summit

The conference proceedings are published into 9 issues of the 2011 Ceramic Engineering & Science Proceedings (CESP); Volume 32, Issues 2-10, 2011 as outlined below:

- Mechanical Properties and Performance of Engineering Ceramics and Composites VI, CESP Volume 32, Issue 2 (includes papers from Symposium 1)
- Advanced Ceramic Coatings and Materials for Extreme Environments, Volume 32, Issue 3 (includes papers from Symposia 2 and 12)
- Advances in Solid Oxide Fuel Cells VI, CESP Volume 32, Issue 4 (includes papers from Symposium 3)
- Advances in Ceramic Armor VII, CESP Volume 32, Issue 5 (includes papers from Symposium 4)
- Advances in Bioceramics and Porous Ceramics IV, CESP Volume 32, Issue 6 (includes papers from Symposia 5 and 9)
- Nanostructured Materials and Nanotechnology V, CESP Volume 32, Issue 7 (includes papers from Symposium 7)
- Advanced Processing and Manufacturing Technologies for Structural and Multifunctional Materials V, CESP Volume 32, Issue 8 (includes papers from Symposium 8)
- Ceramic Materials for Energy Applications, CESP Volume 32, Issue 9 (includes papers from Symposia 6, 13, and 14)
- Developments in Strategic Materials and Computational Design II, CESP Volume 32, Issue 10 (includes papers from Symposium 10 and 11 and from Focused Sessions 1, and 2)

The organization of the Daytona Beach meeting and the publication of these proceedings were possible thanks to the professional staff of ACerS and the tireless dedication of many ECD members. We would especially like to express our sincere thanks to the symposia organizers, session chairs, presenters and conference attendees, for their efforts and enthusiastic participation in the vibrant and cutting-edge conference.

ACerS and the ECD invite you to attend the 36th International Conference on Advanced Ceramics and Composites (http://www.ceramics.org/daytona2012) January 22–27, 2012 in Daytona Beach, Florida.

SUJANTO WIDJAJA AND DILEEP SINGH
Volume Editors
June 2011

Composites: Fibers, Matrices, Interfaces, and Applications

OXIDE FIBER COATINGS FOR SILICON CARBIDE CERAMIC MATRIX COMPOSITES

Emmanuel E. Boakye, Pavel S. Mogilevsky, T. A. Parthasarathy
Air Force Research Laboratory
UES, Inc., Dayton, OH

Randall S. Hay, Michael K. Cinibulk
Air Force Research Laboratory
Materials and Manufacturing Directorate
WPAFB, OH

M. Ahrens
Wright State University
Fairborn, OH *

ABSTRACT

The γ-polymorph of the rare-earth disilicates ($RE_2Si_2O_7$) is a potential oxidation-resistant alternative to carbon or BN for CMC fiber-matrix interphases. The formation of γ-$Y_2Si_2O_7$ and γ-$Ho_2Si_2O_7$ at different temperatures and processing environments was investigated. Silica - yttrium hydroxide and silica - holmium hydroxide dispersions were made and heat-treated at 1200° - 1400°C for 8 h in air. Dense pellets of $Y_2Si_2O_7$ and $Ho_2Si_2O_7$ were made and indented. Sections below Vickers indents were analyzed for evidence of plastic deformation. SCS-0 fibers were put in γ-$RE_2Si_2O_7$ matrix (RE = Y and Ho) and densified at 1200°C/1 h using the field assisted sintering technique (FAST). Fiber push-out experiments were then conducted on these fibers. Preliminary results for characterization of deformation of γ-$RE_2Si_2O_7$ after indentation, and sliding stress values from fiber push-out are reported.

1. INTRODUCTION

SiC/SiC ceramic matrix composites (CMCs) are the most mature CMC system, but their use is often still limited by mechanical property degradation in oxidizing environments. Several methods are used to minimize the oxidation of BN or carbon fiber-matrix interphases in SiC/SiC CMCs;[1] however, an ideal solution would replace BN or carbon with a material that does not oxidize. γ-rare-earth disilicates are candidates.

The processing and stability of γ-rare-earth disilicates (γ-$RE_2Si_2O_7$) in air and in argon has been reported.[2] Rare-earth disilicates ($RE_2Si_2O_7$) exhibit up to seven polymorphs (Fig. 1). Yttrium disilicate ($Y_2Si_2O_7$) is the most thoroughly studied. It has five polymorphs (Fig. 2), and is refractory, melting at 1775°C.[3-10] The mechanical properties of γ-$Y_2Si_2O_7$ are comparable to $LaPO_4$ (monazite),[6,11] which has been demonstrated to function as a weak fiber-matrix interphase in oxide-oxide CMCs.[12] However, $LaPO_4$ monazite decomposes in the reducing atmospheres typical for processing SiC/SiC CMCs, and the rare earth disilicate $La_2Si_2O_7$ forms by reaction between SiO_2 and $LaPO_4$ under reducing atmosphere.[13-15] Of the various rare-earth disilicate polymorphs, γ-$RE_2Si_2O_7$ is a "quasi-ductile" ceramic.[6,11] It is soft (Vickers hardness ~6 GPa) and machineable, and has a low shear modulus which originates from bond strength differences between strong Si-O bonds and weak Y-O chemical bonds.[6,11] γ-$Y_2Si_2O_7$ is thermochemically compatible with SiC and SiO_2,[16] and it has a thermal expansion coefficient (~4×10^{-6}°C^{-1}) similar to SiC.[17] This combination of properties is intriguing and suggests that γ-RE$_2$Si$_2$O$_7$ may function as a fiber-matrix interphase for SiC/SiC CMCs. However, synthesizing stoichiometric, single-phase γ-RE$_2$Si$_2$O$_7$ is challenging.

The rare-earth disilicates are commonly prepared by a variety of methods which include conventional solid-state reaction of the mixed oxides RE$_2$O$_3$ and SiO$_2$, calcination of rare-earth disilicate sol-gel precursors, and hydrothermal processing.[5,18-23] An attempt to make the γ-phase by calcining SiO$_2$ and Y$_2$O$_3$ powders to temperatures as high as 1600°C was not successful.[24] Sol-gel nanoparticle precursors have been used to synthesize α- and δ-Y$_2$Si$_2$O$_7$, but similar work on the synthesis of γ-phase has not been reported.[11,20,25] Felsche[8-9] investigated phase equilibria of rare-earth disilicates (La-Lu) at 1000°-1800°C, and showed that γ-RE$_2$Si$_2$O$_7$ can also be synthesized for rare-earth cations Ho^{3+} and Er^{3+} (Fig. 1). Recently, Sun et al. showed that addition of 3 mol % LiYO$_2$ to Y$_2$O$_3$/SiO$_2$ particles enables formation of γ-phase at 1400°C.[7] LiYO$_2$ promotes the reaction of Y$_2$O$_3$ and SiO$_2$ by formation of a LiYO$_2$-SiO$_2$ liquid phase at 1225°C that enhances transport at grain boundaries to lower the reaction temperature and time. We build on this work to optimize processing methods of γ-RE$_2$Si$_2$O$_7$ (RE = Y, Ho) fiber-matrix interphases on SiC fibers.

The phase transformation temperatures of the different Y$_2$Si$_2$O$_7$ polymorphs most commonly referenced in the literature are those given by Ito and Johnson (Equation 1):[26]

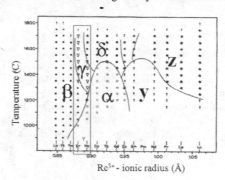

Fig. 1 Rare-earth disilicate polymorphs (after Felsche[8-9])

$$\alpha \xrightarrow{1225°C} \beta \xrightarrow{1445°C} \gamma \xrightarrow{1535°C} \delta \qquad [1]$$

However, other transformation temperatures have also been reported,[4,23,27-28] and a low temperature y-polymorph that formed below 1000°C was recently reported.[5] For the lower temperature polymorph α-Y$_2$Si$_2$O$_7$, different formation temperatures were observed for different synthesis methods.[5,20-21,28-29] The α-Y$_2$Si$_2$O$_7$ forms at temperatures as high as 1300°C by solid-state reaction,[30] and as low as 1000°C by sol-gel methods.[20-21]

The formation of γ-Y$_2$Si$_2$O$_7$ and γ-Ho$_2$Si$_2$O$_7$ using mixtures of rare-earth nitrates and silica sol-gel precursors is reported in this work. Possible plastic deformation of γ-Y$_2$Si$_2$O$_7$ beneath indented γ-$Y_2Si_2O_7$ pellets and sliding stress values from fiber push-out in micro-composites are reported.

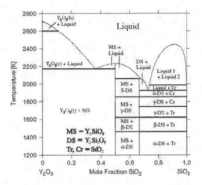

Fig. 2. Phase diagram for the Y_2O_3-SiO_2 system, showing the $Y_2Si_2O_7$ polymorphs.[10]

2. EXPERIMENTAL

2.1 Precursor Synthesis and Characterization

Precursors to $Y_2Si_2O_7$ and $Ho_2Si_2O_7$ were made by adding colloidal silica to solutions of yttrium and holmium nitrate as reported previously.[2] For the $Y_2Si_2O_7$ precursor, 12.7 g of $Y(NO_3)\cdot 6H_2O$ and 0.12g of $LiNO_3$ was dissolved in 50 mL deionized water to make a solution of pH ~1. The pH of sols was determined with a pH/ion meter (Corning Inc., Corning NY). Silica (2.0 – 2.8 g) was added in the form of a sol with a pH of 10. For the $Y_2Si_2O_7$ precursor, it was found that excess silica was essential to form the γ-$Y_2Si_2O_7$ phase, so sols corresponding to Si:Y atomic ratios from 50:50 to 58.8:41.2 were prepared (Table 1). For the $Ho_2Si_2O_7$ precursor, 14.7 g of $Ho(NO_3)\cdot 5H_2O$ and 0.11g of $LiNO_3$ was added to 50 mL of deionized water and 2 g of silica was added. The Si:Ho atomic ratio was 50:50 and the solution pH was ~1. Only stoichiometric Si:Ho ratio was investigated in this case since it formed the required γ-$Ho_2Si_2O_7$ phase.. The mixtures were dried in an oven for 72 h to form $SiO_2/Y(OH)_3$ and $SiO_2/Ho(OH)_3$ and heated at 1200° - 1400°C for 8 h in air and then characterized by X-ray diffraction (XRD) with a Cu-Kα radiation (Model Rotaflex, Rigaku Co., Tokyo, Japan).

Table 1. Y:Si atomic fraction in $Y_2Si_2O_7$

SiO_2 (g)	SiO_2 (mol) x10^{-2}	$Y(NO_3)\cdot 6H_2O$ (g)	$Y(NO_3)\cdot 6H_2O$ (mol) x10^{-2}	at.% Si	at.% Y
2.0	3.3	12.7	3.3	50.0	50.0
2.1	3.5	12.7	3.3	51.5	48.5
2.2	3.7	12.7	3.3	52.9	47.1
2.8	4.7	12.7	3.3	58.8	41.2

2.2 Formation and Densification of Pellets

$SiO_2/Y(OH)_3$ dispersions with Li dopant were dried, heat treatment at 1400°C/8 h to form γ-$Y_2Si_2O_7$. γ-$Y_2Si_2O_7$ powder was ball milled in isopropanol using alumina balls. A 3 vol% polyvinyl butryal resin was added as a binder. Prior to milling, the powder sample was characterized by XRD to confirm that the γ-$Y_2Si_2O_7$ was formed. The milled slurry was separated and dried at 120°C for 18 h. The dried powder was uniaxially pressed into pellets and isostatically cold pressed at ~300 MPa. Pellets were ~10 mm in diameter and ~4 mm in height, with relative densities of ~60%. The pellets were then heat-treated in air at a rate of 1°C/min to 700°C and held for 1 h to remove organics from the binder, and then finally heated at 3°C/min to 1200° or 1400°C for 8 h. After sintering, samples were cooled at 3°C/min to room temperature and polished prior to indentation. Density measurements were done using the Archimedes method.

2.3. Indentation and Characterization

The hardness of sintered pellets was measured by Vickers indentation at loads in the range of 50 g to 1000 g using a Buehler (Lake Bluff, IL) Hardness Tester 1600-2007. Indented samples were examined with FEI (Hillsboro, Oregon) scanning electron microscopes (SEM, Models Sirion and Quanta) with an X-ray energy dispersive spectroscopy (EDS) systems operating at 5-20 kV. The indentation modulus was determined from load displacement curves (Oliver-Pharr method)[31] to the maximum penetration of about 1500 nm using a Nanoindenter II (Nano Instruments, Inc., Oak Ridge, TN) with a Berkovich indenter. Foils for transmission electron microscopy (TEM, Philips CM200, FEI, Hillsboro, Oregon) were cut from areas below the Vickers indentations using FEI (Hillsboro, Oregon) Focused Ion Beam Dual Beam DB 235 workstation equipped with Omniprobe (Dallas, Texas) AutoProbe™ 200 micromanipulator.

2.4 Fiber Push-Out

SCS-0 fibers were sandwiched between two γ-$Ho_2Si_2O_7$ (Re= Y, Ho) green pellets and densified at 1200 °C/1 h using the field assisted sintering technique (FAST) with a force of 20 kN in vacuum. Prior to composite processing, some fibers were heat treated at 600°C for 30 min. in air to oxidize any residual carbon on the fibers that otherwise may enhance debonding and sliding. A ~0.4 mm thick cross sectional specimen with the fibers running perpendicular to the polished surface was prepared for push-out studies. A fiber push-out testing apparatus (Process Equipment, Inc., Troy, OH) was used to obtain load displacement curves. The sliding stress was calculated based on a fiber diameter of 140 μm and using the minimum load corresponding to the load where fiber sliding stopped.

3. RESULTS AND DISCUSSION
3.1 Phase Characterization

3.1.1 Effect of Y to Si atomic ratio, temperature and $LiNO_3$ dopant on phase formation of γ-$Y_2Si_2O_7$:

It has been previously reported that addition of 3 mol% $LiYO_2$ improves formation of the γ-$Y_2Si_2O_7$ polymorph.[4,27] Figure 3 presents the effects of $LiNO_3$ addition, Si to Y atomic ratio, and temperature on the phase presence of γ-$Y_2Si_2O_7$ in this study. Both excess silica and Li-doping enhanced formation of γ-$Y_2Si_2O_7$ over other polymorphs. At 1400°C, both excess silica and LiNO$_3$ doping were required to form single phase γ-Y$_2$Si$_2$O$_7$. Even with 5 mol% of Li, stoichiometric Si:Y compositions formed a mixture of Y$_2$SiO$_5$ (yttrium monosilicate) and Y$_2$O$_3$. Gradual increase of excess silica shifted formation towards β- and γ-Y$_2$Si$_2$O$_7$. Compositions with Si to Y atomic ratios of 52.9:47.1 and 58.9 to 41.2 with 5 mol% of Li formed single phase γ-Y$_2$Si$_2$O$_7$ (Fig. 3a).

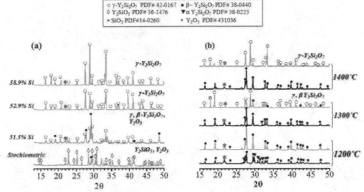

Fig. 3. X-ray diffraction patterns of γ-Y$_2$Si$_2$O$_7$ showing effects of excess silica with 5 mole % of LiNO$_3$ dopant at 1400°C/8 h (a); and temperature and LiNO$_3$ dopant for Si:Y atomic ratio 58.9 to 41.2 (b) on the formation of γ-Y$_2$Si$_2$O$_7$ phase. Gray - with LiNO$_3$, black - without LiNO$_3$.

However, even with 5 mol% Li and the highest amount of excess silica (Si:Y atomic ratio 58.9 to 41.2), 1400°C or higher heat-treatments were necessary to form single phase γ-Y$_2$Si$_2$O$_7$, while at 1300°C traces of β-Y$_2$Si$_2$O$_7$ were still present (Fig. 3b). Without LiNO$_3$, SiO$_2$/Y(OH)$_3$ powder heat-treated at 1200°C formed a mixture of α-Y$_2$Si$_2$O$_7$ and β-Y$_2$Si$_2$O$_7$ (Fig. 3 and Table 2). Powders heat treated at 1300-1400°C formed, β-Y$_2$Si$_2$O$_7$, in agreement with data from Escudero et al.[4,27]

3.1.2 Effect of temperature and LiNO$_3$ dopant on formation of γ-Ho$_2$Si$_2$O$_7$

In contrast to Y$_2$Si$_2$O$_7$, excess silica was not necessary for the formation of γ-Ho$_2$Si$_2$O$_7$. Single phase γ-Ho$_2$Si$_2$O$_7$ was formed from stoichiometric SiO$_2$/Ho(OH)$_3$ powder. However, similar to Y$_2$Si$_2$O$_7$, Li- dopant was necessary for the formation of single phase γ-Ho$_2$Si$_2$O$_7$. Figure 4 summarizes the XRD results of the effect of Li-dopant on formation of γ-Ho$_2$Si$_2$O$_7$ for stoichiometric SiO$_2$/Ho(OH)$_3$ mixtures heat-treated at 1200°C-1400°C/8 h in air. LiNO$_3$ doping reduced the formation temperatures of γ-Ho$_2$Si$_2$O$_7$. In the absence of Li, SiO$_2$/Ho(OH)$_3$ mixtures heat-treated at 1300°C/8 h formed a mixture of α-Ho$_2$Si$_2$O$_7$ and β-Ho$_2$Si$_2$O$_7$. With Li-doping, SiO$_2$/Ho(OH)$_3$ mixtures given the same heat-treatment formed single phase γ-Ho$_2$Si$_2$O$_7$ (Fig. 4).

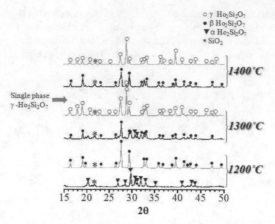

Fig. 4. X-ray diffraction patterns of γ-Ho₂Si₂O₇ showing effects of temperature and LiNO₃ dopant on the formation of the γ-Ho₂Si₂O₇ phase. Gray - with LiNO₃, black - without LiNO₃.

Table 2. Effects of the heat treatment temperature on the phase presence in Re₂Si₂O₇ (Re=Y, Ho) systems with and without Li addition. All powders were heat treated for 8 h.

RE element	Li added	Heat treatment temperature		
		1200°C	1300°C	1400°C
Y	No	α+β	α (trace)+β	β
	Yes	β	β (trace)+γ	γ
Ho	No	α	α+β	β+γ (trace)
	Yes	β	γ	γ

According to published phase diagrams, γ-Ho₂Si₂O₇ is stable at lower temperatures than γ-Y₂Si₂O₇.[8-9,26,32] Our results indicate that regardless of the presence of Li-dopant, the formation of γ-Ho₂Si₂O₇ was thermodynamically and/or kinetically favored over γ-Y₂Si₂O₇ at lower temperatures. Li-doped SiO₂/Ho(OH)₃ mixtures heat-treated at 1300°C/8h completely formed the γ-Ho₂Si₂O₇ phase, whereas Li doped SiO₂/ Y(OH)₃ powder given the same heat-treatment formed mixtures of the γ and β phases (Fig. 3, Table 2). Without LiNO₃, SiO₂/Ho(OH)₃ powder heat-treated at 1400°C formed β-Ho₂Si₂O₇ with a trace of γ-Ho₂Si₂O₇. SiO₂/Y(OH)₃ powder given the same heat treatment formed pure β-Y₂Si₂O₇ without a trace of the γ-polymorph (Fig. 3, Table 2).

3.2 Hardness and Modulus

The relative densities, hardnesses, and moduli of γ-Y₂Si₂O₇ pellets isothermally sintered at 1400°C/8 h are presented in Table 3 for compositions Si to Y of 52.9 to 47.1 and 58.8 to 41.2. Relative densities were determined from the theoretical density of γ-Y₂Si₂O₇ of 4.04 g/cm³. The corresponding SEM micrographs of polished surfaces are shown in Fig. 5. γ-Y₂Si₂O₇ formed from higher silicon to yttrium atomic ratio (Si:Y = 58.8:41.2) had lower relative density of 87%

and was softer, with a Vicker's hardness of 3.7 GPa (Table 4). γ-$Y_2Si_2O_7$ formed with lower silicon to yttrium atomic ratio (Si:Y = 52.9:47.1) had a higher relative density of 94%, which is consistent with reported data.[7] The Vickers hardness of 6.1 GPa and indentation modulus of 165 GPa were also consistent with reported values (Vickers hardness 6.2 GPa and Young's modulus 155 GPa).[7] SEM/EDS shows silica-rich regions with almost no yttrium in the EDS spectrum (Fig. 5). Quantitative image analysis gave an estimate of 18 vol% of excess silica for γ-$Y_2Si_2O_7$ formed with high Si:Y ratio and ~5 vol% excess silica for material formed with lower Si:Y ratio. This was in reasonable agreement with amount of the excess silica, 21.4 vol% and 7.3 vol%, respectively, calculated from Si:Y atomic ratios in the starting compositions (Table 4). Using the rule of mixtures, the calculated excess silica volume fraction, and the densities of silica and γ-$Y_2Si_2O_7$, the expected relative densities are 90% and 97%, respectively. The experimental values were somewhat lower, 87% and 94%, respectively. Therefore, both materials may have about 3% porosity after sintering.

Table 3. Hardness, density and modulus of γ-$Y_2Si_2O_7$ sintered at 1400°C

Si:Y ratio	Density g/cm^3	Relative Density, %	Hardness, GPa	Modulus, GPa
52.9:47.1	3.8	94	6.1±0.2	165± 10
58.8:41.2	3.5	87	3.7± 0.3	-

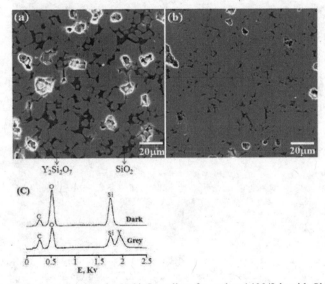

Fig 5. SEM micrographs of γ-$Y_2Si_2O_7$ pellets formed at 1400/8 h with Si:Y ratios of 58.8:41.2 (a), and 52.9:47.1 (b). (c) Representative EDS spectra from sintered γ-$Y_2Si_2O_7$ pellets showing excess silica (dark) and γ-$Y_2Si_2O_7$ (light gray) regions.

3.4 Deformation Behavior under Indentation

As mentioned previously, the mechanical properties of γ-Y$_2$Si$_2$O$_7$ are comparable to monazite, which is soft and deforms plastically by multiple dislocation glide and twinning systems.[33-35] To study deformation mechanisms that may operate in γ-Y$_2$Si$_2$O$_7$, TEM foils were machined beneath indented regions of sintered γ-Y$_2$Si$_2$O$_7$ using FIB. A cross-sectional TEM image of an indent made on sintered γ-Y$_2$Si$_2$O$_7$ (52.9:47.1 Si to Y ratio) is shown in Fig. 6. It shows extensive fracture and plastic deformation in the left side of the indent, and only plastic deformation with slip bands on the right hand side (Fig. 6). Preliminary study also indicated the presence of deformation induced (11$\bar{1}$) stacking faults, likely with R=[101] (Fig. 6). More detailed studies of these features is in progress.

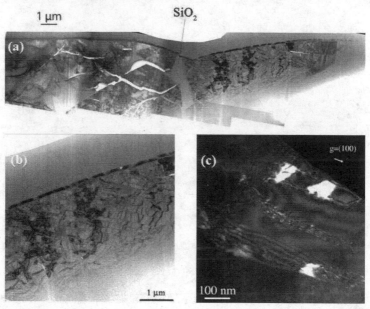

Fig. 6. Plastic deformation under an indent made on sintered γ-Y$_2$Si$_2$O$_7$ (cross-sectional TEM image): (a) overview, bright field; (b) multiple slip bands/active dislocation glide, bright field; (c) (11$\bar{1}$) staking faults, dark field.

3.5 Fiber Push-Out

Preliminary results showed SCS-0 fibers in dense Ho$_2$Si$_2$O$_7$ and Y$_2$Si$_2$O$_7$ matrix to debond and push-out with sliding stresses values of 30-60 MPa,[36-37] within the range reported for C, BN, and LaPO$_4$ coatings. Figure 7 is the sliding stress of SCS-0/Ho$_2$Si$_2$O$_7$ composites made with a) virgin SCS-0 fibers and b) SCS-0 fibers heat treated at 600°C/30 min/air. Due to the removal of interfacial carbon, the average sliding stress of the composite made with heat treated fibers was slightly higher (~45 MPa). The corresponding sliding stress value for the composite formed with virgin fibers was 31 MPa. This value is similar to the corresponding composite formed with $Y_2Si_2O_7$ matrix and virgin SCS fiber (27 MPa) (Fig. 7). Smearing of γ-$Y_2Si_2O_7$ matrix was similar to that observed for $LaPO_4$/alumina composites (Figs. 7c and 7d). Preliminary TEM observation was consistent with intense deformation at the SCS-0 fiber/matrix interface. More detailed study of these features is in progress.

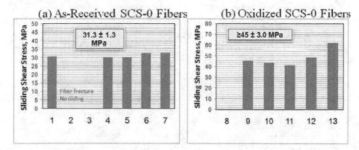

Fig. 7. Sliding stress for push-out SCS in fibers in $Ho_2Si_2O_7$ matrix (a) virgin fibers and (b) heat treated fibers

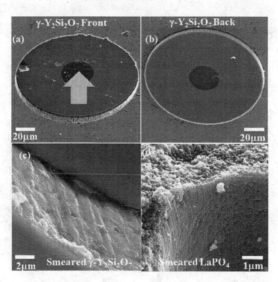

Fig 8 SEM micrographs of pushed-out SCS-0 fibers in $Ho_2Si_2O_7$ and $Y_2Si_2O_7$ matrix; (a) front, (b) back (c) and (d) showing plastic deformation in $Y_2Si_2O_7$ matrix, similar to deformation of monazite.[38]

4. SUMMARY AND CONCLUSIONS

Stoichiometric single phase γ-$Y_2Si_2O_7$ and γ-$Ho_2Si_2O_7$ were made at 1400° and 1300°C, respectively, in air using SiO_2 -$Y(OH)_3$ and SiO_2 -$Ho(OH)_3$ sol-gel precursors and $LiNO_3$ dopant. The addition of $LiNO_3$ enhanced the formation of both γ-$Y_2Si_2O_7$ and γ-$Ho_2Si_2O_7$ at lower temperatures. In the absence of Li-dopant, SiO_2/$Y(OH)_3$ mixtures heat treated at 1400°C/8 h did not form γ-$Y_2Si_2O_7$ and instead formed β-$Y_2Si_2O_7$, and SiO_2/$Ho(OH)_3$ mixtures heat-treated at 1300°C formed β-$Ho_2Si_2O_7$. Phase formation in γ-$Y_2Si_2O_7$ was found to depend on the Si:Y atomic ratio. SiO_2/$Y(OH)_3$ with ~7 vol.% excess silica was necessary to form the γ-$Y_2Si_2O_7$ phase. γ-$Y_2Si_2O_7$ was stable after prolonged heat-treatment at 1200°C. γ-$Y_2Si_2O_7$ pellets had a Vickers hardness of 6.2 GPa and modulus of 165 GPa. TEM of samples directly beneath indentations suggested both extensive dislocation slip and fracture were active. Preliminary results showed SCS-0 fibers in dense $Ho_2Si_2O_7$ and $Y_2Si_2O_7$ matrix to debonded and push-out with sliding stresses of 30-60 MPa, within the range reported for C, BN, and $LaPO_4$ coatings.

REFERENCES

1 Kerans, R. J., Hay, R. S., Parthasarathy, T. A. & Cinibulk, M. K. Interface Design for Oxidation Resistant Ceramic Composites. *J. Am. Ceram. Soc.* **85**, 2599-2632 (2002).

2 Boakye, E. E., Mogilevsky, P., Hay, R. S. & Cinibulk, M. K. Rare-Earth Disilicates as Oxidation-Resistant Fiber Coatings for Silicon Carbide Ceramic Matrix Composites. *in press, J Am. Ceram. Soc.* (2010).

3 Becerro, A. I. & Escudero, A. Polymorphism in the $Lu_{2-x}Y_xSi_2O_7$ system at high temperatures. *J. Eur. Ceram. Soc.* **26**, 2293 (2006).

4 Escudero, A. & Becerro, A. I. Stability of the Low Temperature Polymorphs (y and α) of Lu-Doped $Y_2Si_2O_7$. *J. Phys. Chem. Solids* **68**, 1348-1353 (2007).

5 Becerro, A. I., Naranjo, M., Perdigon, A. C. & Trillo, J. M. Hydrothermal Chemistry of Silicates: Low Temperature Synthesis of y-Yttrium Disilicate. *J. Am. Ceram. Soc.* **86**, 1592-1594 (2003).

6 Sun, Z., Zhou, Y., Wang, J. & Li, M. g-$Y_2Si_2O_7$, a Machinable Silicate Ceramic: Mechanical Properties and Machinability. *J. Am. Ceram. Soc.* **90**, 2535-2541, doi:doi:10.1111/j.1551-2916.2007.01803.x (2007).

7 Sun, Z., Zhou, Y. & Li, M. Low-Temperature Synthesis and Sintering of γ-$Y_2Si_2O_7$. *J. Mater. Res.* **21**, 1443-1450 (2006).

8 Felsche, J. Polymorphism and Crystal Data of the Rare-Earth Disilicates of Type $RE_2Si_2O_7$. *J. Less-Common Metals* **21**, 1-14 (1970).

9 Felsche, J. The Crystal Chemistry of Rare-Earth Silicates. *Struct. Bonding* **13**, 99-113 (1973).

10 Seifert, H. J. *et al.* Yttrium Silicate Coatings on Chemical Vapor Deposition-SiC-Precoated C/C-SiC: Thermodynamic Assessment and High-Temperature Investigation. *J. Am. Ceram. Soc.* **88**, 424-430 (2005).

11 Wang, J. Y., Zhou, Y. C. & Lin, Z. J. Mechanical Properties and Atomistic Deformation Mechanism of γ-$Y_2Si_2O_7$ from First-Principles Investigations. *Acta mater.* **2007**, 6019-6026 (2007).

12 Keller, K. A. *et al.* Effectiveness of Monazite Coatings in Oxide/Oxide Composites After Long Term Exposure at High Temperature. *J. Am. Ceram. Soc.* **86**, 325-332 (2003).

13   Boakye, E. E., Mogilevsky, P., Welter, J., Hay, R. S. & Kerans, R. J. Monazite Coatings on SiC Fibers I: Fiber Strength and Thermal Stability. *J. Am. Ceram. Soc.* **89**, 3475-3480 (2006).

14   Mogilevsky, P., Boakye, E. E., Hay, R. S. & Kerans, R. J. Monazite Coatings on SiC Fibers II: Oxidation Protection. *J. Am. Ceram. Soc.* **89**, 3481-3490 (2006).

15   Cinibulk, M. K., Fair, G. E. & Kerans, R. J. High-Temperature Stability of Lanthanum Orthophosphate (Monazite) on Silicon Carbide at Low Oxygen Partial Pressure. *Am. Ceram. Soc.* **91**, 2290–2297 (2008).

16   Cupid, D. M. & Seifert, H. J. Thermodynamic Calculations and Phase Stabilities in the Y-Si-C-O System. *J. Phase Equil. Diff.* **28**, 90-100 (2007).

17   Sun, Z., Zhou, Y., Wang, J. & Li, M. Thermal Properties and Thermal Shock Resistance of γ-$Y_2Si_2O_7$. *J. Am. Ceram. Soc.* **91**, 2623-2629 (2008).

18   Ranganathan, V. & Klein, L. C. Sol-gel Synthesis of Erbium-doped Yttrium Silicate Glass-ceramics. *J. Non-Cryst. Solids* **354**, 3567-3571 (2008).

19   Zhou, P., Yu, X., Yang, S. & Gao, W. Synthesis of $Y_2Si_2O_7$:Eu Nanocrystal and its Optical Properties. *J. Lumin.* **124**, 241-244 (2007).

20   Diaz, M. G.-C. I., Mello-Castanho, S., Moya, J. S. & Rodriguez, M. A. Synthesis of Nanocrystalline Yttrium Dislicate Powder by Sol-Gel Method. *J. Non. Cryst. Sol.* **289**, 151-154 (2001).

21   Diaz, M. Synthesis, Thermal Evolution, and Luminescence Properties of Yttrium Disilicate Host Matrix. *Chem. Mater.* **17**, 1774-1782 (2005).

22   Maier, N., Rixecker, G. & Nickel, K. G. Formation and Stability of Gd, Y, Yb and Lu Disilicates and their Solid Solutions. *J. Solid State Chem.* **179**, 1630-1635 (2006).

23   Parmentier, J. Phase Transformations in Gel-Derived and Mixed-Powder-Derived Yttrium Disilicate, $Y_2Si_2O_7$, by X-Ray Diffraction and ^{29}Si MAS NMR. *J. Solid State Chem.* **149**, 16-20 (2000).

24   Giesche, H. & Matijevic, E. Preparation, Characterization, and Sinterability of Well-Defined Silica/Yttria Powders. *J. Mater. Res.* **9**, 436-450 (1994).

25   Moya, J. S., Diaz, M., Serna, C. J. & Mell-Castanho, S. Formation of Nanocrystalline Yttrium Disilicate Powder by an Oxalate Gel Method. *J. Eur. Ceram. Soc.* **18**, 1381-1384 (1998).

26   Ito, J. & Johnson, H. Synthesis and Study of Yttrialite. *Am. Mineralogist* **54**, 1940-1952 (1968).

27   Escudero, A., Alba, M. D. & Becerro, A. I. Polymorphism in the $Sc_2Si_2O_7$-$Y_2Si_2O_7$ System. *J. Solid State Chem.* **180**, 1436-1445 (2007).

28   Becerro, A. I., Naranjo, M., Alba, M. D. & Trillo, J. M. Structure-directing effect of Phyllosilicates on the Synthesis of y-$Y_2Si_2O_7$. Phase Transitions in $Y_2Si_2O_7$. *J. Mater. Chem.* **13**, 1835-1842 (2003).

29   Kahlenberg, V. Rietveld Analysis and Raman Spectrospcopic Investigation on α-$Y_2Si_2O_7$. *Z. Anorg. Allg. Chem.* **634**, 1166-1172 (2007).

30   Toropov, N. A., Bondar, I. A., Sidorenko, G. A. & Koroleva, L. N. *AIzv. Akad. Nauk. SSSR, Neorg. Mater.* **1**, 218 (1965).

31   Oliver, W. C. & Pharr, G. M. An Improved Technique for Determining Hardness and Elastic Modulus Using Load and Displacement Sensing Indentation Experiments. *J. Mat. Res.* **7**, 1564-1583 (1992).

32   Vomacka, P. & Babushkin, O. Yttria-Alumina-Silica Glasses with Addition of Zirconia. *J. Eur. Ceram. Soc.* **15**, 921-928 (1995).

33   Hay, R. S. Monazite and Scheelite Deformation Mechanisms. *Ceram. Eng. Sci. Proc.* **21**, 203-218 (2000).

34   Hay, R. S. (120) and (12$\underline{2}$) Monazite Deformation Twins. *Acta mater.* **51**, 5255-5262 (2003).

35   Hay, R. S. Climb-Dissociated Dislocations in Monazite. *J. Am. Ceram. Soc.* **87**, 1149-1152 (2004).

36   Rebillat, F. *et al.* Interfacial Bond Strength in SiC/C/SiC Composite Materials, as Studied by Single-Fiber Push-Out Tests. *J. Am. Ceram. Soc.* **81**, 965-978 (1998).

37   Morgan, P. E. D. & Marshall, D. B. Ceramic Composites of Monazite and Alumina. *J. Am. Ceram. Soc.* **78**, 1553-1563 (1995).

38   Kerans, R. J. *et al.* in *High Temperature Ceramic Matrix Composites* (ed R. Naslain W. Krenkel, H. Schneider) 129-135 (Wiley-VCH, 2001).

TRANSMISSION ELECTRON MICROSCOPY OF RARE-EARTH ORTHOPHOSPHATE FIBER-MATRIX INTERPHASES THAT DEFORM BY TRANSFORMATION PLASTICITY DURING FIBER PUSH-OUT

R. S. Hay, G. E. Fair
Air Force Research Laboratory
Materials and Manufacturing Directorate
WPAFB, OH

E. E. Boakye, P. Mogilevsky, T. A. Parthasarathy
Air Force Research Laboratory
UES, Inc., Dayton, OH

M. Ahrens, T. J. Godar
Wright State University
Fairborn, OH

ABSTRACT
The mechanical properties of ceramic matrix composites are sensitive to the stresses that develop along the fiber-matrix interface during fiber pullout. Lowering these stresses in rare-earth orthophosphate coated fibers by use of a coating that deforms easily by transformation plasticity is studied. Single-crystal alumina fibers (Saphikon^{TM}) were coated with a $(Gd_{0.4},Dy_{0.6})PO_4$ xenotime slurry. Polycrystalline alumina matrices were densified around the coated fibers. Deformation and the xenotime $\rightarrow$ monazite martensitic phase transformation in the $(Gd_{0.4},Dy_{0.6})PO_4$ fiber-matrix interphase were characterized by TEM after fiber push-out. The coated fibers were pushed out with stresses averaging about 30 MPa, significantly lower than those for fibers coated with orthophosphates that do not undergo a phase transformation.

INTRODUCTION
Oxide-oxide CMCs with rare-earth orthophosphate interphases such as monazite and xenotime have been successfully demonstrated by a number of different research groups.[1-6] One performance limiting factor for oxide-oxide CMCs is high fiber-pullout shear stress (friction) of the rare-earth orthophosphate fiber-matrix interfaces.[7] The high fiber pull-out stresses, typically ~80 - 200 MPa, may be near the borderline of acceptable values.[8-10] These stresses contrast with 5 – 20 MPa typically measured or inferred for carbon and BN interphases.[11-12] High pull-out stress biases fiber failure towards shorter pullout lengths, which may slightly increase strength, but at the expense of toughness (strain to failure) and flaw tolerance, the most desirable CMC attributes.[13-14] For monazite rare-earth orthophosphate fiber-matrix interphases, pull-out friction is governed by the plastic deformation mechanisms of a thin layer (~100 – 300 nm) of interphase material adjacent to the fiber.[15] These mechanisms include dislocation slip, microfracture, deformation twinning, and cataclastic flow of deformed nanoparticles.

An approach to lowering pull-out stress involves use of interphases that undergo transformation plasticity with a – V martensitic phase transformation. Transformation plasticity occurs when the atomic rearrangements during a phase transformation simultaneously accommodate stress and therefore weaken the material. It has been known to metallurgists for over 80 years,[16-20] and mechanistically is closely related to deformation twinning. Transformation plasticity has also been proposed to weaken earth materials.[18,21] The phase transformation can be driven by local high pressures and shear stresses caused by accommodation of fiber roughness during pullout. There are two possible effects; friction reduction by transformation plasticity, or by local reduction of normal stress from contraction of small volumes of the interphase material during the phase transformation.

Large rare-earths, such as La and Ce, form orthophosphates with the monazite structure. Small rare earths, such as Y and Lu, form orthophosphates with the xenotime structure (Fig. 1). The change between the structures occurs between the $GdPO_4$ and $TbPO_4$ compositions.[22] The transformation

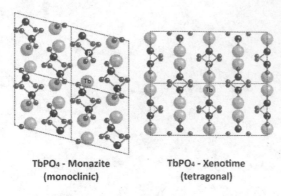

TbPO₄ - Monazite
(monoclinic)

TbPO₄ - Xenotime
(tetragonal)

Fig. 1 *Crystal structures for monazite and xenotime rare-earth orthophosphate polymorphs.*

between the two structures should be martensitic and kinetically facile, since atomic shuffles of only 0.35Å, half that of shuffles required for (100) and (001) monazite deformation twinning, are involved.[23] The shear accompanying the xenotime monazite transformation does not have a shear sense, or sign, unlike deformation twinning,[24] and because of the tetragonal xenotime symmetry, the transformation can occur on (100), (010), and (001) in xenotime. Pressures that induce the transformation can be estimated from thermodynamic data,[25-29] and are ~ 1 GPa for TbPO₄, but should in principle approach zero for $(Dy_x,Gd_{1-x})PO_4$ solid-solutions near monazite-xenotime equilibrium. Thermodynamic calculations suggest monazite-xenotime equilibrium for a $(Dy_{0.8}Gd_{0.2})PO_4$ solid-solution, with little temperature dependence.[30-31] The volume loss of the xenotime monazite transformation is ~5.8 vol%. The ease of inducing the transformation, and the subsequent effect on mechanical properties through transformation plasticity, can be assessed by the stress-strain signature and hysteresis during instrumented nano-indentation, using a method that is widely applied to silicon.[32-35]

Preliminary results on fiber push-out experiments of oxide fibers coated with rare-earth orthophosphates that undergo transformation plasticity through the xenotime monazite phase transformation are presented. Deformation and phase transformation characteristics are characterized by TEM. Preliminary results for indentation of rare-earth orthophosphate compositions that demonstrate significant softening by transformation plasticity are published,[36] and a more thorough presentation, discussion, and analysis of both indentation and fiber push-out is in progress.[37]

EXPERIMENTS

Reagent grade dysprosium nitrate hydrate, gadolinium nitrate hexahydrate, and phosphoric acid, (Aldrich Chemical Co., Milwaukee, WI) were used without further purification. The water of hydration of $Dy(NO_3)_3 \cdot xH_2O$ was determined to be 0.5. Gadolinium nitrate hydrate water of hydration 0.6, $Gd(NO_3)_3 \cdot 0.6H_2O$, was confirmed. Water was purified with the nanopure system (Model D4744, Barnstead /Thermolyne corp., Dubuque, IA) for all experiments.

Rare-earth orthophosphate powders with GdPO₄ and DyPO₄ compositions were made by forced precipitation with phosphoric acid at a pH=~10 as previously reported.[38] Rare-earth solid-solutions of (Gd,Dy)PO₄ corresponding to Gd:Dy molar ratio 40:60 were also formed by mixing portions of Gd and Dy nitrate and subsequently precipitating their phosphates as described previously.[38] The particles were washed with water, ethanol and isopropanol before further processing for fiber coatings.

To retain a xenotime phase coating on the fiber, $(Gd_{0.4}Dy_{0.6})PO_4$ slurry in the xenotime phase was used as the coating sol. Hydrated rare-earth orthophosphate powders ($(Gd_{0.4},Dy_{0.6})PO_4 \cdot H_2O$) were converted to tetragonal xenotime by heat treatment at 1200-1400°C/10h, ball-milled, and redispersed in isopropanol with a concentration of 100g/L. A colloidal dispersion of the $(Gd_{0.4}Dy_{0.6})PO_4$ composition was multiply coated onto Saphikon™ single-crystal alumina fibers 15-20 times. The coatings were heat treated at 900°C/10 minutes between coats in air and were given a final heat-treatment at 1400-1600°C/10h to insure retention of the xenotime, rather than monazite, phase.

Coated fibers were sandwiched between two slip-cast Sumitomo alumina pellets and hot-pressed at 1400°C, 15 MPa for 1 hour to make minicomposites. Two composites specimen were made. In specimen #1, the hot-pressed composite was heat-treated at 1600°C/20h. In specimen #2, the hot-pressed composite was heat-treated at 1600°C/20h and then further densified by hot-isostatic pressing at a temperature of 1400°C/1h with a pressure of 200 MPa. The high processing temperatures were necessary to densify xenotime. A ~0.4 mm thick cross sectional specimen with fibers running perpendicular to the polished surface was prepared for push-out studies. Fiber push-out testing apparatus (Process equipment Inc., Troy, OH) was used to measure load-displacement curves. Interface sliding stress was measured for five fibers for specimen #1 and six fibers for specimen #2. The sliding stress was calculated from a fiber diameter of 120 m using the minimum load at which fiber sliding stopped.

Coated fibers were characterized by SEM and x-ray. As-processed composites were characterized by SEM. Push-out samples were characterized by SEM for push-out displacement and coating deformation. TEM sections were prepared from cross-sections of pushed-out fibers through the middle of the 0.4 mm thick specimens using tripod polishing with diamond films. Thin sections were epoxied to copper grids and washers, followed by ion-milling in a low-angle ion mill. The push-out specimens were infiltrated with low-viscosity epoxy prior to TEM section preparation. TEM observation, including compositional mapping, was done at 200 kV (Phillips CM200) and 300 kV (FEI Titan).

RESULTS AND DISCUSSION

SEM and TEM confirm that the $(Gd_{0.4}Dy_{0.6})PO_4$ coatings were not fully dense after hot-pressing and hot-isostatic pressing. Local relative densities were estimated to be between $70 - 85\%$. The difficulty of densifying rare-earth orthophosphates in the xenotime phase, in contrast with those in the monazite phase, has been previously noted.[39] Sumitomo alumina powder infiltrated the porous $(Gd_{0.4}Dy_{0.6})PO_4$ coatings in small amounts during push-out specimen processing. Infiltration extent was greater in specimen #1 than in #2. X-ray diffraction of the coated fibers heat-treated at 1600°C showed only xenotime phase in the coating. However, the $(Gd_{0.4}Dy_{0.6})PO_4$ coatings show some reaction with the single-crystal alumina fibers; EDX and X-ray diffraction show a xenotime and a garnet phase (Fig. 2). For coatings heat treated at 1400°C/10h, TEM confirms formation of about 0.5

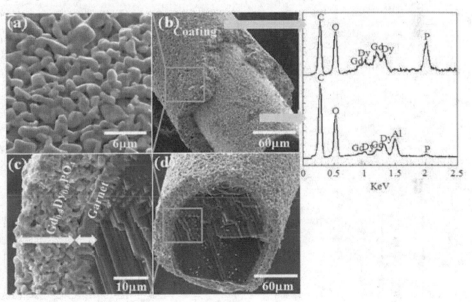

Fig. 2 *SEM micrographs and EDS spectra from coated fibers of specimen 1, showing coating morphology and reaction of the coating with the single-crystal alumina fiber.*

to 1 m of $(Gd,Dy)_3Al_5O_{12}$ garnet at the single-crystal alumina fiber surface in specimen #2 (Fig. 3); SEM characterization suggests a larger reaction extent in specimen #1 heat-treated at 1600°C/10h (Fig. 2). Reaction with the small amount of Sumitomo alumina powder that had infiltrated the porous coatings was not observed (Fig. 3). The reaction has been previously observed between other $(Gd,Dy)PO_4$ compositions and alumina after heat-treatment at 1400°C and higher.[37] Monazite and

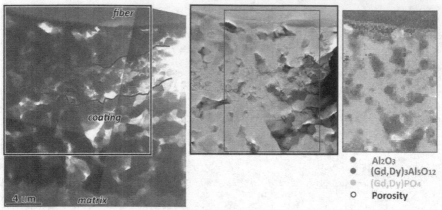

- ● Al_2O_3
- ● $(Gd,Dy)_3Al_5O_{12}$
- ● $(Gd,Dy)PO_4$
- ○ Porosity

Fig. 3 *Low magnification TEM micrograph showing fiber, fiber-matrix interphase (coating) with deformation band outlined in red, and matrix. An STEM image of the outlined area is in the middle, and an x-ray dot map of the area outlined in blue is on the right. The x-ray dot map for composition shows reaction to $(Gd,Dy)_3Al_5O_{12}$ garnet at the fiber surface, and some infiltration of alumina in the coating.*

xenotime rare-earth orthophosphates are thermochemically stable with alumina.[40-43] The reaction was suspected to be due to incongruent vaporization of phosphate during the heat-treatment of the fiber coatings, which leads to formation of rare-earth rich phosphates ($RE_4P_2O_{11}$, $RE_6P_4O_{19}$) that will react with alumina to form $RE_3Al_5O_{12}$ garnets and $REPO_4$. Presence of a dense alumina matrix around the coated fibers should suppress phosphate vaporization, and limit reaction with infiltrated alumina powder or alumina matrix.

The sliding stresses for five fibers from specimen #1 and six from #2 are shown in figure 4. The averages for specimen #1 and #2 were 64 MPa and 11 MPa, respectively. These are significantly lower than those found for other orthophosphate fiber-matrix interphases (80-200 MPa).[8-10] Significantly greater alumina infiltration into the relatively more porous xenotime coatings of specimen #1 may be at least a partial cause of both the higher push-out stresses and greater variation in those stresses in this specimen.

Deformation in the fiber-matrix interphase after fiber push-out was concentrated in a broad damage zone several microns wide in the 15 m thick fiber coatings (Fig. 3, 5). This damage zone was often not adjacent to the fiber; there was often undeformed coating next to the fiber. The width of the damage zone was in sharp contrast to that observed after push-out of $LaPO_4$ monazite coated single-crystal alumina fibers in alumina matrices.[15] For $LaPO_4$

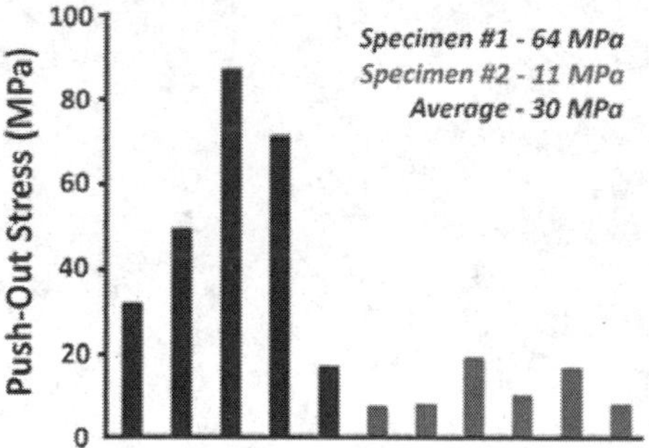

Specimen #1 - 64 MPa
Specimen #2 - 11 MPa
Average - 30 MPa

Fig. 4 *Push-out stresses (friction) for five fibers in specimen #1 and six in specimen #2.*

interphases, intense plastic deformation was concentrated in a band several hundred nanometers in width adjacent to the fiber.[15]

Cataclastic flow best describes the general nature of deformation in the damage zone of the fiber-matrix interphase.[44] Granulation of material under shear, translation and rotation of the granules, accompanied by brittle and ductile deformation of the granules are diagnostic of this mechanism. Higher resolution TEM images show evidence of extensive microcracking and plastic deformation in individual grains (Fig. 5). Thin regions (100 – 200 nm) with fine-grained microstructure between granules, diagnostic of dynamic recrystallization, are also present. These are similar to intensely

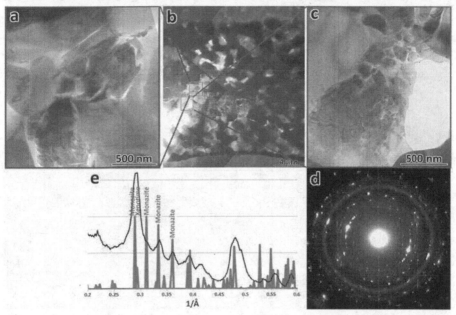

Fig. 5 a,c) High magnification TEM micrographs of regions in low magnification micrograph b), which shows the matrix, $Gd_{0.4}Dy_{0.6}PO_4$ interphase with a cataclastic deformation band, and the fiber with a thin surface layer of garnet. d) Selected area diffraction pattern from region in micrograph "c". e) Digital summation of the radial intensity distribution of diffraction pattern in "d", with peaks indexed to xenotime and monazite.

deformed and recrystallized areas observed after push-out of $LaPO_4$-coated fibers, although those were more extensive and better developed.

Selected area electron diffraction patterns of deformed areas find some whole grains have transformed to monazite. Digital summation of the radial intensity distribution from diffraction patterns taken in heavily deformed, fine-grained areas also confirm the presence of monazite (Fig. 5). Sample stage tilting in the TEM finds a fine-scale lamellar structure in many of the grains in the deformed zone to be prevalent (Fig. 6). It was suspected that these lamellar structures are a fine mixture of monazite and xenotime lamellae that form during martensitic transformation. These structures were observed previously in indented $TbPO_4$ xenotime samples that had partially transformed to monazite underneath the indents, and may be metastable.[36] Selected area electron diffraction patterns of these lamellar phases show extensive streaking in xenotime-indexed patterns,

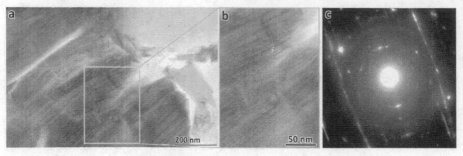

Fig. 6 *a) TEM micrograph of deformed material with a fine-scale lamellae, b) Higher magnification micrograph of region in "a". c) Selected area diffraction pattern from "b".*

but with satellite reflections that cannot be indexed as monazite. Analysis of the patterns and further observation of the structures by high resolution TEM is underway.

SUMMARY AND CONCLUSIONS

Low fiber push-out friction attributed to transformation plasticity was demonstrated for $(Gd_{0.4}Dy_{0.6})PO_4$ xenotime fiber-matrix interphases on single-crystal alumina fibers. Push-out stresses as low as ~10 MPa were measured in one specimen. Transformation of $(Gd_{0.4}Dy_{0.6})PO_4$ xenotime to monazite was characterized by TEM in the deformed interphase after push-out. Deformation of the interphase by cataclastic flow occurred in bands several microns wide that were not necessarily adjacent to the fiber, which contrasts with the much narrower bands adjacent to the fiber previously observed for $LaPO_4$ monazite interphases. A phase with fine-scale lamellae forms profusely in the deformed region, and requires further characterization.

Several issues remain with these fiber-matrix interphases. Porosity in the interphase, and partial infiltration of that porosity by alumina matrix, may affect push-out and the associated interphase deformation mechanisms. High processing temperatures of at least 1400°C are necessary to either form xenotime or densify coatings applied as xenotime slurries, and are not compatible with existing commercially available polycrystalline oxide fibers such as Nextel[TM] 610 and Nextel[TM] 720. These high processing temperatures also cause some incongruent vaporization of phosphate that make the remaining rare-earth rich coating reactive with alumina. The xenotime transformation plasticity mechanism for interphase friction reduction therefore requires either improved low-temperature processing methods or improved fibers for further use and development.

REFERENCES
1   Keller, K. A., Mah, T., Parthasarathy, T. A., Boakye, E. E., Mogilevsky, P., & Cinibulk, M. K. Effectiveness of Monazite Coatings in Oxide/Oxide Composites After Long Term Exposure at High Temperature. *J. Am. Ceram. Soc.* **86**, 325-332 (2003).
2   Lee, P.-Y., Imai, M. & Yano, T. Fracture Behavior of Monazite-Coated Alumina Fiber-Reinforced Alumina-Matrix Composites at Elevated Temperature. *J. Ceram. Soc. Japan* **112**, 628-633 (2004).
3   Marshall, D. B. & Davis, J. B. Ceramics for Future Power Generation Technology: Fiber Reinforced Oxide Composites. *Curr. Opin. Solid State Mater. Sci.* **5**, 283-289 (2001).
4   Kaya, C., Butler, E. G., Selcuk, A., Boccaccini, A. R. & Lewis, M. H. Mullite (Nextel[TM] 720) Fibre-Reinforced Mullite Matrix Composites Exhibiting Favourable Thermomechanical Properties. *J. Eur. Ceram. Soc.* **22**, 2333-2342 (2002).
5   Davis, J. B., Marshall, D. B. & Morgan, P. E. D. Oxide Composites of $LaPO_4$ and Al_2O_3. *J. Eur. Ceram. Soc.* **19**, 2421-2426 (1999).
6   Davis, J. B., Marshall, D. B. & Morgan, P. E. D. Monazite Containing Oxide-Oxide Composites. *J. Eur. Ceram. Soc.* **20**, 583-587 (2000).

7   Zok, F. W. Developments in Oxide Fiber Composites. *J. Am. Ceram. Soc.* **89**, 3309-3324 (2006).
8   Morgan, P. E. D. & Marshall, D. B. Ceramic Composites of Monazite and Alumina. *J. Am. Ceram. Soc.* **78**, 1553-1563 (1995).
9   Kuo, D.-H., Kriven, W. M. & Mackin, T. J. Control of Interfacial Properties through Fiber Coatings: Monazite Coatings in Oxide-Oxide Composites. *J. Am. Ceram. Soc.* **80**, 2987-2996 (1997).
10   Chawla, K. K., Liu, H., Janczak-Rusch, J. & Sambasivan, S. Microstructure and Properties of Monazite (LaPO$_4$) Coated Saphikon Fiber/Alumina Matrix Composites. *J. Eur. Ceram. Soc.* **20**, 551-559 (2000).
11   Cao, H. C., Bischoff, E., Sbaizero, O., Ruhle, M., Evans, A. G., Marshall, D. B. & Brennan, J. J. Effect of Interfaces on the Properties of Fiber-Reinforced Ceramics. *J. Am. Ceram. Soc.* **73**, 1691-1699 (1990).
12   Curtin, W. A., Eldredge, J. I. & Srinivasan, G. V. Push-Out Test on a New Silicon Carbide/Reaction Bonded Silicon Carbide Ceramic Matrix Composite. *J. Am. Ceram. Soc.* **76**, 2300-2304 (1993).
13   Kerans, R. J., Hay, R. S., Parthasarathy, T. A. & Cinibulk, M. K. Interface Design for Oxidation Resistant Ceramic Composites. *J. Am. Ceram. Soc.* **85**, 2599-2632 (2002).
14   Curtin, W. A., Ahn, B. K. & Takeda, N. Modeling Brittle and Tough Stress-Strain Behavior in Unidirectional Ceramic Matrix Composites. *Acta mater.* **46**, 3409-3420 (1998).
15   Davis, J. B., Hay, R. S., Marshall, D. B., Morgan, P. E. D. & Sayir, A. The Influence of Interfacial Roughness on Fiber Sliding in Oxide Composites with La-Monazite Interphases. *J. Am. Ceram. Soc.* **86**, 305-316 (2003).
16   Sauveur, A. What is Steel? Another Answer. *The Iron Age* **113**, 581-583 (1924).
17   Wassermann, G. Untersuchungen an einer Eisen-Nickel Legierung uber die Verformbarkeit Wahrend der g-a Unwandlung. *Archiv Fur der Eisenhutt* **7**, 321-325 (1937).
18   Poirier, J.-P. *Creep of Crystals*. (Cambridge University Press, 1985).
19   Stringfellow, R. G., Parks, D. M. & Olson, G. B. A Constitutive Model for Transformation Plasticity Accompanying Strain-Induced Martensitic Transformations in Metastable Austenitic Steels. *Acta metall. mater.* **40**, 1703-1716 (1992).
20   Fischer, F. D. A Micromechanical Model for Transformation Plasticity in Steels. *Acta metall. mater.* **38**, 1535-1546 (1990).
21   Poirier, J. P. On Transformation Plasticity. *J. Geophys. Res.* **87**, 6791-6798 (1982).
22   Kolitsch, U. & Holtsam, D. Crystal Chemistry of REEXO$_4$ Compounds (X = P, As, V). II. Review of REEXO$_4$ Compounds and their Stability Fields. *Eur. J. Mineral.* **16**, 117-126 (2004).
23   Hay, R. S. & Marshall, D. B. Deformation Twinning in Monazite. *Acta mater.* **51**, 5235-5254 (2003).
24   Wenk, H.-R. in *Texture and Anisotropy* eds U. F. Kocks, C. N. Tome, & H-R. Wenk) 561-596 (Cambridge University Press, 1998).
25   Ushakov, S. V., Helean, K. B., Navrotsky, A. & Boatner, L. A. Thermochemistry of Rare-Earth Orthophosphates. *J. Mater. Res.* **16**, 2623-2633 (2001).
26   Thiriet, C., Konings, R. J. M., Javorsky, P. & Wastin, F. The Heat Capacity of Cerium Orthophosphate CePO$_4$, the Synthetic Analogue of Monazite. *Phys. Chem. Minerals* **31**, 347-352 (2004).
27   Popa, K., Sedmidubsky, D., Benes, O., Thiriet, C. & Konings, R. J. M. The High Temperature Heat Capacity of LnPO$_4$ (Ln = La, Ce, Gd) by Drop Calorimetry. *J. Chem. Thermo.* (2005).
28   Thiriet, C., Konings, R. J. M., Javorsky, P., Magnani, N. & Wastin, F. The Low Temperature Heat Capacity of LaPO$_4$ and GdPO$_4$, the Thermodynamic Functions of the Monazite-Type LnPO$_4$ Series. *J. Chem. Thermo.* **37**, 131-139 (2005).
29   Dorogova, M., Navrotsky, A. & Boatner, L. A. Enthalpies of Formation of Rare Earth Orthovanadates, REVO$_4$. *J. Solid State Chem.* **180**, 847-851 (2007).
30   Mogilevsky, P., Boakye, E. E. & Hay, R. S. Solid Solubility and Thermal Expansion in LaPO$_4$-YPO$_4$ System. *J. Am. Ceram. Soc.* **90**, 1899-1907 (2007).
31   Keller, K. A., Mogilevsky, P., Parthasarathy, T. A., Lee, H. D. & Mah, T.-I. Monazite coatings in dense (≥90%) alumina-chromia minicomposites. *J. Am. Ceram. Soc.* (2007).

32   Bradby, J. E., Williams, J. S., Wong-Leung, J., Swain, M. V. & Munroe, P. Mechanical Deformation in Silicon by Micro-Indentation. *J. Mater. Res.* **16**, 1500-1507 (2001).

33   Domnich, V., Gogotsi, Y. & Dub, S. Effect of Phase Transformations on the Shape of the Unloading Curve in the Nanoindentation of Silicon. *Appl. Phys. Lett.* **76**, 2214-2217 (2000).

34   Zarudi, I., Zhang, L. C. & Swain, M. V. Behavior of Monocrystalline Silicon Under Cyclic Microindentations with a Spherical Indenter. *Appl. Phys. Lett.* **82**, 1027-1029 (2003).

35   Zhang, L. & Zarudi, I. Towards a Deeper Understanding of Plastic Deformation in Mono-crystalline Silicon. *Int. J. Mech. Sci.* **43**, 1985-1996 (2001).

36 . Hay, R. S., Fair, G. E., Boakye, E. E., Mogilevsky, P., Parthasarathy, T. A. & Davis, J., Softening of Rare Earth Orthophosphates by Transformation Plasticity: Possible Applications to Fiber-Matrix Interphases in Ceramic Composites, Design, Development, and Applications of Engineering Ceramics and Composites: *Ceramic Transactions*, **215**, Singh, Zhu, Zhou, Singh (eds.) (2010).

37   Hay, R. S., Boakye, E. E., Mogilevsky, P., Fair, G. E. & Parthasarathy, T. A. Transformation Plasticity in $TbPO_4$ and $(Gd_{0.4}Dy_{0.6})PO_4$ Orthophosphates: Indentation of Bulk Specimens and Pushout of Coated Fibers. *J. Am. Ceram. Soc.* (in preparation).

38   Boakye, E. E., Fair, G. E., Mogilevsky, P. & Hay, R. S. Synthesis and Phase Composition of Lanthanide Phosphate Nanoparticles $LnPO_4$ (Ln = La, Gd, Tb, Dy, Y) and Solid Solutions for Fiber Coatings. *J. Am. Ceram. Soc.* **91**, 3841-3849 (2008).

39   Hikichi, Y., Ota, T., Daimon, K., Hattori, T. & Mizuno, M. Thermal, Mechanical, and Chemical Properties of Sintered Xenotime-Type RPO_4 (R = Y, Er, Yb, or Lu). *J. Am. Ceram. Soc.* **81**, 2216-2218 (1998).

40   Morgan, P. E. D., Marshall, D. B. & Housley, R. M. High Temperature Stability of Monazite-Alumina Composites. *Mater. Sci. Eng.* **A195**, 215-222 (1995).

41   Marshall, D. B., Morgan, P. E. D., Housley, R. M. & Cheung, J. T. High-Temperature Stability of the Al_2O_3-$LaPO_4$ System. *J. Am. Ceram. Soc.* **81**, 951-956 (1998).

42   Morgan, P. E. D. & Marshall, D. B. Functional Interfaces for Oxide/Oxide Composites. *Mater. Sci. Eng.* **A162**, 15-25 (1993).

43   Lewis, M. H., Tye, A., Butler, E. G. & Doleman, P. A. Oxide CMCs: Interphase Synthesis and Novel Fiber Development. *J. Eur. Ceram. Soc.* **20**, 639-644 (2000).

44   Paterson, M. S. *Experimental Rock Deformation - The Brittle Field*. 1 edn, Vol. 13 (Springer Verlag, 1978).

PROCESSING OF OXIDE/OXIDE COMPOSITES FOR GAS TURBINE APPLICATIONS BASED ON BRAIDING TECHNIQUE (OXITEX™)

Christian Wilhelmi, EADS Innovation Works, Munich, Germany
Thays Machry, EADS Innovation Works, Munich, Germany
Ralf Knoche, ASTRIUM Space Transportation, Bremen, Germany
Dietmar Koch, Advanced Ceramics, University of Bremen, Bremen, Germany

ABSTRACT

New concepts for aerospace applications such as gas turbine aero-engines require advanced lightweight structural materials with superior high temperature environmental stability in order to increase operation efficiency and to reduce CO_2 emission. Beside others, oxide/oxide CMC can fulfil these demands for high performance turbines due to their inherent oxidation resistance, good thermo-mechanical properties at elevated temperatures and low density.

For some decades now EADS Innovation Works uses the filament winding technique and PIP process (Polymer Infiltration Pyrolysis) for the manufacture of a specific oxide/oxide CMC termed UMOX™. Continuous oxide based Nextel™ fibers (3M™, USA) such as Nextel™610 alumina fibers are impregnated with pre-ceramic slurries via filament winding and further processed to a green body following by high temperature treatment where the matrix is converted to a ceramic state. Besides specific advantages such as excellent reproducibility the drawback of this process is based on high manufacturing costs due to long process times and pyrolysis steps.

To overcome this disadvantage and due to the strong demand for high performance materials a new process for the fabrication of an oxide/oxide CMC type, termed OXITEX™, using the tailored circular braiding technique was successfully investigated and developed. The development approach and exemplary CMC material characteristics are presented. The paper concludes with the next steps together with the demonstration of a preliminary combustor liner structure for gas turbine engine application since part of the presented work was performed within the German BMBF HiPOC program (**Hi**gh **P**erformance **O**xide **C**eramics). The main focus of the project is the development of oxide/oxide CMC materials and structures for the envisaged use in gas turbine engine applications.

INTRODUCTION

Oxide based Ceramic Matrix Composites (CMC) are highly promising materials for aerospace applications due to their high strength, toughness, notch insensitivity, refractoriness and inherent oxidation resistance at high temperatures where metals are usually limited by their low melting temperature and monolithic ceramics are limited due to their very poor damage tolerance. Furthermore the introduction of oxide/oxide CMC as structural components for e.g. combustor liners in gas turbine engines appear to be very promising due to their elevated thermo-mechanical performance at high temperatures and their low density compared to metal alloys or inter-metallic materials. These CMC materials with tailored thermal management by appropriate cooling schemes show high potential to significantly increase operating temperatures. In doing so the efficiency of future gas turbine engines can be improved and CO_2 emission significantly reduced. Thus the investigation of state of the art oxide based CMC with a high Technology Readiness Level (TRL) as well as the new development of suchlike CMC materials is one of the main tasks in the near future.

This scope has led to the German BMBF (German Federal Ministry of Education Research) HiPOC program (**Hi**gh **P**erformance **O**xide **C**eramics) which started in February 2009 and consists of 3 companies and 4 research institutes. Main objectives of the project are based on the development of different oxide based CMC materials which are candidates to be used in gas turbines for power generation and aerospace propulsion, or as spin-off in space applications such as for Thermal Protection Systems and Hot Structures. In combination with an improved thermal management the main objectives are to minimise the fuel consumption and thereby reduce the CO_2 emission from the gas turbine. To achieve this goal, high performance oxide based CMC material concepts have to be developed and investigated. Furthermore design concepts for selected CMC components such as combustor liner or turbine seal segment have to be developed with focus on material and manufacturing relevant aspects. Beside others, issues like tailored material properties, variation of the micro-structural design of the CMC materials in terms of different fibre architecture and processing of matrix, structural feasibility with respect to manufacturing process as well as material testing in various loading modes (tension, compression, shear, off-axis loading) from room temperature to maximum application temperature are addressed. These studies indicate the high temperature potential of the CMC materials under investigation

Within the HiPOC program EADS Innovation Works intensively focuses on the further development and investigation of a well established oxide based CMC material termed UMOX™ fabricated via standard CMC manufacturing route PIP (Polymer Infiltration Pyrolysis) process as well as on the development of new oxide based CMC material concepts and processes with respect to the fabrication of structural components for e.g. gas turbine engines [1-4].

UMOX™ is the standard oxide based CMC material developed, manufactured and used at EADS Innovation Works and within other companies of the EADS group such as Astrium Space Transportation. It was developed and continuously improved during the last 20 years and already successfully flight tested with a DO228 aircraft jet engine equipped with exhaust components. Beside others the matrix for one type of the UMOX™ material is based on a commercial micron-sized mullite powder and polysiloxane precursor. For instance continuous alumina fibres of type Nextel™610 (3M™, USA) are used as reinforcement fibres. In order to conform to criterion for a weak fibre-matrix interface preventing propagation of cracks through the fibres, a gap between fibre and matrix is generated by using an organic fibre coating which is removed after composite manufacture (fugitive interface). The CMC is manufactured by the PIP process. The coated continuous fibres are infiltrated with liquid pre-ceramic matrix slurry (Liquid Polymer Infiltration, LPI) and wound onto a constantly rotating drum or mandrel (up to 1.5 m in diameter). The unidirectional lay-up of the impregnated fibre bundles in each layer is realized by fully automated 6+2 axis robot controlled filament winding process, as shown in Figure 1, allowing high flexibility in fibre architectures with various fibre orientations and complex geometrical shapes with the feasibility of integral structure design [5-7].

Figure 1 – Liquid Polymer Infiltration (LPI) of continuous fibres via filament winding: pre-impregnated alumina fibres (Nextel™610, 3M™, USA) on drum (left) and robot controlled filament winding facility at EADS Innovation Works, Munich, Germany (right)

After a drying step the impregnated oxide fibres (pre-pregs) are vacuum packed and consolidated in an autoclave process with pressure of p>10 bar and temperature T>150°C. The polymer in the matrix forms a cross-linked network at this temperature to bond the laminate together and lead to a green body that is stable enough for handling and further processing. Following, the pre-ceramic matrix is converted to a ceramic phase by high temperature treatment in an inert atmosphere at temperatures T>1000 °C. The change of matrix density (organic/inorganic) and shrinkage of the pre-ceramic matrix during pyrolysis leaves some porosity in the final ceramic matrix. Re-infiltration of the composite with a pre-ceramic polymer followed by further high temperature treatment reduces the open porosity. The number of re-infiltration cycles depends on the desired material microstructure, porosity and properties [8]. Figure 2 exemplarily shows the microstructure (left) and SEM image of a fracture surface (right) of such an oxide based CMC material UMOX^{TM} with alumina fibre (Nextel^{TM}610, 3M^{TM}) and mullite based matrix manufactured via filament winding and subsequent PIP process.

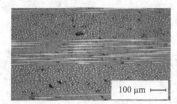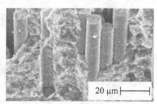

Figure 2 – Microstructure (left) and SEM image of a fracture surface (right) of UMOX^{TM} with alumina fibre (Nextel^{TM}610, 3M^{TM}) and mullite based matrix, fabricated via filament winding & PIP process

The UMOX^{TM} material is characterized by outstanding properties and high reproducibility. Besides the good performance of this oxide based CMC material, the main drawback encountered nowadays is the high production cost. The use of alternative textile technologies such as circular braiding for the fabrication of oxide based CMC can be more time and cost efficient compared to the described conventional PIP process.

Braiding of continuous oxide based ceramic fibers in combination with the manufacture of oxide based CMC is a field in which nearly no developments have been done until today due to the difficulties encountered. For instance ceramic fibers tend to be more brittle than carbon fibers which are commonly braided, leading (consequently) to a high probability of fiber damage and rupture during the braiding process. The braiding technique offers the possibility of manufacturing fiber preforms in 2D or 3D fiber architecture, depending on the braider used. The crossing of the yarns in the preform increase the stability and bending strength. The main advantages and disadvantages of the braiding technique are summarized in Table 1 [9, 10].

Table 1 - Advantages and disadvantages of the braiding technique

Advantages	Disadvantages
Allows fabrication of preform with different fiber architecture, fiber angles, different geometries and fiber types	Limited by no availability of peripheral positions, sliding or folding wear of the fibers is too large in cross-sections
Fast manufacture times	High investing costs
Low production costs	Fiber damage when using ceramic fibers
High preform stability & reduction of delamination	
Excellent reproducibility	

The braiding machine has multiple horn gears containing bobbins with the fiber bundles. These bobbins follow an alternating direction of rotation as shown schematically in Figure 3a. The objective of these horn gears is to transport and transfer the bobbins by a sinusoidal movement. The crossover is done by two opposing, intersecting track patterns on which the individual threads are guided through the process. As seen in Figure 3b showing the braiding machine the two tracks on the lace weaver are marked by coloured lines. The bobbins describe a circular motion through the superposition of a sinusoidal motion of the bobbins [9-12].

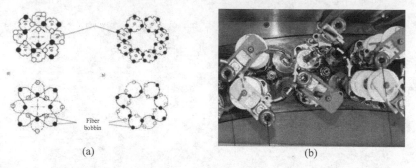

(a)   (b)

Figure 3 – Schematic illustration of rotating bobbins in a braiding machine (a) and bobbins in a braiding machine at EADS Innovation Works, Munich, Germany, showing the circular motion

At EADS Innovation Works a circular braiding machine from Herzog Machine Factory GmbH & Co KG, Germany, with 144 bobbins (as shown in Figure 4) with a fully automated handling robot from KUKA Robots GmbH, Germany, is available. This facility enables the fabrication of textile preforms with different lengths, diameters and also with different fiber architectures by adjusting the braiding parameters.

Figure 4 - Circular braiding machine at EADS Innovation Works, Munich, Germany

EXPERIMENTAL SET-UP

In the first step investigations on the braiding process were performed with different oxide based Nextel™ fiber types from 3M™, USA. These fibers and corresponding characteristics are listed in Table 2.

Table 2 - Properties of Nextel™ fibers from 3M™, USA, used for braiding experiments [13]

Properties		Nextel™440	Nextel™610	Nextel™720
Tensile Strength	[MPa]	2000	3100	2100
E-Modulus	[GPa]	190	380	260
Melting Point	[°C]	1800	2000	1800
Density	[g/cm³]	3,05	3,9	3,4
Deformation	[%]	1,1	0,8	0,8

In a second step tensile tests with fiber bundles after braiding were conducted (according to DIN EN 1007-5) in order to quantify fiber damage during the braiding process. Here, Nextel™610 fibers (alumina) with coating, with and without finishing were tested and, for comparison, the fibers as received from 3M™, were also investigated.

For the fabrication of 2D CMC material sample plates, Nextel™610 fibers were used in combination with a specific oxide matrix system manufactured via the sol-gel technique. The ceramic matrix is derived from a water based sol with dispersed silica nanoparticles, mullite powder and additive agents. After impregnation of the fibers with the pre-ceramic slurry system, the material is frozen for gelation. The nanoparticles form a three-dimensional network due to the sol-gel transition which is forced by the freezing of the solvent. With growth of the ice crystals the solid phase is condensed simultaneously. At moderate temperatures, just above room temperature, the green body is then dried and ice crystals are evaporated or sublimated. The resulting porosity corresponds to the space occupied by the former ice crystals [6].

The CMC material sample plates were braided using the circular braiding facility at EADS Innovation Works (see Figure 4) with a braiding angle of ± 45°. Hence the samples for mechanical tests could be prepared with fiber architecture of 0°/90°. Furthermore, 3 different CMC material sample plates were fabricated and characterized concerning microstructure, tensile strength and bending strength. The mechanical tests were performed according to DIN EN 658. Table 3 gives an overview of the CMC material samples fabricated.

Table 3 - Parameters used for the fabrication of 3 oxide/oxide CMC material sample plates via braiding

Parameter	CMC 1	CMC 2	CMC 3
Fiber type/Coating	Nextel™610/Coating A	Nextel™610/Coating A	Nextel™610/Uncoated
Fiber Denier	A (3000)	A (3000)	B (10000)
Fiber Finishing	A	A + oxidized	A + oxidized
N° of Layers	8	8	3
Technique	Freeze Gelation	Compacting + Freeze Gelation	Compacting + Freeze Gelation

RESULTS AND DISCUSSIONS

Since until today only very few experiments with braiding of oxide ceramic fibers are described in literature a general assessment of the workability of the oxide based Nextel™ fibers via preliminary tests on a single bobbin was performed. The fibers were periodically tensioned through the corresponding thread guiding element at an angle of 90° until rupture of the roving or 100 times in maximum (see Figure 5a). All tests were conducted with and without fiber finishing. The results are presented in Figure 5b, showing that the use of a specific fiber finishing significantly improves the workability of the fibers. The fiber finishing enables the braiding of Nextel[TM]440, 610 and 720 fiber types with coating A and also uncoated. The use of coating B leads to fiber rupture after a few periodically tensioned cycles. Due to this significant fiber damage which would represent a severe drop in mechanical performance of the composite, braiding appears not possible when oxide based fibers without finishing or especially with coating B are used.

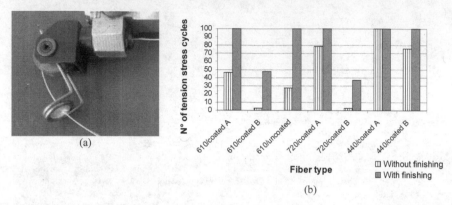

(a)

(b)

Figure 5 - Thread guiding element of braiding machine for tension test of the fibers (a); results of testing different fiber types with and without fiber finishing (b)

In order to evaluate the influence of the braiding process and its parameters on the fiber damage more precisely, Nextel[TM]610/coating A (with and without finishing) fiber bundles were taken directly from the braiding tool after braiding and were mechanically tested (tensile test). Additionally, the Nextel[TM]610 fiber type as received was tested for comparison. Figure 6, 7 and 8 show the load-strain graphs.

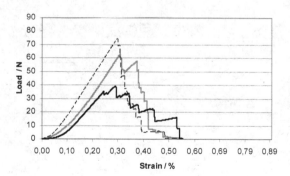

Figure 6 - Load-strain plot of Nextel[TM]610 roving (3M[TM]) as received, after braiding process

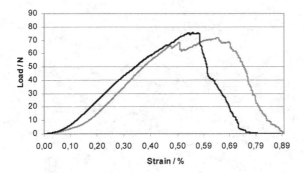

Figure 7 - Load-strain plot of Nextel^{TM}610 roving (3M^{TM}) with coating A and without finishing, after braiding process

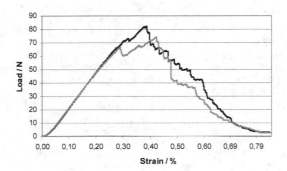

Figure 8 - Load-strain plot of Nextel^{TM}610 roving (3M^{TM}) with coating A and with finishing, after braiding process

Since only a limited number of fibers could be tested and during sample preparation some of the bundles were broken, the results show a relatively large scatter. The roving as received shows an average strength of 58.8 N, the roving without finishing an average strength of 73.8 N and the one with finishing reached an average strength of 77.8 N. At the beginning of the test, not all individual fiber filaments are equally tensioned, which can be seen at the diagrams as a non-linearity. The diagrams of the roving as received (Figure 6) and with coating A (without or with finishing, Figure 7 and 8) show distinct linear regions (constant stiffness), meaning that no filament breakage has occurred. In the roving without finishing the declining force curve of the continuously breaking single filaments is observable (Figure 7).

After the tensile tests the roving as received and especially the roving without finishing the individual filaments do not form a fiber bundle anymore. The fibers with finishing, however, show practically the same appearance as before tensile testing. Figure 9 exemplarily shows each roving type after tensile test.

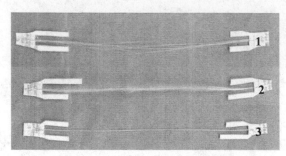

Figure 9 - Nextel™610 roving after tensile test: (1) roving as received; (2) roving with coating A and without finishing; (3) roving with coating A and finishing

Because of the small number of roving samples tested and therefore the large scatter in the test results a comparison of the braided fibers with the fiber as received is difficult, especially due to the unexpected fact that the fiber as received shows a marginal lower strength than the braided fibers. Furthermore the fiber bundle tests bring out that the use of fiber finishing leads to a certain fiber protection which decreases the damage of the fiber filaments. Fibers without finishing contain unprotected filaments. Thus fiber failure occurring during processing is more likely even when the fibers are already coated.

After these investigations, CMC material sample plates were fabricated with Nextel™610 uncoated, coated and with fiber finishing. The preference for the fiber Nextel™610 is based on its properties which are more suitable for potential high temperature application such as future gas turbines in aero-engines. As previously shown in Table 3 the parameters for the fabrication of 3 different CMC material sample plates were varied. Braiding parameters, sol-gel based ceramic slurry system, freeze gelation process for consolidation as well as drying and sintering parameters were kept constant for the three CMC material sample plates. Only the number of fiber filaments, fiber coating and fiber finishing were varied. The fiber preforms were infiltrated with the ceramic slurry, frozen, dried and sintered at 1200°C for 5 hours (at ambient pressure in air). No re-infiltration was conducted. Figure 10a shows the 3 different CMC material sample plates after sintering step and Figure 10b illustrates the fabrication of material sample plate CMC 3 during circular braiding process. After the fabrication process the CMC material samples were characterized concerning their microstructure, tensile strength and bending strength.

In Figure 11 micrographs of the 3 different CMC material modifications are presented. The microstructures are characterized by the individual braided and infiltrated CMC layers. With the micrographs in Figure 12 the homogenous infiltration of the single filaments with the ceramic matrix can be seen in more detail. It can be observed (Figure 11) that CMC 2 shows better interlaminar bonding, whereas the microstructure of the other two composite materials CMC 1 and 3 appear with small gaps (light grey) between the layers. This gap normally appears along one fiber layer but it can also be seen along two layers. This indicates that partially the adjacent layers are not connected with matrix. Homogeneous fiber filament infiltration could be achieved in all three CMC material variations as it can be observed in Figure 12, even for sample CMC 3 where the fiber filaments show a slightly different shape (10000 denier roving). The porosities of the CMC material samples are 62.5 Vol.-% (CMC 1), 32.6 Vol.-% (CMC 2) and 32.0 Vol.-% (CMC 3). The very high porosity of CMC 1 is the result of the missing compacting step after preform impregnation. Furthermore the fiber finishing was not removed (in contrast to CMC 2 and 3).

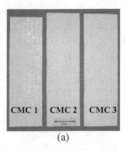

CMC 1   CMC 2   CMC 3

(a)

(b)

Figure 10 – Three different CMC material sample plates fabricated with Nextel^{TM}610 roving via circular braiding process, fiber impregnation, freezing, drying and sintering (a); fabrication of preform CMC 3 with Nextel^{TM}610 roving and coating A with finishing, during circular braiding

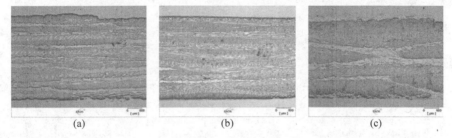

(a)   (b)   (c)

Figure 11 - Microstructure of the material samples CMC 1 (a), CMC 2 (b) and CMC 3 (c)

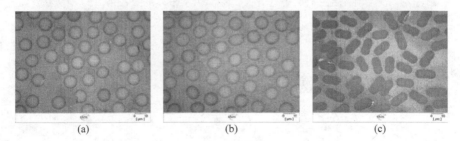

(a)   (b)   (c)

Figure 12 - Microstructure of the material samples CMC 1 (a), CMC 2 (b) and CMC 3 (c)

As a consequence from the high porosity of CMC 1 it was not possible to perform tensile tests since the clamping pressure damaged the samples. The stress-displacement curves from tensile tests for CMC 2 and CMC 3 are presented in Figure 13.

CMC 2 (see Figure 13a) shows very similar results and stress-displacement behaviour for the 3 samples tested indicating homogeneous material. The curves show quasi-elastic behaviour up to displacements of approximately 0.05 mm. All samples fail abrupt after reaching their maximum strength (average strength of $\sigma_{1M} = 35.8$ MPa and standard deviation s = 3.7 MPa). Since all samples failed at the edge of the clamping jaw the test is formally not valid. However this also indicates that the maximum values are below the materials potential.

CMC 3 stress-displacement curves (see Figure 13b) also indicate homogeneous material but with a slightly lower strength of $\sigma_{1M} = 30.6$ MPa in average and a standard deviation of s = 2.0 MPa. Similar to CMC 2 the plot reveal a quasi-elastic behaviour up to approximately 0.05 mm displacement. At higher stresses the behaviour changes towards quasi-plastic characteristics until the maximum stress is reached.

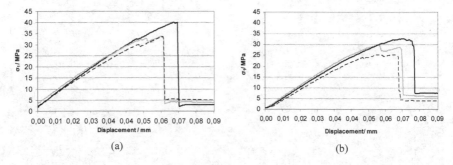

(a)   (b)

Figure 13 - Stress-displacement graphs from tensile tests of CMC 2 (a) and CMC 3 (b)

3-point bending tests were performed for all three CMC materials, the stress-displacement graphs are presented in Figure 14. Due to the high porosity the CMC 1 samples were slightly compressed at the supports leading occasionally to interference as seen in the stress decrease of one sample in Figure 14a. All CMC 1 samples are characterized by brittle fracture behaviour, showing a steep drop in stress. The average bending strength is $\sigma_B = 77.3$ MPa with a standard deviation of s = 8.3 MPa.

The stress-displacement curves of CMC 2 samples indicate an increased damage tolerant behaviour compared to CMC 1. Furthermore in comparison to CMC 1 the stress level is similar. The average bending strength is $\sigma_B = 75.7$ MPa with a standard deviation of s = 9.5 MPa.

The bending strengths of material samples CMC 3 are, in comparison to CMC 1 and CMC 2, significantly lower (values of 40-60 MPa). The average bending strength of CMC 3 is $\sigma_B = 48.7$ MPa with a standard deviation of s = 8.2 MPa.

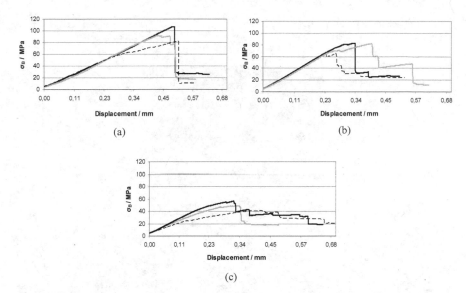

(a)

(b)

(c)

Figure 14 - Stress-displacement graphs from 3-point bending tests for CMC 1 (a), CMC 2 (b) and CMC 3 (c)

In order to demonstrate the feasibility of braiding preforms with Nextel™610 fibers in complex shapes a manufacturing demonstrator was fabricated in a first step. This preform was braided near net-shape according to a preliminary combustor design for gas turbine engines, as shown in Figure 15.

Figure 15 – Braided preform demonstrator with a preliminary combustor design for gas turbine engines, using Nextel™610 fibers

In a second step the same technology demonstrator with a preliminary combustor design for gas turbine engines was fabricated with oxide/oxide CMC, termed OXITEX™, using Nextel™610 fibers. The final technology demonstrator is illustrated in Figure 16.

Figure 16 –OXITEX™ technology demonstrator (oxide/oxide CMC) with a preliminary combustor design for gas turbine engines, fabricated via braiding technique and using Nextel™610 fibers

CONCLUSIONS

The use of Nextel™440, 610 and 720 fiber types for the fabrication of oxide/oxide CMC using the braiding technique was proved to be feasible. In order to improve textile workability of the oxide ceramic fibers and to avoid fiber damage during braiding it appeared to be profitable to apply coating A and fiber finishing. Fiber damage was investigated by fiber bundle tensile test in braided preform samples. Although the amount of tested fiber bundle samples is not enough for representative conclusions, the different behaviour of the roving under tension load with and without finishing could be demonstrated. The use of a specific fiber finishing seems to protect the fiber filaments against the high stresses during the braiding process. In order to achieve more representative results these experiments will be continued.

Three different 2D CMC material set-ups (CMC 1-3) were fabricated. The microstructure of material sample CMC 2 shows a closer bonding between the single layers as a consequence of compacting before the consolidation step by freezing. CMC 1, which was not compacted, hence shows weaker interlayer connection and higher porosity. According to the fiber bundle infiltration in between the single filaments all three CMC materials present a good and homogeneous impregnation, even for CMC 3 containing a fiber type with 10000 denier and different filament shape (bigger diameter and so-called "dog-bones" shape).

Mechanical testing of CMC 1 shows inhomogeneous results. Tensile tests could not be carried out due to the very large material porosity and, consequently, low pressure resistance leading to damage during clamping in the test machine. Results of 3-point-bending test show brittle fracture behaviour but homogeneous stress-displacement curve progression, the average bending strength is $\sigma_B = 77.3$ MPa. Due to the homogeneous microstructure and good fiber impregnation material sample CMC 2 is characterized by consistent mechanical testing results. The stress-displacement curves of tensile tests are similar in the elastic range, cracking does not occur. The reduction of tension stress proceeds stepwise, thus showing a damage-tolerant behaviour. It seems that the additional compacting step and oxidizing of the fiber finishing lead to improved mechanical strength and homogeneity of the composite. CMC 3, on the other hand, shows a microstructure with large voids due to the thicker roving with a higher denier number used. During mechanical testing failure occurs probably caused by defects and large pores in the composite.

Comparing these first results with properties of UMOX^{TM} material, which is fabricated by filament winding and PIP process, it can be concluded that UMOX^{TM} material still presents superior mechanical properties as expected since this material was developed and constantly improved over 15 years leading to high TRL. The two material types are fabricated by different processing routes, contain different matrix systems and therefore show different material characteristics such as porosity, density or mechanical properties. As an example, the tensile strength of UMOX^{TM} material with 0/90° fiber architecture is approximately 180 MPa and of the braided OXITEX^{TM} material roughly 36 MPa. Although currently characterized by lower mechanical properties the present study reveals the feasibility of braiding oxide ceramic fiber preforms with Nextel^{TM}610 fibers and the processing of corresponding oxide/oxide CMC.

Moreover a manufacturing demonstrator with a preliminary combustor design for gas turbine engines was fabricated for both textile preform and OXITEX^{IM} CMC material, again indicating the high potential of this textile technique with superior near net-shaping capabilities.

However braiding of oxide ceramic fibers and the subsequent fabrication of oxide/oxide CMC is in an early stage and thus the promising work on OXITEX^{TM} will be continued.

ACKNOWLEDGMENTS

Part of the presented work was performed within the HiPOC project (**Hi**gh **P**erformance **O**xide **C**eramics) started in February 2009 (contract number 03X3528) within the WING framework. It is funded in part by the German Federal Ministry of Education and Research (BMBF) and administered by Projektträger Jülich (PTJ). Their support is gratefully acknowledged.

REFERENCES

[1] T. Machry, D. Koch and C. Wilhelmi, Development of a new Oxide Ceramic Matrix Composite, *Proceedings of the 7th International Conference of High-Temperature Ceramic Matrix Composites*, Bayreuth, Germany, 435-445 (2010).

[2] T. Machry, C. Wilhelmi and D. Koch, Novel High Temperature Wound Oxide Ceramic Matrix Composites Manufactured via Freeze Gelation, in press: *Proceedings of the 34th International Conference on Advanced Ceramics and Composites (ICACC)*, Florida, USA (2011).

[3] R. Knoche, E. Werth, M. Weth, J. Gómez García, C. Wilhelmi and M. Gerendás, Design and Development Approach for Gas Turbine Combustion Chambers made of Oxide Ceramic Matrix Composites, in press: *Proceedings of the 34th International Conference on Advanced Ceramics and Composites (ICACC)*, Florida, USA (2011).

[4] M. Gerendás, Y. Cadoret, C. Wilhelmi, T. Machry, R. Knoche, T. Behrendt, T. Aumeier, S. Denis, J. Göring, D. Koch, and K. Tushtev, Improvements of Oxide/Oxide CMC and Development of Combustor and Turbine Components in the HiPOC Program, in press: *Proceedings of the ASME Turbo Expo 2011: Power for Land, Sea Air*, GT2011-45460, Canada (2011).

[5] B. Clauß, B., Fibres for Ceramic Matrix Composites, In: Krenkel, W. *Ceramic Matrix Composites*, Germany: Wiley-VCH, 1-20 (2008).

[6] B. Newman, B. and W. Schäfer, Processing and Properties of Oxide/Oxide Composites for Industrial Applications, In: Krenkel, W.; Naslain, R.; Schneider, H. *High Temperature Ceramic Matrix Composites* Germany: Wiley-VCH, 600-609 (2001)

[7] G. Motz, S. Schmidt and S. Beyer, The PIP-process: Precursor Properties and Applications, In: Krenkel, W. *Ceramic Matrix Composites*, Germany: Wiley-VCH, 165-186 (2008)

[8] W.D. Vogel and U. Spelz, Cost effective production techniques for continuous fibre reinforced ceramic matrix composites, In: *Ceramic Processing Science and Technology*, 51, 225-259 (1995)

[9] M. Stobbe (Supervisor C. Wilhelmi and T. Machry), Entwicklung eines Verfahrens zur Herstellung von oxidkeramischen Verbundwerkstoffen (CMC) über die Flechttechnik, Diploma Thesis, Technical University Darmstadt / EADS Innovation Works, Munich, Germany (2010)

[10] R. Groß (Supervisor C. Wilhelmi), Untersuchungen zu Optimierungsmöglichkeiten bei der textiltechnischen Verarbeitung von $SiBN_3C$-Rovings, Diploma Thesis, Technical University Munich / EADS Innovation Works, Munich, Germany (2008)

[11] M. D. Mello and R.A. Florentine, 3-D Braided Continuous Fiber Ceramic Matrix Composites Produced by Chemical Vapor Infiltration, SBIR-Report, US Army Research Laboratory (1993)

[12] 3TEX Inc., 3Braid™ Product Data sheet (2005)

[13] 3M Nextel Keramische Textilien und Werkstoffe, Nextel-Brochure, 3M Deutschland GmbH (2009)

Environmental Effects of Ceramics and Composites

RELATIONSHIPS BETWEEN FIBER STRENGTH , PASSIVE OXIDATION AND SCALE CRYSTALLIZATION KINETICS OF HI-NICALON[TM]-S SIC FIBERS

R. S. Hay, G. E. Fair
Air Force Research Laboratory
Materials and Manufacturing Directorate, WPAFB, OH

R. Bouffioux
New Mexico Tech. U., Socorro, NM

E. Urban
Appalachian State University, Boone, NC

J. Morrow, A. Hart
U. Cincinnati, Cincinnati, OH

M. Wilson
Ohio State University, Columbus, OH

ABSTRACT
    The strengths of Hi-Nicalon[TM]-S SiC fibers were measured after oxidation in dry air between 700° and 1400°C. The oxidation and scale crystallization kinetics were also measured. Scale thickness was measured by TEM so that amorphous scale could be clearly distinguished from crystalline scale. Oxidation initially produces an amorphous scale that has significant carbon. These scales begin to crystallize to cristobalite and tridymite in 100 hours at 1000°C or in one hour at 1300°C. Crystallization nucleates at the scale surface with rapid growth parallel to the surface relative to through thickness. The activation energy for parabolic oxidation for uncrystallized SiO_2 scale was 248 kJ/mol. The fiber strength increased by approximately 10% for SiO_2 scale thickness up to ~100 nm, and decreased for thicker scales. The onset of strength degradation was strongly correlated with crystallization of the scale. Possible mechanisms for strength increases and decreases with scale thickness are discussed.

INTRODUCTION
    SiC fiber strength is affected by oxidation. Fiber strength defines the ultimate attainable CMC strength.[1] Complications include fiber impurities, particularly alkali and alkali earths, that increase oxidation rates, reduce scale viscosity, and lower temperatures for scale crystallization.[2-3] Moisture has similar effects.[4-9] The general consensus is that oxidation of SiC fibers reduces fiber strength.[10-17] However, recent work has shown that thin silica scales (< 100 nm) actually increase SiC fiber strength,[18-19] as might be expected from residual compressive stress in the scales and surface flaw healing. Ambiguous effects of SiC oxidation on strength have also been observed for bulk material.[20-21]
    Preliminary data and analysis for the strength of oxidized Hi-Nicalon[TM]-S SiC (β-SiC) fiber as a function of oxidation product thickness are presented. Oxidation and scale crystallization kinetics are also presented and relationships with fiber strength are discussed. This paper complements a more thorough description and analysis of fiber oxidation and scale crystallization kinetics.[22] Hi-Nicalon[TM]-S fiber was chosen because it has near-stoichiometric SiC composition (~1 at% oxygen and ~2 at% carbon),[23] and the smoothest surface of currently available SiC fibers.[24] The properties of Hi-Nicalon[TM]-S fiber are described in several publications.[11,23,25-29] It is hoped that the data will be useful in designing processing methods and developing life prediction methods for SiC-SiC CMCs.

EXPERIMENTS
    Hi-Nicalon[TM]-S fibers have a PVA (polyvinyl alcohol) sizing. If this sizing was burned off at 800 - 1000°C, the thickness of SiO_2 scales formed during subsequent oxidation was non-uniform (Fig. 1). Contamination of the scale by inorganic decomposition products from the sizing were the suspected cause. Consequently, the fiber sizing was removed by dissolution from 50 inches of Hi-Nicalon[TM]-S

Fig. 1 *Optical micrographs (reflected light) of non-uniform oxidation after 1 hour at 1000°C after burning off the sizing, rather than dissolving it off in hot distilled water.*

fiber tow in 175 cc of boiling distilled deionized water in a Pyrex glass beaker for one hour. This process was repeated with fresh distilled deionized water. Fibers were then dried in a drying oven for 20 minutes at 120°C. The water was collected for inductively coupled plasma (ICP) chemical analysis of sizing impurities, and compared to control experiments done without fibers. The desized SiC fibers were also analyzed for impurities by ICP analysis. The main sizing contaminants were Ca and Na. There was enough Ca in the sizing to yield ~1 monolayer of CaO on the fiber surface after sizing burn-off. The main fiber contaminants were Cl (620 ppm), S (52 ppm), Ca (45 ppm), Na (35 ppm), and Fe (27 ppm).

The desized fiber tows were oxidized in flowing dry air (< 10 ppm H_2O) using an alumina muffle tube furnace and an alumina boat dedicated to these experiments. Both the muffle tube and boat were baked-out at 1540°C for 4 hours in laboratory air prior to use. These bake-outs have been shown to be necessary to prevent contamination by alumina impurities.[30] Fiber oxidation was done at 700 to 1400°C for times up to 100 hours. A total of 46 different heat-treatments were done (Table I). Heat-up and cool-down rates of 10°C/minute were used.

The strengths of the oxidized fibers were measured by tensile testing of at least 30 filaments using published methods.[31] The average and Weibull characteristic value for failure stress were calculated, along with the Weibull modulus. Strengths were calculated using both the original (r_i) and final SiC radius after oxidation (r_f) (Fig. 2). The average fiber diameter measured by optical microscopy and SEM was 12.1 μm.

The uniformity SiO_2 scale uniformity was characterized using reflected light interference fringes. Surface morphology and cracking of SiO_2 were characterized by SEM. Cross-section TEM specimens were prepared from the oxidized fibers by published methods.[32-33] TEM sections were ion-milled at 5 kV and examined using a 200 kV Phillips LaB_6-filament TEM and a 300 kV FEI Titan TEM. SiO_2 oxidation product thickness and cracking, crystallization, and microstructures of the SiO_2 scale were characterized. SiO_2 thickness and the relative abundance of crystallized and amorphous SiO_2 were measured for a minimum of five filaments, and in many cases more than twenty. Where possible, the crystallized SiO_2 phase (tridymite or cristobalite) was identified from selected area electron diffraction patterns. Oxide scale composition, with particular attention to carbon, was characterized by analytical TEM with EDS.

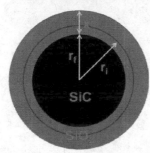

Fig. 2. *Schematic diagram showing initial (r_i) and final (r_f) SiC thickness after oxidation and SiO_2 thickness x.*

RESULTS AND DISCUSSION
Fiber Strength

The relationship between oxide scale thickness and Weibull characteristic strength of the fibers is shown in figure 3. As-received fibers had average strengths of 2.85 GPa and Weibull characteristic strengths of 3.0 GPa. Strengths after oxidation at various times at 700 - 1300°C are plotted. Crystallized and partially crystallized scales are denoted by open circles. The strengths are plotted for the final SiC fiber radius (r_f) after oxidation, which can be calculated from the original fiber radius (r_i) and the oxide scale thickness (x) from:

$$r_f = [x^2 (\Omega_{SiC}^2/\Omega_{SiO2}^2 - \Omega_{SiC}/\Omega_{SiO2}) + r_i^2]^{1/2} - (\Omega_{SiC}/\Omega_{SiO2})x \qquad [1]$$

where Ω_{SiC} and Ω_{SiO2} are the molar volumes for SiC and SiO_2, respectively.

Fiber strength increased about 10% with a weakly defined maximum near SiO_2 thicknesses of

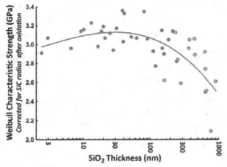

Fig. 3 *Weibull characteristic strength vs. oxide scale thickness. Open circles are scales that are partly or fully crystallized.*

Fig. 4. *TEM image of a partially crystallized scale formed after oxidation for 10 h at 1200°C. A dot map of the carbon/oxygen ratio (white is high ratio) is shown above the image. Carbon is depleted in the crystallized scale.*

products.[4,9,37-40]

SiO$_2$ crystallized to cristobalite (minor phase) and various tridymite (major phase) polymorphs (Fig. 5) at 1000 and 1400°C, but at 1200°C the phase abundance was roughly evenly split (Table I). Most SiO$_2$ crystallization studies find that cristobalite predominates at higher temperatures.[41-42] Both tridymite and cristobalite were extensively twinned; cristobalite twinning on (101) was particularly common (Fig. 5). Uncracked crystalline silica scales were always thinner than amorphous scales under the same conditions, as observed previously (Fig. 6).[43] Thicknesses of crystalline scales varied more than those of amorphous scales (Table I). Presumably the crystalline scale variation was caused by the

about 50 nm, and decreased for thicker scales. A relationship between oxidation temperature and fiber strength was not observed; a 100 nm thick scale that formed in a short time at high temperature had about the same effect on fiber strength as a 100 nm scale that formed in a long time at low temperature. However, fiber strength decreases were correlated with scale crystallization (Fig. 3). Both scale crystallization and growth kinetics are related to strength changes after oxidation.

Crystallization

Crystallization always nucleated at the scale surface. Growth was more rapid parallel to the surface than through the scale thickness (Fig. 4). Compositions of adjacent crystalline and amorphous areas measured using analytical TEM show small and varying carbon concentrations in the amorphous scale, but little or no carbon in the adjacent crystalline scale (Fig. 4). Measured carbon concentrations in amorphous scales varied from sample to sample. Problems with quantification of the "soft" carbon x-rays by EDS are well known,[34-36] and difficult to ameliorate. Qualitatively, there is less carbon in crystallized silica scales than in adjacent amorphous scale, but the data is not of sufficient quality to be quantified. Rejection of carbon during crystallization is consistent with surface nucleation and more rapid growth parallel to the surface. There are many other reports of carbon in SiC oxidation

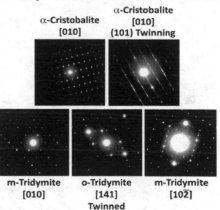

Fig. 5 *Selected area electron diffraction patterns of crystalline SiO$_2$ in Hi-Nicalon^{TM}-S oxidation scales.*

Table I. *Hi-Nicalon*^{TM}*-S oxidation experiments. f is the fraction of scale that is crystallized*

#	# of Obs.	T (°C)	t (hrs)	x (nm) amorphous	x (nm) crystalline	f	Comments
1	24	700	100	9.3 ± 3.3	—	0	
2	9	800	0	2.4 ± 0.3	—	0	
3	7	800	1	6.2 ± 1.1	—	0	
4	29	800	10	18.8 ± 2.4	—	0	
5	22	800	100	91.6 ± 5.5	—	0	
6	7	900	0	3.0 ± 0.6	—	0	
7	8	900	2	16.8 ± 2.0	—	0	
8	18	900	4	31.7 ± 2.7	—	0	
9	48	900	16	81.9 ± 5.9	—	0	
10	43	900	49	150 ± 10	—	0	
11	28	900	100	218 ± 6.	—	0	
12	8	1000	0	13.3 ± 2.5	—	0	
13	18	1000	0.067	10.3 ± 1.9	—	0	
14	28	1000	0.15	20.1 ± 3.5	—	0	
15	30	1000	0.25	21.3 ± 2.9	—	0	
16	30	1000	0.39	25.2 ± 3.4	—	0	
17	6	1000	0.56	38.8 ± 5.0	—	0	
18	16	1000	0.67	40.0 ± 3.2	—	0	
19	11	1000	1	52.7 ± 3.6	—	0	
20	37	1000	1	63.4 ± 11.5	—	0	Not Desized
21	55	1000	1.56	41.3 ± 3.5	—	0	
22	10	1000	4	153 ± 6	—	0	
23	15	1000	16	249 ± 9	—	0	
24	27	1000	100	703 ± 41	538. ± 39	0.26	3 Tridymite, 1 Cristobalite
25	13	1050	100	1009 ± 7	792 ± 70	0.77	6 Tridymite, 0 Cristobalite
26	25	1100	0	28.3 ± 3.7	—	0	
27	32	1100	1	122 ± 8	—	0	
28	49	1100	4	258 ± 15		0.07	
29	29	1100	10	402 ± 14	368 ± 11	0.53	2 Tridymite, 1 Cristobalite
30	14	1100	30	609 ± 16	510 ± 85	1	3 Tridymite, 1 Cristobalite
31	16	1100	81	—.	1034 ± 170	1	
32	20	1100	100	—	933 ± 104.	1	9 Tridymite, 4 Cristobalite
33	21	1200	0	112 ± 4	—	0	
34	24	1200	0.33	190 ± 3	—	0	
35	19	1200	1	239 ± 12	225 ± 12	0.05	1 Tridymite
36	34	1200	3	356 ± 29	381 ± 58	0.77	1 Tridymite, 2 Cristobalite
37	26	1200	10	535 ± 30	454 ± 74	0.89	4 Tridymite, 7 Cristobalite
38	34	1200	30	—	687 ± 63	1	11 Tridymite, 9 Cristobalite
39	5	1300	0	171 ± 12	—	0	
40	41	1300	0.33	—	154 ± 17	0.42	
41	21	1300	1	472±20	466 ± 50	0.96	5 Tridymite, 1 Cristobalite
42	35	1300	3	—	633 ± 48	1	1 Tridymite, 3 Cristobalite
43	15	1400	0	186±18	143±38	0.18	
44	6	1400	0.33	—	507 ± 47	1	4 Tridymite
45	2	1400	1	—	—	1	
46	8	1400	5	—	2380 ± 390	1	8 Tridymite

different times a particular scale section spent in the amorphous and crystalline states.

Crystallized silica was often cracked (Fig. 7). Through thickness cracks were often along twin boundaries. Debond cracks between SiO$_2$ scale and SiC were also common (Fig. 7c). Cracked crystalline scales were previously observed for Hi-Nicalon^{TM} and Hi-Nicalon^{TM}-S, and were suggested to cause lower fiber strengths.[10,44] Unlike amorphous SiO$_2$, cristobalite and tridymite have much larger coefficients of thermal expansion than SiC. Cracks are assumed to form from thermal stress during cooling, and from volume contraction during the $\alpha \rightarrow \beta$ cristobalite phase transformation.

Thicker crystalline scales often had growth cracks that were parallel with the fiber axis. These had wide apertures and erratic morphology; they clearly formed during oxidation (Fig. 7a). They are assumed to be caused by hoop tensile growth stresses that develop during oxidation in cylindrical

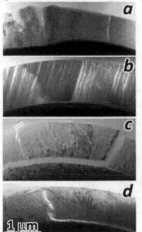

Fig. 6 *TEM images of scale formed in 100h at 1050°C. Scale is amorphous in (a) and crystalline in (b). Dislocations in SiO₂ near SiC in (b) are from plastic deformation driven by growth stresses.*

geometry caused by the large volume expansion for SiC oxidation.[45-50] There were dimples in SiC under the growth cracks, and crystalline scale was thicker in areas adjacent to these cracks. When these cracks form the scale is no longer passivating they are a short-circuit path for O_2 ingress.[48]

Kolmogorov-Johnson-Mehl-Avrami (KJMA) kinetics were used to analyze silica crystallization kinetics:[51-52]

$$f = 1 - \exp[-Kt^n] \qquad [2]$$
$$K = K_o\exp[-Q/RT] \qquad [3]$$

where f is the fraction crystallized (Table I), K is a rate constant, t is time, n is a the time growth exponent, K_o is a pre-exponential factor and Q an activation energy for growth, and RT has the usual meaning in an Arrhenius expression. The analysis gives a parameter best fit for t, T, and f of:

$$Q = 514 \text{ kJ/mol} \qquad [4a]$$
$$K_o = 2 \times 10^{12} \qquad [4b]$$
$$n = 1.5 \qquad [4c]$$

Best fits of the crystallization fraction data to the KJMA model [2-3] converged strongly on a growth exponent n of 1.5, which is diagnostic of three-dimensional growth from site-saturated nucleation.[52] A more thorough analysis and discussion of these results is given elsewhere.[22]

Oxidation Kinetics
Amorphous Scale

Oxidized fibers with amorphous scales usually had uniform interference color in reflected light microscopy, indicating uniform scale thickness. However, for thick scales, filaments facing the alumina boat during oxidation sometimes had a slightly different interference color than those facing away from the alumina boat at the top of a fiber tow. This suggests that there was a slight pO_2 gradient through the tow during oxidation, and/or slight contamination from impurities in the alumina boat. There were also always a few "rogue" filaments in each tow bundle that had much thicker scales, and were in some cases prematurely crystallized. These were ignored for measurement, but suggest that there is some non-uniformity in filament composition and impurity distribution.

SiO_2 scales formed after oxidation at low temperatures and short times were amorphous. Thinner scales (< 50 nm) were more variable in thickness, in some cases varying with the SiC surface orientation. One sample (1h, 1000°C) was oxidized after the sizing was burned off, rather than dissolved off. This sample had a slightly thicker scale, with much greater thickness variation, than a comparable sample with sizing dissolved off in hot distilled water (Table I).

An impurity phase that oxidized at a slower rate than SiC is visible in a scale formed in 100 h at 800°C (Fig. 8). This was uncommon, and unfortunately this particular feature could not be relocated and similar features could not be found to measure

Fig. 7 *TEM micrographs of cracks from thermal stress (b-d) and growth stress (a) after oxidation for 100 h at 1100°C.*

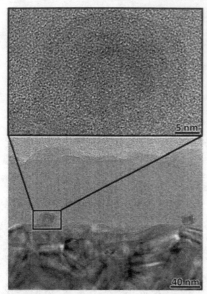

Fig. 8 *TEM images of SiC oxidation scale after 100 h at 800°C, showing a second-phase at the SiC-SiO₂ interface that oxidized at a different rate than SiC.*

composition by EDS. No cracks were observed in amorphous scales, even in scales a micron thick (100 h at 1050°C).

Models

Passive oxidation kinetics for SiC are usually described as parabolic (diffusion controlled) for thick SiO₂ scales and linear (interface controlled) for thin scales. They usually obey Deal-Grove kinetics developed for silicon oxidation, with diffusion of molecular O_2 through SiO_2 as the rate limiting step in the parabolic regime.[9,53]

The thickness of the SiO_2 scale ($\mathbf{x}$) described by Deal-Grove kinetics for the flat plate geometry is:

$$d\mathbf{x}/dt = \mathbf{B}/(\mathbf{A} + 2\mathbf{x}) \qquad [5]$$
$$\mathbf{A} = \mathbf{A_o} \exp[-\mathbf{Q_a}/\mathbf{RT}] \qquad [6]$$
$$\mathbf{B} = \mathbf{B_o} \exp[-\mathbf{Q_b}/\mathbf{RT}] \qquad [7]$$

where $\mathbf{A_o}$ and $\mathbf{B_o}$ are constants and $\mathbf{Q_a}$ and $\mathbf{Q_b}$ are activation energies. Deal-Grove kinetics was reformulated for SiC; the mathematical description is the same, but the definition of $\mathbf{A}$ and $\mathbf{B}$ are different and incorporate CO diffusion coefficients and concentrations.[54] $\mathbf{B}$ is the parabolic rate constant and $\mathbf{B}/\mathbf{A}$ is the linear rate constant. For an initial SiO_2 thickness of $\mathbf{x_i}$, the solution to [5] is:

$$\mathbf{x} = \tfrac{1}{2}\mathbf{A}\{[1 + (t + \tau)/(\mathbf{A}^2/4\mathbf{B})]^{1/2} - 1\} \qquad [8]$$
$$\tau = (\mathbf{x_i}^2 + \mathbf{A}\mathbf{x_i})/\mathbf{B} \qquad [9]$$

where τ is a time shift that corrects for the presence of an initial oxide layer. For long times, [8] becomes the simple expression for parabolic oxidation kinetics:

$$\mathbf{x}^2 = \mathbf{B}\,t \qquad [10]$$

The initial SiO_2 thickness $\mathbf{x_i}$ is often formed during heat-up to the heat-treatment temperature. For a known heat-up rate $\mathbf{T(t)}$, $\mathbf{x_i}$ can be found from equations [5-9] by solution of the non-linear differential equation:

$$d\mathbf{x[t]}/dt = \mathbf{B_o}\,e^{-\mathbf{Qb}/\mathbf{RT[t]}}/(\mathbf{A_o}\,e^{-\mathbf{Qa}/\mathbf{RT[t]}} + 2\mathbf{x[t]}),\ \mathbf{x[t=0]} = 0 \qquad [11]$$

An analytical solution for $\mathbf{x[t]}$ cannot be found for a constant heat-up rate $\mathbf{T(t)}$. However, [11] can be solved numerically by use of a program like *Mathematica*[TM]. For a constant heat-up and cool-down rate, without any hold time, the ratios for SiO_2 thicknesses formed on heat-up to those formed during cool-down vary from 1 : 1 for the linear regime to 1 : ($\sqrt{2} - 1$) for the parabolic regime.

Deal-Grove kinetics were reformulated for the cylindrical geometry of Nicalon[TM] fibers.[55] Unfortunately this treatment assumed no volume change during oxidation; this is clearly not correct for near-stoichiometric SiC fibers such as Hi-Nicalon[TM]-S. The much higher molar volume of amorphous silica ($\Omega_{SiO_2} = 27.34\ \text{cm}^3$) causes the oxidized fiber to expand much more than the remaining SiC ($\Omega_{SiC} = 12.46\ \text{cm}^3$ for β-SiC) contracts. However, Deal-Grove kinetics have been solved for cylinders with oxidation volume expansion using silicon as an example (Fig. 2).[46,56]

$$d\mathbf{r_f}/dt = \mathbf{B}/\{(1+\sigma)[\mathbf{A} + \mathbf{r_f}\,\text{Ln}(-\sigma + \mathbf{r_i}^2/\mathbf{r_f}^2 + \mathbf{r_i}^2\,\sigma/\mathbf{r_f}^2)]\} \qquad [12]$$
$$\sigma = \Omega_{SiO2}(1 - \Omega_{SiC}/\Omega_{SiO2})/\Omega_{SiC} \qquad [13]$$

$$r_f = [x^2 (\Omega_{SiC}^2/\Omega_{SiO2}^2 - \Omega_{SiC}/\Omega_{SiO2}) + r_i^2]^{1/2} - (\Omega_{SiC}/\Omega_{SiO2})x \qquad [14]$$

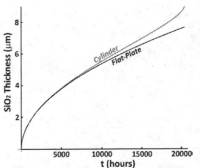

Fig. 9. *Deal-Grove oxidation kinetics for a cylinder and flat-plate geometry. Plot uses parameters for Hi-Nicalon[TM]-S: 6.1 μm radius fiber, and Deal-Grove kinetic parameters calculated for 1000°C*

where Ω_{SiC} and Ω_{SiO2} are the molar volumes of SiC and SiO_2, respectively. A complex analytical solution was reported for [12-14], but they can also be numerically solved. Numerical analysis confirms that fiber oxidation kinetics do not deviate significantly from those for flat plate geometry until the oxidation products for 12 μm diameter fibers are several microns thick (Fig. 9), as found previously for Nicalon[TM] fiber[55] and silicon cylinders.[46,56] The SiO_2 scales generally crystallized when they were greater than 1 μm thick, so the flat-plate geometry is adequate for analysis of data in Table I.

Analysis

The scale thickness – time – temperature data was first analyzed assuming parabolic kinetics [7,10] for amorphous scales. Crystalline scales were not analyzed because there was no way of knowing how long they had been growing in a crystalline state and in an amorphous state. Two fitting methods were used. The usual method of finding **B** for each temperature from **x** and **t** (Table I) using [8] with linear regression to fit B_o and Q_b to a ln(**B**) versus 1/**T** plot was employed. **B** values

calculated in this manner are plotted in figure 10, along with other SiC oxidation data for comparison. The B_o and Q_b parameters were also directly fit to [7] and [10] for all data without calculation of separate **B** values for each temperature. At higher temperatures **B** for Hi-Nicalon[TM]-S tended to be higher than that reported in other studies. Reasons for this are discussed extensively in another publication,[22] but clearly one major source of discrepancy is mixing data for crystallized and amorphous SiO_2 – oxidation rates are much slower if the scale is crystallized.[43]

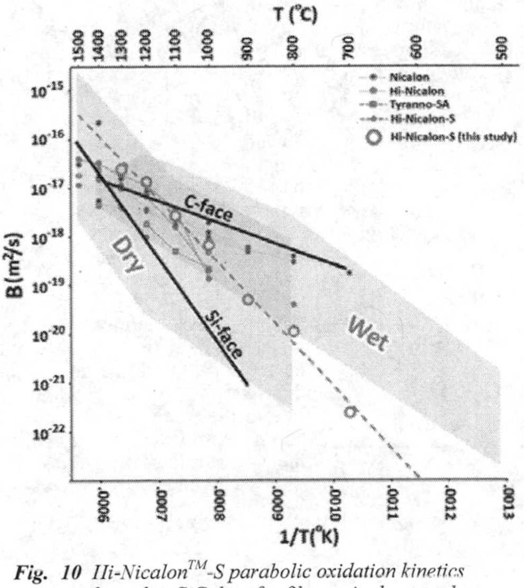

The scale thickness – time – temperature data was analyzed using Deal-Grove kinetics [5-9] for amorphous scales only, using methods described elsewhere.[22] The best fit for "Deal-Grove" parameters for oxidation kinetics for Hi-Nicalon[TM]-S fiber are:

$$A_o = 4.7 \times 10^{-4} \text{ m} \qquad [15a]$$
$$Q_a = 110 \text{ kJ/mol} \qquad [15b]$$
$$B_o = 1.2 \times 10^{-8} \text{ m}^2/\text{s} \qquad [15c]$$
$$Q_b \text{ (parabolic)} - 248 \text{ kJ/mol} \qquad [15d]$$
$$Q_{b/a} \text{ (linear)} = 138 \text{ kJ/mol} \qquad [15e]$$

Oxidation times were corrected by τ [9] for scale formation that occurred during heat-

Fig. 10 *Hi-Nicalon[TM]-S parabolic oxidation kinetics compared to other SiC data for fibers, single crystals, polycrystalline, and powder. Data sources: Wet[57-58]; Single Crystal[37-39,43,54,59-72]; Polycrystalline[60,73-80]; Powder, whiskers[81-83]; CVD[37,77,84-89]; Nicalon[TM], other SiC fibers[10,89-93]; Hi-Nicalon[TM 10,14,91,94-96]; Tyranno[TM]-SA[19]; Hi-Nicalon[TM]-S[10,91].*

up using solution to numerical solutions to [11].

The fit had strong convergence to the Q_b value, but was less sensitive to the other parameters, particularly Q_a and A_o.

Literature values for activation energies for parabolic oxidation kinetics (Q_b) vary widely. Reviews of oxidation data find that Q_b can vary from 80 to 612 kJ/mol,[9] from 134 to 498 kJ/mol,[97] or from 69 to 498 kJ/mol,[94] depending on the material studied and the methods used. Q_b can vary from ~100 to ~270 kJ/mol°K for C-face (fast oxidation) and Si-face (slow oxidation) surfaces, respectively, in dry air.[9] The accepted values for B_o and Q_b for O_2 diffusion in amorphous silica are 2.7×10^{-8} m^2/s and 113 kJ/mol.[98] The reported values for various SiC fibers are in Table II. For fibers, our Q_b of 248 kJ/mol is the highest reported, although it is very similar to one reported for Hi-Nicalon[TM,44] and 30% larger than one reported for Tyranno[TM] SA.[19] For most other fiber data (Table II), crystallization is

Table II. *Activation energies (Q_b) for the parabolic oxidation rate constant reported for SiC fibers.*

SiC Fiber Type	Q_b (kJ/mol)	T range	Comments	Reference
Hi-Nicalon[TM]-S	248	700-1300	For amorphous scales	this study
Hi-Nicalon[TM]-S	79	1000-1500	Crystallized at high T	11
Tyranno[TM]-SA	189	1000-1200		19
PCS derived	69 - 77	800-1200		95
PCS derived –C-rich	96	1000-1400	Crystallized at high T	95
Nicalon[TM]	102	800-1500	Crystallized at high T	94,101
Nicalon[TM]	137	1100-1500	Crystallized at high T	44
Hi-Nicalon[TM]	109	1200-1500	Crystallized at high T	94
Hi-Nicalon[TM] SiC$_{1.4}$O$_{0.05}$	234	1100-1500	Crystallized at high T	44,94
Hi-Nicalon[TM]	107	1000-1400	Crystallized at high T	48,94

reported at high temperatures and is probably a significant source of the discrepancy. Some data was measured by weight gain for carbon rich fibers, another possible source of discrepancy.

Literature values for activation energies for linear oxidation kinetics ($Q_{b/a}$) also vary widely. Almost all reported linear oxidation data for SiC is for single crystal polymorphs with specific surface orientation. Some idea of how the $Q_{b/a}$ and Q_b values calculated for Hi-Nicalon[TM]-S compare with those measured for single crystal SiC can be gained from figure 11.

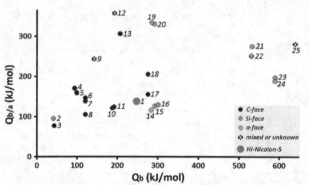

As noted, the C-face for SiC has low Q_b, and Si-face is high. Hi-Nicalon[TM]-S Q_b is close to and roughly intermediate in value and between at least some values reported for 4H-C[54,72] and 4H-a,[54] 6H-Si,[99] and 4H-Si[54] SiC surfaces. Corresponding $Q_{b/a}$ values are very similar.[54,72,99] This is qualitatively consistent with expectations; Hi-Nicalon[TM]-S surface β-SiC grain orientations should be random in distribution and therefore intermediate in oxidation kinetic parameters. Possible reasons for the wide range of values are discussed elsewhere,[22] and in several review papers.[9,100]

Fig. 11 *SiC oxidation studies where both activation energies for linear ($Q_{b/a}$) and parabolic (Q_b) oxidation kinetics are reported. Sources: 1. This Study 2. 3C-Si[57,102] 3. 4H & 6H-C[57,102] 4. β-C (CVD)[37] 5. 6H-C[37] 6. 6H-C[54,99] 7. 3C-C[103] 8. 6H-C[37,103] 9. 3C[60,104] 10. 4H-C[72] 11. 4H-C[54,74] 12. 6H-C & 6H-Si[66] 13. 3C[61,76] 14. 4H-a[54,60] 15. 6H-Si[99] 16. 4H-Si[54] 17. 6H-C[103] 18. 3C-C[103] 19. β-Si (CVD)[37] 20. 6H-Si[37] 21. 6H-Si[87,100] 22. 3C[37,104] 23. 3C-Si[105] 24. 6H-Si[103] 25. 3C[104]*

SUMMARY AND CONCLUSIONS
The strength of Hi-Nicalon™-S fiber increased slightly after oxidation if the SiO_2 scales were not crystallized. Fiber strengths decreased when the scale crystallizes, which typically began at scale thicknesses greater than 100 nm. Thicker crystallized scales correlated with greater fiber strength degradation. Scale crystallization was described by KJMA kinetics with an activation energy of 514 kJ/mol. Fiber oxidation for amorphous SiO_2 scale obeys Deal-Grove kinetics with an activation energy (Q_b) for the parabolic rate constant B of 248 kJ/mol. Amorphous scales were uncracked. Crystalline scales clearly formed at lower rates, but these scales were cracked from thermal stress, polymorphic phase transformations, and tensile hoop growth stress. These scale cracks are suggested to be responsible for lower fiber strength. At higher temperatures the measured oxidation rates were often higher than those reported for other SiC fibers and other forms of SiC; in many cases it was clear that lower rates could be attributed to crystallized scale. Future work will try to establish quantitative relationships between fiber strength, oxidation kinetics, crystallization kinetics, and scale residual stress. Quantitative measurements of C in the oxide scales, and the influence of C on O_2 permeability and crystallization of SiO_2 will also be required. The influence of impurities on oxidation and crystallization kinetics, particularly those like boron common in SiC-SiC fiber coatings and those like H_2O found in application environments will also require assessment.

REFERENCES
1   Curtin, W. A., Ahn, B. K. & Takeda, N. Modeling Brittle and Tough Stress-Strain Behavior in Unidirectional Ceramic Matrix Composites. *Acta mater.* **46**, 3409-3420 (1998).
2   Doremus, R. H. Viscosity of Silica. *J. Appl. Phys.* **92**, 7619-7629 (2002).
3   Pezzotti, G. & Painter, G. S. Mechanisms of Dopant-Induced Changes in Intergranular SiO_2 Viscosity in Polycrystalline Silicon Nitride. *J. Am. Ceram. Soc.* **85**, 91-96 (2002).
4   Akashi, T., Kasajima, M., Kiyono, H. & Shimada, S. SIMS Study of SiC Single Crystal Oxidized in Atmosphere Containg Isotopic Water Vapor. *J. Ceram. Soc. Japan* **116**, 960-964 (2008).
5   Opila, E. J. & Hann, R. E. Paralinear Oxidation of CVD SiC in Water Vapor. *J. Am. Ceram. Soc.* **80**, 197-205 (1997).
6   Opila, E. J. Oxidation Kinetics of Chemically Vapor-Deposited Silicon Carbide in Wet Oxygen. *J. Am. Ceram. Soc.* **77**, 730-736 (1994).
7   Narushima, T., Goto, T. & Hirai, T. High-Temperature Passive Oxidation of Chemically Vapor deposited Silicon Carbide. *J. Am. Ceram. Soc.* **72**, 1386-1390 (1989).
8   Maeda, M., Nakamura, K. & Ohkubo, T. Oxidation of Silicon Carbide in a Wet Atmosphere. *J. Mater. Sci.* **23**, 3933-3938 (1988).
9   Presser, V. & Nickel, K. G. Silica on Silicon Carbide. *Crit. Rev. Solid State Mater. Sci.* **33**, 1-99 (2008).
10 Takeda, M., Urano, A., Sakamoto, J. & Imai, Y. Microstructure and Oxidation Behavior of Silicon Carbide Fibers Derived from Polycarbosilane. *J. Am. Ceram. Soc.* **83**, 1171-1176 (2000).
11 Shimoo, T., Takeuchi, H. & Okamura, K. Oxidation Kinetics and Mechanical Property of Stoichiometric SiC Fibers (Hi-Nicalon-S). *J. Ceram. Soc. Japan* **108**, 1096-1102 (2000).
12 Kim, H.-E. & Moorhead, A. J. Strength of Nicalon Silicon Carbide Fibers Exposed to High-Temperature Gaseous Environments. *J. Am. Ceram. Soc.* **74**, 666-669 (1991).
13 Brennan, J. J. Interfacial Characterization of a Slurry-Cast Melt-Infiltrated SiC/SiC Ceramic-Matrix Composite. *Acta mater.* **48**, 4619-4628 (2000).
14 Gauthier, W., Pailler, F., Lamon, J. & Pailler, R. Oxidation of Silicon Carbide Fibers During Static Fatigue in Air at Intermediate Temperatures. *J. Am. Ceram. Soc.* **92**, 2067-2073 (2009).
15 Gogotsi, Y. & Yoshimura, M. Oxidation and Properties Degradation of SiC Fibres Below 850 C. *J. Mater. Sci. Lett.* **13**, 680-683 (1994).
16 Lara-Curzio, E. Stress-Rupture of Nicalon/SiC Continuous Fiber Ceramic Composites in Air at 950 C. *J. Am. Ceram. Soc.* **80**, 3268-3272 (1997).
17 Lara-Curzio, E. Oxidation Induced Stress-Rupture of Fiber Bundles. *J. Eng. Mater. Tech* **120**, 105-109 (1998).

18 Hay, R. S. *et al.* Oxidation Kinetics and Strength Versus Scale Thickness for Hi-Nicalon[TM]-S Fiber. *PACRIM 2009 Proceedings* **in press** (2010).

19 Mogilevsky, P., Boakye, E. E., Hay, R. S. & Kerans, R. J. Monazite Coatings on SiC Fibers II: Oxidation Protection. *J. Am. Ceram. Soc.* **89**, 3481-3490 (2006).

20 Easler, T. E., Bradt, R. C. & Tressler, R. E. Strength Distributions of SiC Ceramics After Oxidation and Oxidation Under Load. *J. Am. Ceram. Soc.* **64**, 731-734 (1981).

21 Badini, C., Fino, P., Ortona, A. & Amelio, C. High Temperature Oxidation of Multilayered SiC Processed by Tape Casting and Sintering. *J. Eur. Ceram. Soc.* **22**, 2017-2079 (2002).

22 Hay, R. S. *et al.* Oxidation and Scale Crystallization Kinetics of SiC Fiber. *J. Am. Ceram. Soc.* (submitted).

23 Dong, S. M. *et al.* Characterization of Nearly Stoichiometric SiC Fibres. *J. Mater. Sci.* **36**, 2371-2381 (2001).

24 Hinoki, T., Snead, L. L., Lara-Curzio, E., Park, J. & Kohyama, A. Effect of Fiber/Matrix Interfacial Properties on Mechanical Properties of Unidirectional Crystalline Silicon Carbide Composites. *Ceram. Eng. Sci. Proc.* **23**, 511-518 (2002).

25 Sauder, C. & Lamon, J. Tensile Creep Behavior of SiC-Based Fibers With a Low Oxygen Content. *J. Am. Ceram. Soc.* **90**, 1146-1156, doi:doi:10.1111/j.1551-2916.2007.01535.x (2007).

26 Bunsell, A. R. & Piant, A. A Review of the Development of Three Generations of Small Diameter Silicon Carbide Fibres. *J. Mater. Sci.* **41**, 823-839 (2006).

27 Ishikawa, T. Advances in Inorganic Fibers. *Adv. Polym. Sci.* **178**, 109-144 (2005).

28 Sha, J. J., Nozawa, T., Park, J. S., Katoh, Y. & Kohyaman, A. Effect of Heat-Treatment on the Tensile Strength and Creep Resistance of Advanced SiC Fibers. *J. Nucl. Mater.* **329-333**, 592-596 (2004).

29 Tanaka, T., Shibayama, S., Takeda, M. & Yokoyama, A. Recent Progress of Hi-Nicalon Type S Development. *Ceram. Eng. Sci. Proc.* **24**, 217-223 (2003).

30 Opila, E. Influence of Alumina Reaction Tube Impurities on the Oxidation of Chemically-Vapor-Deposited Silicon Carbide. *J. Am. Ceram. Soc.* **78**, 1107-1110 (1995).

31 Petry, M. D., Mah, T. & Kerans, R. J. Validity of Using Average Diameter for Determination of Tensile Strength and Weibull Modulus of Ceramic Filaments. *J. Am. Ceram. Soc.* **80**, 2741-2744 (1997).

32 Hay, R. S., Welch, J. R. & Cinibulk, M. K. TEM Specimen Preparation and Characterization of Ceramic Coatings on Fiber Tows. *Thin Solid Films* **308-309**, 389-392 (1997).

33 Cinibulk, M. K., Welch, J. R. & Hay, R. S. Preparation of Thin Sections of Coated Fibers for Characterization by Transmission Electron Microscopy. *J. Am. Ceram. Soc.* **79**, 2481-2484 (1996).

34 Armigliato, A. & Rosa, R. X-Ray Microanalysis Combined with Monte Carlo Simulation for the Analysis of Layered Thin Films: The Case of Carbon Contamination. *Micros. Microanal.* **15**, 99-105 (2009).

35 Isabell, T. C., Fischione, P. E., O'Keefe, C., Guruz, M. U. & Dravid, V. P. Plasma Cleaning and Its Applications for Electron Microscopy. *Micros. Microanal.* **5**, 126-135 (1999).

36 Rolland, P., Carlino, V. L. & Vane, R. Improved Carbon Analysis with Evactron Plasma Cleaning. *Micros. Microanal.* (2004).

37 Ramberg, C. E., Cruciani, G., Spear, K. E., Tressler, R. E. & Ramberg, C. F. Passive-Oxidation Kinetics of High-Purity Silicon Carbide from 800 to 1100 C. *J. Am. Ceram. Soc.* **79**, 2897-2911 (1996).

38 Christiansen, K. & Helbig, R. Anisotropic Oxidation of 6H-SiC. *J. Appl. Phys.* **79**, 3276-3281 (1996).

39 Narushima, T., Kato, M., Murase, S., Ouchi, C. & Iguchi, Y. Oxidation of Silicon and Silicon Carbide in Ozone-Containing Atmospheres at 973K. *J. Am. Ceram. Soc.* **85**, 2049-2055 (2002).

40 Vickridge, I. & et al. Growth of SiO_2 on SiC by dry thermal oxidation: mechanisms. *J. Physics D: Applied Physics* **40**, 6254 (2007).

41 Guinel, M. J. F. & Norton, M. G. Oxidation of Silicon Carbide and the Formation of Silica Polymorphs. *J. Mater. Res.* **21**, 2550-2563 (2006).

42 Horvath, E. *et al.* Microstructural Characterization of the Oxide Scale on Nitride Bonded SiC-Ceramics. *Ceramics Int.* **34**, 151-155 (2008).

43   Presser, V., Loges, A., Hemberger, Y. & Nickel, K. G. Microstructural Evolution of Silica on Single-Crystal Silicon Carbide. Part I: Devitrification and Oxidation Rates. *J. Am. Ceram. Soc.* **92**, 724-731 (2009).

44   Shimoo, T. *et al.* Mechanism of Oxidation of Low-Oxygen SiC Fiber Prepared by Electorn Radiation Curing Method. *J. Ceram. Soc. Japan* **102**, 617-622 (1994).

45   Delph, T. J. Intrinsic strain in SiO_2 thin films. *J. Appl. Phys.* **83**, 786-792 (1998).

46   Kao, D.-B., McVittie, J. P., Nix, W. D. & Saraswat, K. C. Two-Dimensional Thermal Oxidation of Silicon - II. Modeling Stress Effects in Wet Oxides. *IEEE Trans. Electron. Dev.* **35**, 25-37 (1988).

47   Rafferty, C. S., Borucki, L. & Dutton, R. W. Plastic Flow During the Thermal Oxidation of Silicon. *Appl. Phys. Lett.* **54**, 1516-1518 (1989).

48   Chollon, G. *et al.* Thermal Stability of a PCS-Derived SiC Fibre with a Low Oxygen Content (Hi-Nicalon). *J. Mater. Sci.* **32**, 327-347 (1997).

49   Hsueh, C. H. & Evans, A. G. Oxidation Induced Stresses and Some Effects on the Behavior of Oxide Films. *J. Appl. Phys.* **54**, 6672-6686 (1983).

50   Brown, D. K., Hu, S. M. & Morrissey, J. M. Flaws in Sidewall Oxides Grown on Polysilicon Gate. *J. Electrochem. Soc.* **129**, 1084-1089 (1982).

51   Liu, F., Sommer, F. & Mittemeijer, E. J. Analysis of the Kinetics of Phase Transformations; Roles of Nucleation Index and Temperature Dependent Site Saturation, and Recipes for the Extraction of Kinetic Parameters. *J. Mater. Sci.* **42**, 573-587 (2007).

52   Liu, F., Sommer, F., Bos, C. & Mittemeijer, E. J. Analysis of Solid State Phase Transformation Kinetics: Models and Recipes. *Int. Mater. Rev.* **52**, 193-212 (2007).

53   Deal, B. E. & Grove, A. S. General Relationships for the Thermal Oxidation of Silicon. *J. Appl. Phys.* **36**, 3770-3778 (1965).

54   Song, Y., Dhar, S., Feldman, L. C., Chung, G. & Williams, J. R. Modified Deal Grove Model for the Thermal Oxidation of Silicon Carbide. *J. Appl. Phys.* **95**, 4953-4957 (2004).

55   Zhu, Y. T., Taylor, S. T., Stout, M. G., Butt, D. P. & Lowe, T. C. Kinetics of Thermal, Passive Oxidation of Nicalon Fibers. *J. Am. Ceram. Soc.* **81**, 655-660 (1998).

56   Wilson, L. O. & Marcus, R. B. Oxidation of Curved Silicon Surfaces. *J. Electrochem. Soc.* **134**, 481-490 (1987).

57   Yoshimura, M., Kase, J. & Somiya, S. Oxidation of SiC Powder by High-Temperature, High-Pressure H_2O. *J. Mater. Res.* **1**, 100-103 (1986).

58   Opila, E. J. Variation of the Oxidation Rate of Silicon Carbide with Water Vapor Pressure. *J. Am. Ceram. Soc.* **82**, 625-636 (1999).

59   Hijikata, Y., Yamamoto, T., Yaguchi, H. & Yoshida, S. Model Calculation of SiC Oxidation Rates in the Thin Oxide Regime. *Mater. Sci. Forum* **600-603**, 663-666 (2009).

60   Costello, J. A. & Tressler, R. A. Oxidation Kinetics of Silicon Carbide Crystals and Ceramics: I. In Dry Oxygen. *J. Am. Ceram. Soc.* **69**, 674-681 (1986).

61   Fung, C. D. & Kopanski, J. J. Thermal Oxidation of 3C Silicon Carbide Single-Crystal Layers on Silicon. *Appl. Phys. Lett.* **45**, 757-759 (1984).

62   Shenoy, J. N., Cooper, J. A. & Melloch, M. R. Comparison of Thermally Oxidized Metal-Oxide-Semiconductor Interfaces on 4H and 6H Polytypes of Silicon Carbide. *Appl. Phys. Lett.* **68**, 803-805 (1996).

63   Zheng, Z., Tressler, R. E. & Spear, K. E. The effects of Cl_2 on the oxidation of single crystal silicon carbide. *Corrosion Science* **33**, 557-567 (1992).

64   Zheng, Z., Tressler, R. E. & Spear, K. E. The effect of sodium contamination on the oxidation of single crystal silicon carbide. *Corrosion Science* **33**, 545-556 (1992).

65   Chatterjee, A. N. & Matocha, K. S. Multiscale modeling and simulation of silicon carbide oxidation. *Phys. Stat. Sol. C* **6**, 1476-1479 (2009).

66   Harris, R. C. A. Oxidation of 6-H SiC Platelets. *J. Am. Ceram. Soc.* **58**, 7-9 (1975).

67   Munch, W. V. & Pfaffeneder, I. Thermal Oxidation and Electrolytic Etching of Silicon Carbide. *J. Electrochem. Soc.* **122**, 642-643 (1975).

68   Powell, J. A. *et al.* Application of Oxidation to the Structural Characterization of SiC Epitaxial Films. *Appl. Phys. Lett.* **59**, 183-185 (1991).

69   Schmitt, J. & Halbig, R. Oxidation of 6H-SiC as Function of Doping Concentration and Temperatrure. *J. Electrochem. Soc.* **141**, 2262-2265 (1994).

70   Ray, D. A., Kaur, S., Cutler, R. A. & Shetty, D. K. Effects of Additives on the Pressure-Assisted Densification and Properties of Silicon Carbide. *J. Am. Ceram. Soc.* **91**, 2163-2169 (2008).
71   Szilagyi, E. *et al.* Oxidation of SiC Investigated by Ellipsometry and Rutherford Backscattering Spectrometry. *J. Appl. Phys.* **104**, 104903 (2008).
72   Kakubari, K., Kubaki, R., Hijikata, Y., Yaguchi, H. & Yoshida, S. Real Time Observations of SiC Oxidation Using an In Situ Ellipsometer. *Mater. Sci. Forum* **527-529**, 1031-1034 (2006).
73   Miller, W. J. High Temperature Oxidation of Silicon Carbide. *AFIT Thesis*, 1-55 (1972).
74   Costello, J. A. & Tressler, R. A. Oxidation Kinetics of Hot-Pressed and Sintered a-SiC. *J. Am. Ceram. Soc.* **69**, 674-681 (1981).
75   Ogbuji, L. J. T. & Singh, M. High-Temperature Oxidation Behavior of Reaction-Formed Silicon Carbide Ceramics. *J. Mater. Res.* **10**, 3232-3240 (1995).
76   Gomez, E., Iturriza, I., Echeberria, J. & Castro, F. Oxidation Resistance of SiC Ceramics SIntered in the Solid State or in the Presence of a Liquid Phase. *Scripta metall. mater.* **33**, 491-496 (1995).
77   Ramberg, C. E. & Worrell, W. L. Oxygen Transport in Silica at High Temperatures: Implications of Oxidation Kinetics. *J. Am. Ceram. Soc.* **84**, 2607-2616 (2001).
78   Singhal, S. C. Effect of Water Vapor on the Oxidation of Hot-Pressed Silicon Nitride and Silicon Carbide. *J. Am. Ceram. Soc.* **59**, 81-82 (1976).
79   Rodríguez-Rojas, F., Ortiz, A. L., Guiberteau, F. & Nygren, M. Oxidation behaviour of pressureless liquid-phase-sintered a-SiC with additions of $5Al_2O_3 + 3RE_2O_3$ (RE = La, Nd, Y, Er, Tm, or Yb). *J. Eur. Ceram. Soc.* **30**, 3209-3217 (2010).
80   Lu, W. J., Steckl, A. J., Chow, T. P. & Katz, W. Thermal Oxidation of Sputtered Silicon Carbide Thin Films. *J. Electrochem. Soc.* **131**, 1907-1914 (1984).
81   Das, D., Farjas, J. & Roura, P. Passive-Oxidation Kinetics of SiC Microparticles. *J. Am. Ceram. Soc.* **87**, 1301-1305 (2004).
82   Adamsky, R. F. Oxidation of Silicon Carbide in the Temperature Range 1200-1500C. *J. Phys. Chem.* **63**, 305-307 (1959).
83   Wang, P. S., Hsu, S. M. & Wittberg, T. N. Oxidation Kinetics of Silicon Carbide Whiskers Studied by X-ray Photoelectron Spectroscopy. *J. Mater. Sci.* **26**, 1655-1658 (1991).
84   Fox, D. S. Oxidation Behavior of Chemically-Vapor-Deposited Silicon Carbide and Silicon Nitride from 1200 to 1600 C. *J. Am. Ceram. Soc.* **81**, 945-950 (1998).
85   Goto, T. & Homma, H. High-Temperature Active/Passive Oxidation and Bubble Formation of CVD SiC in O_2 and CO_2 Atmospheres. *J. Eur. Ceram. Soc.* **22**, 2749-2756 (2002).
86   Ogbuji, U. J. T. & Opila, E. J. A Comparison of the Oxidation Kinetics of SiC and Si_3N_4. *J. Electrochem. Soc.* **142**, 925-930 (1995).
87   Ogbuji, L. U. J. T. Effect of Oxide Devitrification on Oxidation Kinetics of SiC. *J. Am. Ceram. Soc.* **80**, 1544-1550 (1997).
88   Sibieude, F., Rodríguez, J. & Clavaguera-Mora, M. T. Kinetics and crystallization studies by in situ X-ray diffraction of the oxidation of chemically vapour deposited SiC. *Thin Solid Films* **204**, 217-227 (1991).
89   Naslain, R. *et al.* Oxidation mechanisms and kinetics of SiC-matrix composites and their constituents. *J. Mater. Sci.* **39**, 7303-7316 (2004).
90   Mah, T. *et al.* Thermal Stability of SiC Fibres (Nicalon). *J. Mater. Sci.* **19**, 1191-1201 (1984).
91   Shimoo, T., Morisada, Y. & Okamura, K. Suppression of Active Oxidation of Polycarbosilane-Derived Silicon Carbide Fibers by Preoxidation at High Oxygen Pressure. *J. Am. Ceram. Soc.* **86**, 838-845 (2003).
92   Huger, M., Souchard, S. & Gault, C. Oxidation of Nicalon SiC fibres. *J. Mater. Sci. Lett.* **12**, 414-416 (1993).
93   Filipuzzi, L. & Naslain, R. in *Advanced Structural Inorganic Composites* (ed P. Vincenzini) 35-46 (Elsevier Science Publishers, 1991).
94   Shimoo, T., Toyoda, F. & Okamura, K. Oxidation Kinetics of Low-Oxygen Silicon Carbide Fiber. *J. Mater. Sci.* **35**, 3301-3305 (2000).
95   Chollon, G. *et al.* A Model SiC-Based Fibre with a Low Oxygen Content Prepared from a Polycarbosilane Precursor. *J. Mater. Sci.* **32**, 893-911 (1997).

96   Rebillat, F., Garitte, E. & Guette, A. in *Design, Development, and Applications of Engineering Ceramics and Composites* Vol. 215 *Ceramic Transactions* eds Dileep Singh, Dongming Zhu, Yanchun Zhou, & Mrityunjay Singh) 350 (Wiley, 2010).

97   Luthra, K. L. Some New Perspectives on Oxidation of Silicon Carbide and Silicon Nitride. *J. Am. Ceram. Soc.* **74**, 1095-1103 (1991).

98   Norton, F. J. Permeation of Gaseous Oxygen through Vitreous Silica. *Nature* **191**, 701 (1961).

99   Akita, H., Kimoto, T., Inoue, N. & Matsunami, H. Proceedings of the International Conference on SiC and Related Materials, Kyoto, Japan. *Institute of Physics Publishing* **142**, 725 (1996).

100   Raynaud, C. Silica films on silicon carbide: a review of electrical properties and device applications. *J. Non. Cryst. Sol.* **280**, 1-31 (2001).

101   Shimoo, T., Toyoda, F. & Okamura, K. Effect of Oxygen Partial Pressure on Oxidation Rate of Si-C-O Fiber. *J. Ceram. Soc. Japan* **106**, 447-451 (1998).

102   Golz, A., Horstmann, G., Stein-von-Kamienski, E. & Kurz, H. Silicon Carbide and Related Materials. *Inst. Phys. Conf. Series* **142**, 633-636 (1996).

103   Petit, J. B., Neudeck, P. G., Matus, L. G. & Powell, J. A. Thermal Oxidation of Single-Crystal Silicon Carbide: Kinetics, Electrical, and Chemical Studies. *Springer Proceedings in Physics* **71**, 190 (1992).

104   Palmour, J. W., Kim, H. J. & Davis, R. F. *MRS Symp. Proc.* **54**, 553 (1986).

Fracture Mechanics, Modeling, and Mechanical Testing

SPECIMEN STRESS EQUILIBRIUM IN SPLIT HOPKINSON PRESSURE BAR TESTS OF CERAMICS AT HIGH STRAIN RATE

Jianming Yuan[1], Jan Ma[1,2], and Geoffrey E.B. Tan[2]
[1] Temasek Laboratories, Nanyang Technological University, 50 Nanyang Drive, Singapore 637553, Singapore
[2] School of Materials Science and Engineering, Nanyang Technological University, Singapore 639798, Singapore

ABSTRACT

This paper provides an analytical method and guidelines for achieving valid Split Hopkinson Pressure Bar (SHPB) tests of ceramics at high strain rate. A one dimensional model of a SHPB test of an elastic specimen is expressed by the wave equation with the initial and boundary conditions. For any incident stress pulse, the analytical expression of the specimen stress is derived via Fourier transform analysis. Consequently, the general expressions for the bar stress at the two ends of the specimen, the specimen stress uniformity and the strain rate are established. For linear ramp loading, which is usually generated using pulse shaping technique in SHPB tests, the threshold strain rate to achieve stress equilibrium and constant strain rate is determined. In addition, the strain rate when the specimen fractures is obtained. The effects of the bar-specimen mechanical impedance and the rising rate of the incident pulse on the stress uniformity and the strain rate are investigated. The analytical results are applied to determine the optimal test conditions for accurate and reliable SHPB experiments on ceramics at several thousand strain per-second.

INTRODUCTION

Split Hopkinson Pressure Bar (SHPB) testing of ceramics present unique challenges. Ceramics exhibit no more than 1 to 2% strain before fracture. Even within this small strain, most of the assumptions used in the traditional SHPB technique are violated, thus proper modifications must be made in the testing procedure to ensure the accuracy and reliability of the stress and strain measurements [1]. Using the pulse shaping technique [2], SHPB tests of ceramics at low dynamic strain rate, i.e. a few hundred strain per-second, were carried out successfully. The validity of the obtained test data was examined in terms of both specimen stress equilibrium and constant strain rate [3]. At high dynamic strain rates of several thousand strain per-second, constant strain rate and stress equilibrium, which are two of the assumptions used in SHPB technique, may not be achieved simultaneously. The discrepancy and scatter in the measured ceramics failure strength were reported [4-5]. Loading rate, instead of strain rate, was introduced to address ceramics strength to avoid the issue of constant strain rate [6]. Upper limits of strain rate in SHPB tests of ceramics were originally investigated by Ravichandran et al [7] and subsequently improved by Pan et al [8] which provided experimental verification to their proposed analytical models [8]. Specimen stress equilibrium in SHPB tests of ceramics using pulse shaping technique was studied by applying the method of characteristics in wave propagation to obtain numerical results [7].

In this paper, one-dimensional wave equation with initial and boundary conditions for the specimen stress is formulated and solved by using the Fourier transform. The general analytical expression of the specimen stress obtained is applied to derive the expression for the bar stress at the two ends of the specimen, the stress equilibrium and the strain rate under general incident stress pulse. Moreover, for linear ramp loading (pulse shaping technique in SHPB tests), it is applied to determine optimal test conditions in accurate and reliable SHPB test of ceramics at a few thousand strain rate.

SPECIMEN STRESS FUNCTION, STRESS UNIFORMITY AND STRAIN RATE

In Fig. 1, an incident longitudinal elastic stress wave, σ_i, reaches the interface between the incident bar and the specimen, a reflected wave, σ_r, is generated in the incident bar. After a delay of certain amount of time, a transmitted wave, σ_t, is produced in the transmitted bar. A ratio between the mechanical impedances of the bars and the specimen is defined as

$$r = \frac{\rho_0 c_0 A_0}{\rho c A} \quad (1)$$

where ρ_0, c_0 and A_0 are the density, the wave velocity and the cross-sectional area of the bars; ρ, c and A are, the density, the wave velocity and the cross-sectional area of the specimen, respectively. Let L be the length of the specimen. A reference time τ required for the wave to travel through the specimen once is defined as

$$\tau = \frac{L}{c} \quad (2)$$

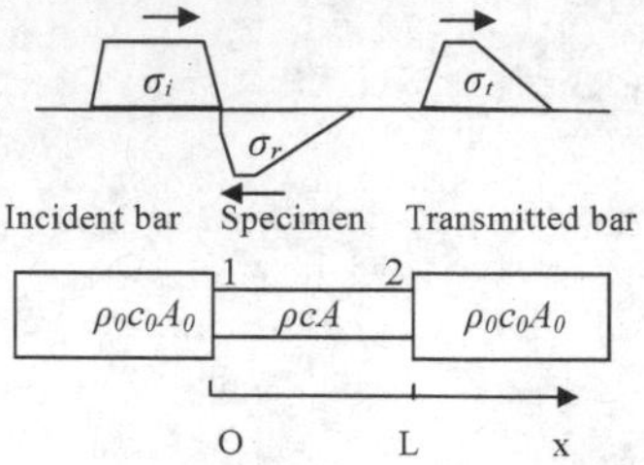

Figure 1. Analytical model of SHPB test of an elastic specimen.

The one-dimensional wave equation with the initial and boundary conditions for the specimen stress for SHPB test of an elastic specimen is formulated and solved by using Fourier transform (see Appendix A). The specimen stress function of location and time is obtained as (see Eq. (A12) in Appendix A)

$$\sigma(x, t+2t) = \frac{(1-r)^2}{(1+r)^2}\sigma(x,t) + \frac{2A_0}{A}\left[\frac{1}{1+r}\sigma_i\left(t - \frac{x}{c} + 2\tau\right) - \frac{(1-r)}{(1+r)^2}\sigma_i\left(t - \frac{x}{c}\right)\right] \quad (3)$$

Let $\sigma_1(t)$ and $\sigma_2(t)$ denote the bar stress at the bar-specimen interfaces shown in Fig. 1. Subscripts 1 and 2 represent the locations of the front and rear ends of the specimen. The difference and average of the two bar stress can be written as

$$\Delta\sigma = \sigma_1(t) - \sigma_2(t)$$

$$\sigma_a = \frac{\sigma_1(t) + \sigma_2(t)}{2} \quad (4)$$

The degree of specimen stress uniformity is able to be quantified by

$$\frac{\Delta\sigma}{\sigma_a} = \frac{2[\sigma_1(t) - \sigma_2(t)]}{\sigma_1(t) + \sigma_2(t)} \quad (5)$$

where $\sigma_1(t)$ and $\sigma_2(t)$ are obtained as (see Eq. (A14) in Appendix A)

$$\sigma_1(t + 2\tau) = \frac{(1-r)^2}{(1+r)^2}\sigma_1(t) + \frac{2}{(1+r)^2}[(1+r)\sigma_i(t+2\tau) - (1-r)\sigma_i(t)]$$

$$\sigma_2(t + 2\tau) = \frac{(1-r)^2}{(1+r)^2}\sigma_2(t) + \frac{4r}{(1+r)^2}\sigma_i(t+\tau)$$

$$(6)$$

Let $v_1(t)$ and $v_2(t)$ denote, respectively, the particle velocities at interfaces 1 and 2. The strain rate of the specimen is (see Eq. (A17) in Appendix A)

$$\dot{\varepsilon}_1(t) = \frac{1}{L}[v_2(t) - v_1(t)] = \frac{1}{\rho_0 c_0 L}[2\sigma_i(t) - \sigma_1(t) - \sigma_2(t)] \tag{7}$$

Given any incident stress pulse, $\sigma_i(t)$, in a SHPB test, Eq.(3) describes the variation of the stress distribution along the specimen, Eq. (6) is the analytical expression of the bar stress at the two ends of the specimen; while Eq. (5) and Eq. (7) are the general expressions of the specimen stress uniformity and the strain rate, respectively. In this paper, Eqs. (5) - (7) are applied to analyze stress equilibrium and constant strain rate under linear ramp loading, which is usually generated by using pulse shaper technique in SHPB tests, so as to investigate optimal test conditions for SHPB tests on ceramics at high strain rate.

STRESS EQUILIBRIUM TIME, CONSTANT STRAIN RATE TIME AND SPECIMEN FRACTURE STRAIN RATE UNDER LINEAR RAMP LOADING

A linear ramp stress incident pulse in Fig. 2 is usually applied in SHPB tests with using pulse shaping technique [2]. The analytical expressions for the stress uniformity, the strain and strain can be derived as follows (see Eqs. (B5) - (B9) in Appendix B):

$$\frac{\Delta\sigma}{\sigma_a} = \frac{\dfrac{1}{r}\left\{1 - \exp\left[-\dfrac{2rt}{\tau(1+r^2)}\right]\right\}}{\dfrac{t}{\tau} - \dfrac{r}{2}\left\{1 - \exp\left[-\dfrac{2rt}{\tau(1+r^2)}\right]\right\}} \tag{8}$$

$$\frac{\dot{\varepsilon}_1(t)}{\dot{\varepsilon}_{max}} = 1 - \exp\left[-\frac{2rt}{\tau(1+r^2)}\right] \tag{9}$$

$$\varepsilon(t) = \dot{\varepsilon}_{max}t - \dot{\varepsilon}_{max}\frac{(1+r^2)}{2r}\left\{1 - \exp\left[-\frac{2rt}{\tau(1+r^2)}\right]\right\} \tag{10}$$

$$\dot{\varepsilon}_{max} = \frac{r}{\rho_0 c_0 c}\frac{\sigma_0}{t_0} \tag{11}$$

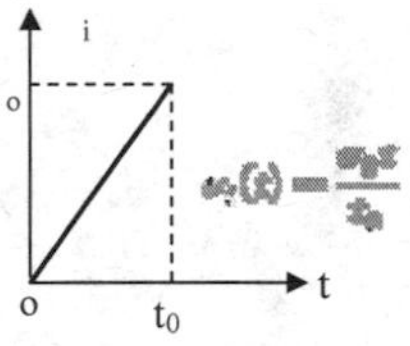

Figure 2. Linear ramp incident wave.

Figure 3a shows the variation of the specimen stress uniformity with the dimensionless time t/τ based on Eqs. (5) and (6). The dimensionless time can be interpreted as the number of wave transits that the pulse is reflected back and forth within the specimen. Mathematically, it takes an infinite amount of time to achieve complete stress equilibrium in Fig. 3a. Therefore, it is reasonable to define a threshold value, Δ, and its corresponding stress equilibrium time, t_Δ, such that the stress within the specimen is considered uniform if $|\Delta\sigma/\sigma_a| \leq \Delta$ is hold for $t \geq t_\Delta$. Considering the accuracy in strength measurement in SHPB tests, it is proposed that $\Delta = 3\%$ is used to determine the stress equilibrium time. Figure 3b plots variation of t_Δ/τ versus r for $\Delta = \Delta\sigma/\sigma_a = 3\%$ and $\Delta\sigma/\sigma_a = 5\%$ based on Eq. (8). In SHPB test of ceramics-like elastic specimen, the duration of the linear ramp incident pulse (t_0 in Fig. 2) must be longer than the stress equilibrium time (t_Δ in Fig. 3b), thus it is to ensure the specimen is under stress equilibrium before fracture is initiated. Figure 3a also shows comparison of the stress uniformity obtained by between this paper and reference [7] which used numerical analysis based on the method of characteristics of wave propagation. Both methods produce exactly the same results. Reference [7] suggested

$\Delta\sigma/\sigma_a$=5% to determine the stress equilibrium time. However, the current analysis suggests that using $\Delta\sigma/\sigma_a$=3% for determining more accurately the stress equilibrium time, as in Figure 3b.

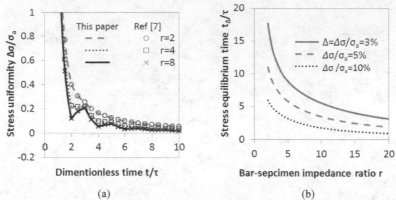

(a)   (b)

Figure 3. Stress uniformity and stress equilibrium time under linear ramp loading.

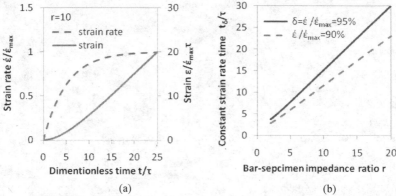

(a)   (b)

Figure 4. Strain rate, strain and constant strain rate time under linear ramp loading.

Figure 4a is the typical strain rate history under a linear ramp loading, in which the specimen strain rate increases until it reaches the maximum strain rate, $\dot{\varepsilon}_{max}$, as predicted by Eq. (9). Because a pulse shaper in SHPB tests is able to provide a nearly linear ramp loading, the constant strain rate can be achieved. Reference [8] initially investigated the variation of strain rate history and derived a similar expression to Eq. (9):

$$\frac{\dot{\varepsilon}(t)}{\dot{\varepsilon}_{max}} = 1 - \exp\left(-\frac{2t}{\tau r}\right) \qquad (12)$$

Experimental results for the strain rate with using pulse shaping technique and Eq. (12) can be found in literatures [2, 8]. However, it must be pointed out that Eq. (12) is only valid for the special case when the impedance ratio r>>1, whereas Eq. (9) derived in this paper is valid for all general conditions of r.

Mathematically, it takes an infinite amount of time to reach constant strain rate in Fig. 4a. It is thus reasonable to define a threshold value, δ, and its corresponding constant strain rate time, t_δ, such that the strain rate is considered constant if $\dot{\varepsilon}/\dot{\varepsilon}_{max} \geq \delta$ is hold for $t \geq t_\delta$. Because strain rate effect on material property is usually evaluated in logarithm scale, it is proposed that $\delta=95\%$ is used to determine the constant strain rate time. Figure 4b plots the variation of t_δ/τ versus r for $\delta=\dot{\varepsilon}/\dot{\varepsilon}_{max}=95\%$ and $\dot{\varepsilon}/\dot{\varepsilon}_{max}=90\%$. In SHPB test of ceramics-like elastic specimen, the duration of the linear ramp incident pulse (t_0 in Fig. 2) must be greater than the constant strain rate time (t_δ in Fig. 4b) to ensure that the specimen is loaded under constant strain rate before fracture is initiated.

Ceramics exhibit no more than 1 to 2% strain before fracture. Before the strain rate reaches the constant strain rate plateau in Fig. 4a, the specimen may have already reached its fracture strain, ε_f. This means the specimen may actually reach the specimen fracture strain rate, $\dot{\varepsilon}_f$, rather than the maximum strain rate, $\dot{\varepsilon}_{max}$ in SHPB tests. The specimen fracture strain rate is determined as (see Eq. (B12) in Appendix B):

$$\frac{\dot{\varepsilon}_f}{\dot{\varepsilon}_{max}} = 1 + lamberW\left[-\exp\left(-\frac{\varepsilon_f}{\tau\dot{\varepsilon}_{max}}\frac{2r}{1+r^2}-1\right)\right] \quad (13)$$

where $lambW$ is the Lambert W function [9]. Figure 5 shows the variation of the fracture strain rate versus the fracture strain. Defining the reference strain as $\varepsilon_{fmin}=\dot{\varepsilon}_{max}\tau(1+r^2)/r$, then when $\varepsilon_f \geq \varepsilon_{fmin}$, see Fig.5, we have $\dot{\varepsilon}_f \geq 0.95\dot{\varepsilon}_{max}$. The specimen is thus effectively loaded under constant strain rate before fracture occurs. The specimen fracture strain rate here refers the strain rate the specimen is experiencing when the specimen fractures. The original works by Ravichandran et al[7] and Pan et al [8] investigated the same parameter, calling it the limiting strain rate. However, this was obtained by numerical methods, as opposed to the current analytical derivation.

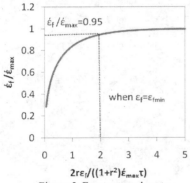

Figure 5. Fracture strain rate under linear ramp loading.

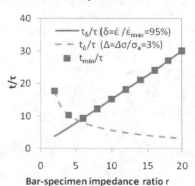

Figure 6. Variation of minimum loading duration with bar-specimen impedance ratio.

OPTIMAL TEST CONDITIONS FOR VALID SHPB TESTS

To conduct SHPB tests of ceramics-like elastic materials at several thousand strain per second, test conditions should be optimized based on the criteria below.

a) For constant σ_0 in Fig. 2 (e.g. the elastic limit of the steel bars used in the SHPB tester), minimize the duration of the linear ramp loading, t_0, in Fig. 2.

b) The loading duration, t_0, in Fig. 2 must be longer than both the stress equilibrium time, t_Δ ($\Delta=\Delta\sigma/\sigma_a=5\%$), in Fig. 3b and the constant strain rate time, t_δ ($\delta=\dot\varepsilon/\dot\varepsilon_{max}=95\%$), in Fig. 4b.

c) The fracture strain must meet the condition, $\varepsilon_f\geq\varepsilon_{fmin}$.

Criterion (a) requires the selection of a short loading time to achieve high strain rate. It is noted that $\dot\varepsilon_{max}$ in Eq. (11) is proportional to the gradient of the linear ramp loading, σ_0/t_0, in Fig. 2. σ_0 is chosen to be constant, for example, the elastic limit of the steel bar of a SHPB tester. Criterion (b) ensures the loading time chosen is sufficient to ensure valid test data when the specimen fractures, i.e. the specimen is under the conditions of stress uniformity and constant strain rate. Based on Criterion (b), the minimum loading duration, t_{min}, at different bar-specimen impedance ratios is listed in Table 1 and in Fig. 6.

Table 1. Minimum loading duration for different bar-specimen impedance ratios.

r	2	4	5.4	5.5	6	8	10	12	14	16	18	20
t_{min}/τ	17.7	10.3	8.44	8.51	9.24	12.2	15.1	18.1	21.1	24.1	27.0	30.0
region	$t_{min}=t_\Delta$,	$(t_\Delta>t_\delta)$					$t_{min}=t_\delta$,		$(t_\delta\geq t_\Delta)$			

The selection of t_{min} is dependent on the bar-specimen impedance r.

(1) When $r<5.5$, $t_\Delta>t_\delta$. The loading history can be divided into 3 stages: (i) from 0 to t_δ, the strain rate is rising fast (not constant) meanwhile the specimen stress is non-uniform; (ii) from t_δ to t_Δ, the strain rate is constant but the specimen stress is non-uniform; (iii) from t_Δ to specimen fracturing, the strain rate is constant meanwhile the specimen stress is uniform. Consequently, $t_{min}=t_\Delta$, resulting the maximum strain rate in valid SHPB tests.

(2) When $r\geq5.5$, $t_\delta\geq t_\Delta$. The loading history can also be divided into 3 stages: (i) from 0 to t_Δ, the specimen stress is non-uniform meanwhile the strain rate is rising fast; (ii) from t_Δ to t_δ, the specimen stress is uniform but the strain rate is still rising fast; (iii) from t_δ to specimen fracturing, the specimen stress is uniform meanwhile the strain rate is constant. Consequently, $t_{min}=t_\delta$, resulting the maximum strain rate in valid SHPB tests.

Table 2. Material parameters used in analysis.

Parameter	c_0	ρ_0	D_0^*	σ_0	t_0^{**}
Unit	m/s	kg/m^3	mm	MPa	μs
Steel bar	5135	7850	20	550	11.6

Parameter	c	ρ	D^*	L	r	τ	t_Δ	$t_{min,,}t_\delta^{***}$	$\dot\varepsilon_{max}$	$\varepsilon_{fmin},\varepsilon_f$
Unit	m/s	kg/m^3	mm	mm		μs	μs	μs	(s^{-1})	
Al$_2$O$_3$	8885	3800	7	7	9.75	0.788	4.47	11.6	1288	1.00%

Note: *D_0 and D are, respectively, the diameters of the steel bar and the alumina specimen.

**To achieve maximum strain rate, we have $t_0=t_{min}$.

***Here $r=9.75$ and $\tau=0.788$. As shown in Fig.6, we have $t_{min}=t_\delta=14.75\tau=11.6$.

To illustrate how the selection of various parameters affects strain rates, consider an alumina specimen with a fracture strain of 1% for a SHPB test. Table 2 lists material properties, dimensions of the steel incident and transmitted bars and the alumina specimen, and SHPB test parameters (loading duration, strain rates) obtained with these bar and specimen conditions.

Figure 7a shows the histories of strain and strain rate for this minimum loading duration. A strain rate of 1288 s^{-1}($\dot{\varepsilon}_{max}$) can be achieved in this condition, with the strain rate becoming constant at the time t_δ just before the specimen fractures (i.e. when the strain accumulated in the specimen reaches 1%). In this case, the specimen stress is non-uniform during the loading history from 0 to 4.47 µs (t_Δ); but becomes uniform after 4.47 µs. Here the loading time has been selected to obtain the maximum strain rate achievable with the tester settings and the specimen parameters given.

In comparison, the histories of strain and strain rate for a longer loading duration ($t_0 = 4t_{min}$ $= 4t_\delta = 46.4$ µs) are plotted in Fig. 7b. In this case, the strain rate achieved is 322 s^{-1}, which is much lower than that (1288 s^{-1}) in Fig. 7a. The specimen strain reaches 1% thus fracturing at the time 35 µs. The strain rate is constant from 11.6 µs (t_δ) until specimen fractures. Here the longer loading results in a lower strain rate.

It should be pointed out that if a smaller specimen is chosen, strain rates of a few thousands per second are achievable. For instance, by using a shorter specimen (L=4 mm) together with a shorter loading time (t_0=6.64 µs), the strain rate will become constant at 2254 s^{-1} when the strain of the specimen reaches 1%.

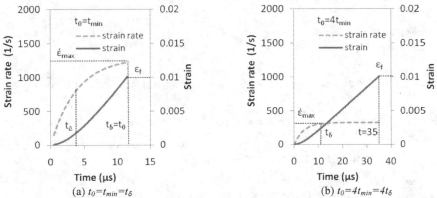

(a) $t_0 = t_{min} = t_\delta$ (b) $t_0 = 4t_{min} = 4t_\delta$

Figure 7. Strain and strain rate histories under linear ramp loading for two different loading time.

CONCLUSIONS

For any incident pulse in SHPB test of an elastic specimen, the analytical expression of the specimen stress can be determined. For linear ramp loading, which is generated by using pulse shaping technique in SHPB test, the stress equilibrium time, the constant strain rate time and the specimen fracture strain rate are investigated, based on the specimen stress expression obtained.

Higher strain rates in SHPB test can be obtained by ensuring a high gradient in the linear ramp loading. The loading duration must be short to achieve the high gradient (when $_0$ is constant); but the chosen loading duration must at the same time be longer than the stress equilibrium time and the constant strain rate time, so that the test data obtained from SHPB tests is valid.

The method and criteria to optimize the test conditions for achieving accurate and reliable SHPB experiment of ceramics at the strain rate region of several thousand per second are proposed. The minimum allowable loading duration is chosen based on the bar-specimen impedance ratio, from which the specimen strain history can be determined. The chosen loading duration should ensure that the ceramics specimen reaches its fracture strain after both stress equilibrium and constant strain rate have been reached.

ACKNOWLEDGEMENT

This work was funded by Defence Research Technology Office, Singapore Ministry of Defence.

REFERENCES

[1] G. Subhash and G. Ravichandran, Split-Hopkinson Pressure Bar Testing of Ceramics, ASM Handbook on Mechanical Testing and Evaluation, Vol. **8**, *ASM International*, 497–504 (2000).
[2] D.J. Frew, M.J. Forrestal and W. Chen, Pulse shaping techniques for testing brittle materials with a split Hopkinson pressure bar, *Exp. Mech.*, **42**(1), 93-106 (2002).
[3] D.J. Frew, M.J. Forrestal and W. Chen, A split Hopkinson bar technique to determine compressive stress-strain data for rock materials, *Exp. Mech.*, **41**(1), 40-46 (2001).
[4] J. M. Staehler, W.W. Predebon, B.J. Pletka and J. Lankford, Testing of High-Strength Ceramics with the Split Hopkinson Pressure Bar, *J. Am. Ceram. Soc.*, **76**(2), 536-538 (1993).
[5] T. Jiao, Y. Li and K.T. Ramesh, High rate response and dynamic failure of structural ceramics, *Int. J. of Applied Ceramic Tech.*, **1**(3), 243-253 (2004).
[6] H. Wang and K.T. Ramesh, Dynamic strength and fragmentation of hot-pressed silicon carbide under uniaxial compression, *Acta Materialia*, **52**, 355–367 (2004).
[7] G. Ravichandran and G. Subhash, Critical appraisal of limiting strain rates for compression testing of ceramics in a split Hopkinson pressure bar, *J. Am. Ceram. Soc.*, **77**(1), 263-267 (1994).
[8] Y. Pan, W. Chen and B. Song, Upper limit of constant strain rates in a split Hopkinson pressure bar experiment with elastic specimens, *J. Exp. Mech.*, **45**(5), 440-446 (2005).
[9] R.M. Corless, G.H. Gonnet, D.E.G. Hare, D.J. Jeffrey, D.E. Knuth, On the Lambert W function, *Adv. Comp. Maths.*, **5**, 329-359 (1996).

APPENDIX A : STRESS ANALYSIS OF ELASTIC SPECIMEN IN SHPB TEST

For SHPB test of an elastic specimen shown in Fig. 1, the one-dimensional wave equation with the boundary conditions at the two ends of the specimen are defined as

$$\frac{\partial^2 \sigma}{\partial t^2} = c^2 \frac{\partial^2 \sigma}{\partial x^2}$$

$$\frac{A}{A_0}\sigma(0,t) - \rho_0 c_0 v(0,t) = 2\sigma_i(t) \tag{A1}$$

$$\frac{A}{A_0}\sigma(L,t) + \rho_0 c_0 v(L,t) = 0$$

where $\sigma = \sigma(x,t)$ and $v=v(x,t)$ are the functions of the specimen stress and the velocity. The boundary conditions in Eq. (A1) represent that the reflected and transmitted waves propagate in the incident and transmitted bars, respectively. $\sigma(0,t)$ and $\sigma(L,t)$ are related to the incident, reflected and transmitted waves in a SHPB test as

$$\frac{A}{A_0}\sigma(0,t) = \sigma_i(t) + \sigma_r(t)$$

$$\frac{A}{A_0}\sigma(L,t) = \sigma_t(t)$$

(A2)

Applying the Fourier transform of Eq. (A1), we have

$$c^2\frac{\partial^2\sigma*}{\partial x^2} + \omega^2\sigma* = 0$$

$$\frac{A}{A_0}\sigma*(0,\omega) - \rho_0 c_0 v*(0,\omega) = 2\sigma_i(\omega)$$

(A3)

$$\frac{A}{A_0}\sigma*(L,\omega) + \rho_0 c_0 v*(L,\omega) = 0$$

where * represents the Fourier transform, i.e.

$$\sigma*(x,\omega) = \frac{1}{\sqrt{2\pi}}\int_{-\infty}^{\infty}\sigma(x,t)e^{i\omega t}dt$$

$$v*(x,\omega) = \frac{1}{\sqrt{2\pi}}\int_{-\infty}^{\infty}v(x,t)e^{i\omega t}dt$$

(A4)

The solution of the Eq. (A3) can be written as

$$\sigma*(x,\omega) = c_1\exp\left(\frac{Ix\omega}{c}\right) + c_2\exp\left(-\frac{Ix\omega}{c}\right)$$

(A5)

where the constants c_1 and c_2 are determined from the boundary conditions in Eq. (A3). For this, $v*$ must be expressed by $\sigma*$. The force equilibrium law referring to the velocity and stress is

$$\rho\frac{\partial v}{\partial t} = \frac{\partial\sigma}{\partial x}$$

(A6)

Applying the Fourier transform of Eq. (A6) and substituting Eqs. (A4) and (A5) into it, we have

$$v*(x,\omega) = \frac{1}{\rho c}\left[c_1\exp\left(\frac{Ix\omega}{c}\right) - c_2\exp\left(-\frac{Ix\omega}{c}\right)\right]$$

(A7)

Eqs. (A5) and (A7) are substituted into the two boundary conditions in Eq. (A3), we have

$$(c_1 + c_2) - \frac{\rho_0 c_0 A_0}{\rho c A}(c_1 - c_2) = \frac{2A_0}{A}\sigma_i^*(\omega)$$

$$c_1 \exp\left(\frac{IL\omega}{c}\right) + c_2 \exp\left(-\frac{IL\omega}{c}\right) + \frac{\rho_0 c_0 A_0}{\rho c A}\left[c_1 \exp\left(\frac{IL\omega}{c}\right) - c_2 \exp\left(-\frac{IL\omega}{c}\right)\right] = 0$$

(A8)

therefore, the constants are determined as

$$c1 = -\frac{2(1-r)A_0}{A}\frac{\exp(-I\omega\tau)\sigma_i^*(\omega)}{(1+r)^2\exp(I\omega\tau) - (1-r)^2\exp(-I\omega\tau)}$$

$$c2 = -\frac{2(1+r)A_0}{A}\frac{\exp(I\omega\tau)\sigma_i^*(\omega)}{(1+r)^2\exp(I\omega\tau) - (1-r)^2\exp(-I\omega\tau)}$$

(A9)

Substituting Eq. (A9) into Eq. (A5), we have

$$\sigma^*(x,\omega) = \frac{2(1+r)A_0}{A}\frac{\exp(I\omega\tau)\exp\left(-\frac{I\omega x}{c}\right)\sigma_i^*(\omega)}{(1+r)^2\exp(I\omega\tau) - (1-r)^2\exp(-I\omega\tau)}$$

$$-\frac{2(1-r)A_0}{A}\frac{\exp(-I\omega\tau)\exp\left(\frac{I\omega x}{c}\right)\sigma_i^*(\omega)}{(1+r)^2\exp(I\omega\tau) - (1-r)^2\exp(-I\omega\tau)}$$

(A10)

The reverse Fourier transform of Eq. (A10) gives

$$\sigma(x,t) = \frac{2A_0}{A(1+r)}\sum_{k=0}^{\infty}\left[\left(\frac{1-r}{1+r}\right)^{2k}\sigma_i\left(t - 2k\tau - \frac{x}{c}\right)\right]$$

$$-\frac{2A_0(1-F)}{A(1+r)^2}\sum_{k=0}^{\infty}\left[\left(\frac{1-r}{1+r}\right)^{2k}\sigma_i\left(t - 2(k+1)\tau + \frac{x}{c}\right)\right]$$

(A11)

or in the iterative equation below, which is more efficient and productive than Eq. (A11).

$$\sigma(x,t+2\tau) = \frac{(1-r)^2}{(1+r)^2}\sigma(x,t) + \frac{2A_0}{A}\left[\frac{1}{1+r}\sigma_i\left(t - \frac{x}{c} + 2\tau\right) - \frac{(1-r)^2}{(1+r)^2}\sigma_i\left(t + \frac{x}{c}\right)\right]$$

(A12)

The bar stress at the two bar-specimen interfaces are

$$\sigma_1(t) = \frac{A}{A_0}\sigma(0,t)$$

$$\sigma_2(t) = \frac{A}{A_0}\sigma(L,t)$$

(A13)

Substituting Eq. (A12) into Eq. (A13), we have

$$\sigma_1(t+2\tau) = \frac{(1-r)^2}{(1+r)^2}\sigma_1(t) + \frac{2}{(1+r)^2}[(1+r)\sigma_i(t+2\tau) - (1-r)\sigma_i(t)]$$

$$\sigma_2(t+2\tau) = \frac{(1-r)^2}{(1+r)^2}\sigma_2(t) + \frac{4r}{(1+r)^2}\sigma_i(t+\tau)$$

(A14)

When τ is small, Eq. (A14) can be written as

$$\frac{d}{dt}\sigma_1(t) = -\frac{2r}{\tau(1+r^2)}\sigma_1(t) + \frac{2}{(1+r^2)}\frac{d}{dt}\sigma_i(t) + \frac{2F}{\tau(1+r^2)}\sigma_i(t)$$

$$\frac{d}{dt}\sigma_2(t) = -\frac{2r}{\tau(1+r^2)}\sigma_2(t) + \frac{2r}{\tau(1+r^2)}\sigma_i(t)$$

(A15)

Derived from Eqs. (A1) and (A13), the velocities of the incident and transmitted bar at the two bar-specimen interfaces, $v_1(t)$ and $v_2(t)$, are

$$v_1(t) = v(0,t) = \frac{A}{\rho_0 c_0 A_0}\sigma(0,t) - \frac{2\sigma_i(t)}{\rho_0 c_0} = \frac{\sigma_2(t) - 2\sigma_i(t)}{\rho_0 c_0}$$

$$v_2(t) = v(L,t) = -\frac{A}{\rho_0 c_0 A_0}\sigma(L,t) = \frac{\sigma_2(t)}{\rho_0 c_0}$$

(A16)

The strain rate of the specimen can be written as

$$\dot{\varepsilon}_1(t) = \frac{1}{L}[v_2(t) - v_1(t)] = \frac{1}{\rho_0 c_0 L}[2\sigma_i(t) - \sigma_1(t) - \sigma_2(t)]$$

(A17)

APPENDIX B : SPECIMEN RESPONSE UNDER RAMP RISE INCIDENT LOADING

For a linear ramp stress incident pulse shown in Fig. 2, Eq. (A15) can be written as

$$\frac{d\sigma_1(t)}{dt} = -\frac{2r\sigma_1(t)}{\tau(1+r^2)} + \frac{2k(rt+\tau)}{\tau(1+r^2)}$$

$$\frac{d\sigma_2(t)}{dt} = -\frac{2r\sigma_2(t)}{\tau(1+r^2)} + \frac{2krt}{\tau(1+r^2)}$$

(B1)

Consequently, the control equations and the initial conditions are

$$\frac{d[\sigma_1(t) - \sigma_2(t)]}{dt} = -\frac{2r[\sigma_1(t) - \sigma_2(t)]}{\tau(1+r^2)} + \frac{2k}{1+r^2}$$

$$\sigma_1(0) - \sigma_2(0) = 0$$

(B2)

and

$$\frac{d[\sigma_1(t) - \sigma_2(t)]}{dt} = -\frac{2r[\sigma_1(t) + \sigma_2(t)]}{2\tau(1+r^2)} + \frac{2k(2rt+\tau)}{\tau(1+r^2)}$$

$$\sigma_1(0) + \sigma_2(0) = 0$$

(B3)

The solutions of Eqs. (B2) and (B3) are determined as

$$\sigma_1(t) - \sigma_2(t) = \frac{\sigma_0 \tau}{t_0 r} \left\{ 1 - \exp\left[-\frac{2rt}{\tau(1+r^2)} \right] \right\}$$

$$\sigma_1(t) + \sigma_2(t) = \frac{2\sigma_0 t}{t_0} - \frac{\sigma_0 \tau r}{t_0} \left\{ 1 - \exp\left[-\frac{2rt}{\tau(1+r^2)} \right] \right\}$$

(B4)

therefore, the stress uniformity is

$$\Delta\sigma = \sigma_1(t) - \sigma_2(t) = \frac{\sigma_0 \tau}{t_0 r} \left\{ 1 - \exp\left[-\frac{2rt}{\tau(1+r^2)} \right] \right\}$$

$$\sigma_a = \frac{\sigma_1(t) + \sigma_2(t)}{2} = \frac{\sigma_0 t}{t_0} - \frac{\sigma_0 \tau r}{2t_0} \left\{ 1 - \exp\left[-\frac{2rt}{\tau(1+r^2)} \right] \right\}$$

$$\frac{\Delta\sigma}{\sigma_a} = \frac{\frac{1}{r}\left\{ 1 - \exp\left[-\frac{2rt}{\tau(1+r^2)} \right] \right\}}{\frac{t}{\tau} - \frac{r}{2}\left\{ 1 - \exp\left[-\frac{2rt}{\tau(1+r^2)} \right] \right\}}$$

(B5)

Substituting Eq. (B4) into Eq. (A17), we have

$$\dot{\varepsilon}(t) = \frac{\sigma_0}{\rho_0 c_0 L} \frac{\tau r}{t_0} \left\{ 1 - \exp\left[-\frac{2rt}{\tau(1+r^2)} \right] \right\}$$

(B6)

Define the maximum strain rate

$$\dot{\varepsilon}_{max} = \frac{\sigma_0}{\rho_0 c_0 L} \frac{\tau r}{t_0} = \frac{r}{\rho_0 c_0 c} \frac{\sigma_0}{t_0}$$

(B7)

Eq. (B6) can be written as

$$\frac{\dot{\varepsilon}(t)}{\dot{\varepsilon}_{max}} = 1 - \exp\left[-\frac{2rt}{\tau(1+r^2)} \right]$$

(B8)

With the initial condition, $\varepsilon(0) = 0$, the strain history is obtained as

$$\varepsilon(t) = \dot{\varepsilon}_{max} t - \dot{\varepsilon}_{max} \frac{(1+r^2)\tau}{2r} \left\{ 1 - \exp\left[-\frac{2rt}{\tau(1+r^2)} \right] \right\}$$

(B9)

It may be expressed as

$$\varepsilon(t) = \dot{\varepsilon}_{max} t - \dot{\varepsilon}_{max} = -\frac{\tau \dot{\varepsilon}_{max}(1+r^2)}{2r} \left\{ \ln\left[1 - \frac{\dot{\varepsilon}(t)}{\dot{\varepsilon}_{max}} \right] + \frac{\dot{\varepsilon}(t)}{\dot{\varepsilon}_{max}} \right\}$$

(B10)

When the specimen fracture, i.e. $\varepsilon = \varepsilon_f$ and $\dot{\varepsilon} = \dot{\varepsilon}_f$, we have

$$\varepsilon_f = -\frac{\tau \dot{\varepsilon}_{max}(1+r^2)}{2r} \left\{ \ln\left[1 - \frac{\dot{\varepsilon}_f}{\dot{\varepsilon}_{max}} \right] + \frac{\dot{\varepsilon}_f}{\dot{\varepsilon}_{max}} \right\}$$

(B11)

The solution of Eq. (B11) is

$$\frac{\dot{\varepsilon}_f}{\dot{\varepsilon}_{max}} = 1 + lamberW\left[-\exp\left(-\frac{\dot{\varepsilon}_f}{\tau \dot{\varepsilon}_{max}} \frac{2r}{1+r^2} - 1 \right) \right]$$

(B12)

where $lambW$ is the Lambert W function [9].

RESIDUAL STRESS IN CERAMIC ZIRCONIA-PORCELAIN CROWNS BY NANOINDENTATION

Y. Zhang, J. C. Hanan
School of Mechanical and Aerospace Engineering
Oklahoma State University, Stillwater OK 74078

ABSTRACT

Residual stress plays a critical role in failure of all ceramic crowns. The magnitude and distribution of residual stress in the crown system are largely unknown. Determining the residual stress quantitatively is challenging since the crown has such complex contours and shapes. This work explored the feasibility and validity of measuring residual stress of zirconia and porcelain in ceramic crowns by nanoindentation. Nanoindentation tests were performed on the cross section of a crown along both porcelain and zirconia. Here, one critical location of the crown was selected. It has the thickest section of porcelain and curvature at the crown surface. A half crown annealed at 400 °C was used as a reference sample. The residual stress was determined by comparing the measured apparent hardness of the stressed sample with that of the reference sample. The nanoindentation impression images were acquired through Scanning Probe Microscope (SPM) equipped with a Hysitron Triboindenter. The derived normal contact area value was corrected using the measured real area. Residual stress is determined along the thickness of crowns at the chosen location for both porcelain and zirconia. Results show residual stress across the thickness for both porcelain and zirconia is significantly different. Porcelain showed more tensile stress closer to the crown surface, while zirconia showed more tensile stress closer to the zirconia-porcelain interface.

INTRODUCTION

Many efforts have been taken to reduce the brittleness of ceramic dental crowns, including veneering porcelain on stronger core materials, such as zirconia and alumina.[1,2] Zirconia has a very high flexural strength of 900~1200MPa and a fracture toughness[3] of 9~10 MPa/m$^{1/2}$. The phase transformation (from tetragonal to monoclinic) behavior of zirconia contributes to its high toughness, but brings other issues like low temperature degradation and sub-cracking.[4,5] Even with such a strong and tough material as zirconia, fracture of the veneer or the core is still the dominant clinical complication.[6] In three published reports, up to 50% of crowns developed crazing or cracking with loss of material after 1-2 years of in service.[7,8,9,10] Crazing and chipping often indicates the presence of tensile stress. Stronger core materials could not bring down the failure rate, suggesting other factors than the mechanical strength played an important role in the failure. Residual stress caused by the mismatch of coefficient of thermal expansion, tempering or grain anisotropy is one of them.[11] Preexisting stress will amplify the applied cycling stress and induce cracks in the region that has preexisting stress, but the magnitude and distribution of residual stress in the crown system is largely unknown. Recently, Bale[12] did measurements on the zirconia core of crown systems using X-ray diffraction (XRD) and found that the magnitude of residual stress in zirconia can be as large as 1 GPa locally after veneering with porcelain, and the magnitude and sign of residual stress can change also. But the magnitude and distribution of residual stress in porcelain has never been measured. This work explores the possibility of measuring the residual stress of both zirconia and porcelain by nanoindentation. It can be a very powerful technique, when proved feasible, due to the relatively simple data analysis, sample preparation, and limited safety requirements. Measuring residual stress with nanoindentation is a new field. The mechanism of the problem was not explored until Tsui et al.[13] pioneered studying the effect of residual stress on nanoindentation hardness using a sharp berkovich indenter. They found that preexisting tensile stress decreases hardness, while compressive residual stress increases hardness. A later numerical simulation[14] suggested variation in hardness is caused by the inability to include the factor of material piling up. Swadener *et al.*[15] proposed a model for a spherical indenter based on the onset of yielding and the contact pressure. Suresh and Giannakopoulos[16] proposed a model using a sharp berkovich indenter which measures the residual stress through comparing the load-depth curves between the reference sample and the stressed sample. Carlsson and Larsson[17] formulated the problem for a sharp indenter using Tabor's relation approach.

This work will explore the feasibility of measuring the residual stress of the zirconia-porcelain crown system by nanoindentation using a berkovich tip.

THEORY
Several researchers had shown that residual stress will affect the measured global parameters, such as hardness, indentation depth, or contact area.[13-17] In reverse, by comparing the difference of the global parameters of stressed and reference sample, residual stress can be extracted. Residual stress acts as the effect of an extra force adding to the applied indentation force, as shown in Figure 1. If the

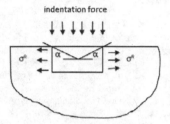
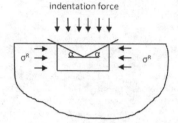

Figure 1 a) Tensile residual stress.   b) compressive residual stress.

residual stress is tensile, it acts the same way as an extra force in the same direction of the indentation force. If the residual stress is compressive, it acts like an extra force in the opposite direction of the indentation. In result, when applying the same indentation force to the stressed and reference material, the measured depth, contact area or hardness will be different. Material with tensile stress will show less resistance to the indentation force and the one with compressive residual stress will show more resistance.

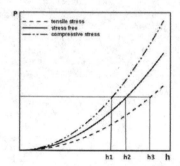

Figure 2 Residual stress will affect the nanoindentation load depth curve.

The residual stress effect on nanoindentation curves will be like that shown in Figure 2. The curve from a sample with compressive residual stress will be above that of the reference sample, while the curve from a sample with tensile residual stress will be below that of the reference sample. The indentation hardness can be extracted from the load-depth curves. Using the differences from the nanoindentation hardness, residual stress can be extracted.

Tabor's relation[18] as used by Swadener[15] and Larson and Carlsson[17] was used to compare hardness from stressed and reference sample.

$$H_0 = C\sigma_Y \tag{1}$$

$$H = C(\sigma_Y - \sigma_R) \tag{2}$$

Where H_o and H are hardness of reference sample and stressed sample, respectively. σ_R is the residual stress. The sign of σ_R is positive for tensile stress and negative for compressive stress. σ_Y is the yield stress. The constrain factor, C, has a value between 0.5 and 3 for elastic-plastic indentation. It was determined using Johnson's "Expanding Cavity" model.[19]

$$C = \frac{2}{3}[2 + \ln\frac{\left(\frac{E\tan\alpha}{\sigma_Y}\right) + 4(1 - 2v)}{6(1 - v^2)}] \tag{3}$$

E and v are Young's Modulus and Poisson's ratio. The angle α, as shown in Figure 3, is angle of the berkovich tip. The hardness was defined as the average pressure,

$$P_{ave} = \frac{P}{A} \tag{4}$$

P_{ave} is the average pressure at the maximum load P, which is directly obtained from nanoindentation. The contact area A having a radius a, as shown in Figure 3, is determined based on the classic Oliver and Pharr method[20], which assumed that the indentation process can be modeled as a rigid axisymmetric indenter into an elastic half space, and at peak load, the edge of the contact is

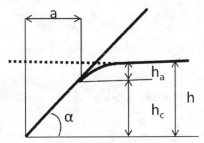

Figure 3 A schematic drawing of nanoindentation half space

vertically displaced relative to the initial position of the surface. The projected area at the contact depth is determined by the area shape function, the relation between contact depth and area was experimentally calibrated using fused silica as,

$$A = \sum_{i=0}^{8} C_i (h_c)^{2-i} \tag{5}$$

Where C_i are a series of constants. The next step is to determine the contact depth h_c, as shown in Figure 3, which is the depth at which the indenter and specimen remain in contact at peak load. According to Sneddon[21]

$$h_c = h - \epsilon \frac{P}{S} \tag{6}$$

The contact stiffness, S, is determined by curve fitting the upper 30% of the unloading load-depth curve as shown in Figure 4. Where $\varepsilon = 0.75$ for a berkovich indenter

$$S = \frac{dP}{dh} \tag{7}$$

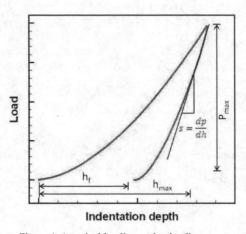

Figure 4 A typical loading and unloading curve.

EXPERIMENT

Nanoindentation was performed on the cross section of an all-ceramic zirconia-porcelain crown using a MTS nanoindenter XP, which has a resolution of 0.2 nm in displacement and 50 nN in load. The nanoindentation impression was examined for pile-up and sink-in by a Scanning Probe Microscope (SPM) equipped with a Hysitron Triboindenter. The Hysitron has a resolution of 0.04 nm in displacement and 1 nN in load. The ceramic crown used have a zirconia core with 0.5 mm thickness and veneered porcelain with varying thickness. Indentation was performed on both stressed and a reference sample. On the stressed sample, the location of indents was where the porcelain was the thickest. Also, this location had a curved surface, shown as dashed line in Figure 5. The thickness of the zirconia and porcelain was 0.5 mm and 2.1 mm. The maximum loads applied on zirconia and porcelain were 200 mN and 30 mN respectively, while the loading time was 40s for both materials. A half crown annealed at 400 °C was used as a reference sample. Randomly chosen, 19 indents were performed on the porcelain and 23 indents were performed on zirconia in the reference half crown. The average hardness value of the indents on the reference crown was used for residual stress computation. Both the reference sample and the stressed sample were polished the same amount to minimize any differences caused by polishing.

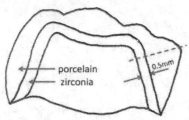

Figure 5 Nanoindentation performed along the dashed line on a zirconia-porcelain crown cross section.

RESULTS AND DISCUSSION
    This study probes the residual stresses in a ceramic crown including both zirconia and porcelain. To compute the value of residual stress, the results from a corresponding reference sample are needed. The samples were assumed to have biaxial stress, and the stress in the direction of thickness was assumed zero due to dimension being much smaller relative to the other two dimensions. Comparing to the reference sample, smaller hardness means tensile stress, while larger hardness means compressive stress. After nanoindentation, the real contact was examined using SPM and compared with the nominal contact area derived from the equation. Zirconia showed significant pile-up, while it was slight for porcelain, as shown in Figure 6 and Figure 7. Due to pile-up, the measured real contact area of zirconia was 12% larger than the nominal area derived mathematically, while it was 5% for porcelain. The effect of pile-up was considered in the following computation of the residual stress.

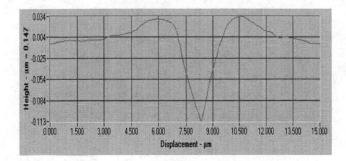

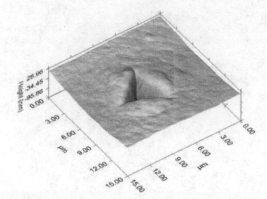

Figure 6 Zirconia nanoindentation impression shows large pile-up.

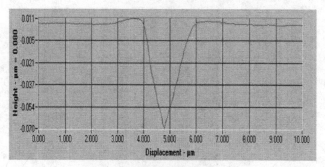

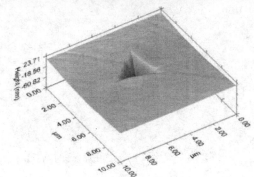

Figure 7 Porcelain nanoindentation impression shows slight pile-up.

At the same indentation load, the indentation load depth curves for the stressed and reference samples will be different because of the different the residual stresses. Figure 8 shows a typical load depth curve of stressed and reference samples for both porcelain and zirconia in case of tensile stress. At the same load, the indentation depth is larger for the stressed sample, resulting smaller hardness. It is opposite for compressive stress.

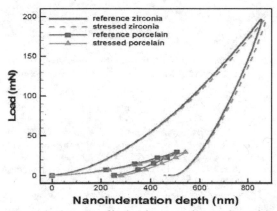

Figure 8 Load depth curves of both reference and stressed porcelain and zirconia

Based on the measured load depth data, hardness is derived. By comparing the hardness of the stressed sample with that of the reference sample, residual stress is determined. Finding the appropriate reference sample is not trivial. A half crown annealed at 400 °C and slowly cooled down to room temperature was used in this case. Fischer *et al.*[22] showed that annealing at about 100°C below the glass transition temperature of porcelain relaxes residual stress while not changing the microstructure. The average hardness for the reference sample was 6.55±0.36 GPa and 13.82±0.77 GPa for porcelain and zirconia respectively.

Figure 9 shows both the hardness and residual stress along the dashed line on the crown for both porcelain and zirconia. Residual stress across thickness for both porcelain and zirconia are significantly different. Porcelain shows more tensile stress as it approaches the crown surface and shows less tensile stress as it gets closer to the zirconia-porcelain interface. Zirconia shows less tensile stress as it approaches the crown surface and shows more tensile stress as it gets closer to the zirconia-porcelain interface. The maximum tensile stress was 380 MPa and 800 MPa for porcelain and zirconia, respectively. This indicates the porcelain thickness affects distribution and magnitude of residual stress, as predicted by Allahkarami *et al.*[23]. For equilibrium we should see both compressive and tensile residual stress, but the experimental results only show tensile stress. One reason could be that the annealed sample was assumed residual stress free, but in reality it must not be completely stress free. Therefore, the measured data are offset to the tensile direction. The local variation of the measurement could be caused by microstructural differences, but the trend in general reflects a reasonable effect from residual stress.

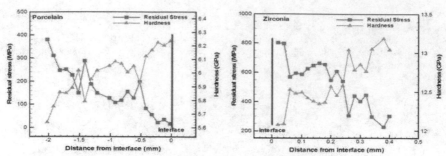

Figure 9 Residual stress and hardness along marked line on crown

Zirconia is known for its phase transformation behavior (from tetragonal to monoclinic). Examining load depth curves is one way to look for signs of phase transformation. The indentation load-depth curves for both porcelain and zirconia shown in Figure 8 are very smooth, and show no apparent phase transformation. However, the more significant pile-up seen in zirconia could be a sign of phase transformation. Gaillard et al.[24] have shown using micro-raman that such a low indentation force as 200 mN will not induce phase transformation. Surface treatment, grinding and polishing, has been shown to slightly alter the surface stress state of zirconia due to stress induced phase transformation.[25,26] Here, to minimize the effect of polishing, both the reference sample and stressed sample were polished the same amount, but further quantitative characterization including polishing effects is needed in the future.

CONCLUSION

A method for measuring residual stress in porcelain layers of dental crowns was identified and the first results have been observed. In this case, stresses were shown to become more tensile for porcelain as a function of the distance from the interface. Similarly, stresses in zirconia were shown to become less tensile moving away from the interface. Residual stress can be extracted by comparing the hardness of a stressed and reference sample. Residual stress along one critical location in a crown across the thickness for both zirconia and porcelain was obtained. While the absolute magnitude of stresses needs confirmation, the results confirm that residual stress can be significantly different across the thickness for both porcelain and zirconia in modern dental crowns.

ACKNOWLEDGEMENT

This research was made possible in part by the Oklahoma Health Research award (project number HR07-134), from the Oklahoma Center for the Advancement of Science and Technology (OCAST). The authors would like to appreciate Drs. P. Coelho, V. Thompson, E. D. Rekow and M. Cabrera at NYU College of Dentistry for providing the samples and valuable discussion.

REFERENCES

[1] E. Tsalouchou, M. J. Cattell, J. C. Knoeles, P. Pittayachawan, A. McDonald, Fatigure and Fracture Properties of Yttria Partially Stabilized Zirconia Crown Systems, *Dental Materials*, **24**, 308-318 (2008)

[2] A. J. Raigrodski, Contemporary Materials and Technologies for All-ceramic Fixed Partial Dentures: a Review of the Literature", *J. Prosthet Dent*, **92**(6) , 557-62 (2004)

[3] P. Christel, A. Munier, M. Heller, J. P. Torre, C. N. Peille, Mechanical Properties and Short-term In-vivo Evaluation of Yttrium-oxide-partially-stablized Zirconia" *J Biomed Mater Res.*, **23**(1), 45-61 (1989)

[4] P.G. Coelhio, N. R. Silva, E. A. Bonfante, P. C. Guess, E. D. Rekow, V. P. Thompson, Fatigue Testing of Two Porcelain-zirconia All-ceramic Crown Systems, *Dental Mater.*, **25**, 1122-1127 (2009)

[5] I. Denry, J.R. Kelly, State of the Art of Zircionia for Dental Application, *Dental Mater.*, **24**, 299-307 (2008)

[6] B.E. Pjetursson, I. Sailer, M. Zwahlen, C.H. Hammerle, A Systematic Review of the Survival and Complication Rates of All-ceramic and Metal-ceramic Reconstructions After an Observation Period of at Least 3 years. Part I: Single Crowns, Clin Oral Inpl Res 18 (suppl3): 73-85: *erratum in Clin Oral Impl Res.*, **19**, 326-328 (2008)

[7] S.P. Von, All-Ceramic Fixed Partial Dentures, Studies on Aluminum Oxide- and Zirconium Dioxide Based Ceramic Systems", *Swed Dent J. Suppl.*, **173**, 1-69 (2005)

[8] S. P. Von, P. Calson, K. Nilner, All-ceramic Fixed Partial Denture Designed According to DC-Zircon® a Two Year Clinical Study." *J. Oral Rehab*, **32**, 180-187 (2005)

[9] C. Larsson, S. P. Von, B. Sunzel, K. Nilner, All-ceramic Two and Five-unit Implant-supported Reconstructions, A Randomized, Prospective Clinical Trial" *Swed Dent J.*, **30**, 45-53 (2006)

[10] A. J. Raigrodski, G. J. Chiche, N. Potiket, J. L. Hochstedler, S. E. Mohamed, S. Billot, The Efficacy of Posterior Three-unit Zirconium-oxide-based Ceramic Fixed Partial Denture Prostheses: a Prospective Clinical Pilot Study", *J Prosthet Dent.*, **96**, 237-244 (2006)

[11] H. A. Bale, J. C. Hanan, Interface Residual Stresses in Dental Zirconia Using Laue Micro Diffraction", *8th International Conference on Residual Stresses*, Denver, CO. (2008)

[12] H. A. Bale, Measurement and Analysis of Residual Stresses in Zirconia Dental Composites Using Micro X-ray *Dissertation, Oklahoma State University*, 2010

[13] T. Y. Tsui, W. C. Oliver, G. M. Pharr, Influence of Stress On the Measurement of Mechanical Properties Using Nanoindentation: Part I. Experimental Studies in an Aluminum Alloy." *J Mater Res.*, **11**(3), 752-759 (1996)

[14] A. Bolshakov, W.C. Oliver, G.M. Pharr, Influence of Stress on the Measurement of Mechanical Properties Using Nanoindentation: Part II. Finite Element Simulations." *J Mterl Res.*, **11**(3), 760-768 (1996)

[15] J. G. Swadner, B. Taljat, G. M. Pharr, Measurement of Residual Stress by Load and Depth Sensing Indentation with Spherical Indenters" *J Mater Res.*, **16**(17), 2091-2102 (2001)

[16] S. Suresh, A. E. Giannakopoulos, A New Method for Estimating Residual Stress by Instrumented Sharp Indentation" *Acta Mater.*, **46**(16), 5755-5767 (1988)

[17] S. Carlsson, P.L. Larsson, On the Determination of Residual Stress and Strain Fields by Sharp Indentation Testing Part II: Experimental Investigation, *Acta Mater.*, **49** 2193-2203 (2001)

[18] D. Tabor, *Hardness of Metals*, Cambridge University Press, Cambridge (1951)

[19] K. L. Johnson, *Contact Mechanics*, Cambridge University Press, Cambridge (1985)

[20] W. C. Oliver and G. M. Pharr, An Improved Technique for Determining Hardness and Elastic Modulus Using Load and Displacement Sensing Indentation Experiments, *J mater Res.*, 7, 1564-1583 (1992)

[21] I. N. Sneddon, The Relation Between Load and Penetration in the Axisymmetrc Boussinnesq Problem for a Punch of Arbitrary Profile, *Int J Eng Sci.*, **3**, 47-571965

[22] H. Fischer, M. Hemelik , R. Telle, R. Mark, Influence of Annealing Temperature on the Strength of Dental Glass Ceramic Materials, *Dental Mater*, **21**, 671-677 (2005)

[23] M. Allahkarami, H. A. Bale, J. C. Hanan, Characterization of Non Uniform Veneer Thickness Distribution on Curved Substrate Zirconia Ceramics Using X-ray Micro-Tomography, ICAAA, Daytona, FL, (2011)

[24] Y. Gaillard, M. Anglada, and E. Jimenez-Pique, Nanoindentation of Yttria-doped Zirconia: Effect of Crystallographyic Structure on the Deformation Mechanisms, *J. Mater. Res.*, **24**(3) 719-727 (2008)

[25] A. Juy, M. Anglada, Strength and Grinding Residual Stresses of Y-TZP with Duplex Microstructures, *Engineering Failure Analysis*, **16**, 2586-2597 (2009)

[26] C. J. Ho, H. C. Liu, W. H. Tuan, Effect of Abrasive Grinding on the Strength of Y-TZP, *J. of European Ceramic Society*, **29**, 2665-2669 (2009)

DESIGN AND DEVELOPMENT APPROACH FOR GAS TURBINE COMBUSTION CHAMBERS MADE OF OXIDE CERAMIC MATRIX COMPOSITES

Ralf Knoche, Erich Werth, Markus Weth, Jesus Gómez García
ASTRIUM Space Transportation, Bremen, Germany

Christian Wilhelmi
EADS Innovation Works, Munich, Germany

Miklós Gerendás
Rolls-Royce Germany, Berlin, Germany

ABSTRACT

Higher efficiencies and reduced emissions in gas turbine engines require the turbine cycle to operate at higher pressures and temperatures in conjunction with improved combustion technologies. A safe operation of high performance gas turbines can only be achieved by using high temperature (HT) materials with appropriate cooling methods. Accordingly the introduction of Ceramic Matrix Composites (CMC) for effusion cooled combustor and turbine liners appears promising due to elevated thermo-mechanical performances and low densities. Especially oxide CMC are of elevated significance since they are inherently resistant to oxidation, which enables an acceptable lifespan, and the required external barrier coatings are of oxide ceramics, too. Thus the development of these CMC and tailored cooling schemes is one of the main tasks for increasing efficiencies in gas turbines.

This scope has led to the cooperative project HiPOC (**H**igh **P**erformance **O**xide Ceramics) consisting of 3 companies and 4 research institutes. In HiPOC ASTRIUM focuses on the structural design of representative gas turbine components made of CMC including attachment systems. The paper describes the design for structural components in gas turbine applications which obviously is influenced by CMC characteristics and its manufacturing. The development approach is presented together with results from the FEM verification of the design. The paper concludes with the next steps in developing CMC components for gas turbines.

INTRODUCTION

Modern day demands for higher efficiencies and reduced emission in gas turbine engines require the gas turbine cycle to operate at higher pressure ratios and at higher turbine entry temperatures in conjunction with improved combustion technologies. Since the current temperature levels in advanced gas turbines can exceed the maximum allowed combustor and turbine liners metal alloy temperature by more than 500 K, a safe operation of high performance gas turbines can only be achieved by using high temperature materials combined with appropriate cooling methods. As a consequence, the introduction of Ceramic Matrix Composites (CMC) as structural components for combustor and turbine liners appears very promising due to their elevated thermo-mechanical performance and their low density compared to metal alloys or inter-metallic materials. Especially oxide CMC are inherent resistant to high temperature oxidation and required external barrier coatings are of oxide ceramics, too, which in concurrence enables an acceptable lifespan of corresponding components. Thus the development of these sophisticated CMC materials and tailored cooling schemes is one of the main focuses for significantly increasing the operating temperatures in future gas turbines.

This scope has led to the German BMBF (German Federal Ministry of Education Research) collaborative HiPOC program started in February 2009 which main objectives is based on the development of three different oxide/oxide CMC materials which are candidates for the use in gas turbines for power generation and aerospace propulsion, or as spin-off in space applications such as for Thermal Protection Systems and Hot Structures. In conjunction with an improved thermal management the main objectives are to minimise the fuel consumption and thereby reduce the emission from the gas turbine. To achieve this goal, design concepts for the attachment of the CMC component to the metal structure of the engine are developed in a first work stream focused on the combustion chamber and the turbine seal segment. Issues like Environmental Barrier Coating (EBC) / Thermal Barrier Coating (TBC) application and volatilisation, allowance for the different thermal expansion and the mechanical fixation are addressed. The design work is accompanied by CFD and FEM simulations. A variation of the microstructural design of the three CMC materials in terms of different fibre architecture and processing of matrix are considered. To enhance the required thermo-mechanical properties and long-term stability of the components, the material concepts are developed further in a second work stream and validated by material testing. By modification of the matrix and the fibre-matrix interface as well as EBC coatings, the high temperature stability and the insulation performance are enhanced. As for precise simulations detailed experimental data are necessary, the oxide/oxide CMC are tested in various loading modes (tension, compression, shear, off-axis loading) from room temperature to maximum application temperature. These studies indicate the high temperature potential of the CMC materials under investigation and support the development tailored design rules.

Within the HiPOC program ASTRIUM Space Transportation focuses on the structural design of representative gas turbine components and demonstrators made of oxide CMC including appropriate attachment systems. These activities strongly interact with the further development of one of the three oxide CMC termed UMOX[TM] which is performed by EADS Innovation Works [1,2]. Within the development ASTRIUM Space Transportation can benefit from many years of experiences in the field of Thermal Protection Systems (TPS) and Hot Structures (HS) during various development programs such as for X-33, X-38, PARES, or IXV [3-8]. Corresponding experiences for instance in CMC adequate design is adapted to the specific gas turbine components and tailored to gas turbine requirements and boundary conditions.

The material UMOX[TM] is currently used for the CMC face sheet within in the Surface Protected Flexible Insulation (SPFI) TPS and was also successfully flight tested for Do 228 Exhaust Nozzle [9]. The oxide CMC Exhaust cone (as manufactured) and exhaust nozzles (after flight testing) are depicted in figure 1.

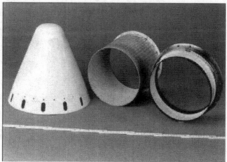

Figure 1. Oxide CMC (UMOX[TM]) exhaust cone for feasibility study (left) and successfully flight tested UMOX[TM] exhaust nozzles for DO 228 airplane (right).

UMOX[TM] - OXIDE CMC MATERIAL AND MANUFACTURING

UMOX[TM] is the standard oxide based CMC material used within ASTRIUM ST / EADS IW and is available in specific material configurations. The matrix of one of these UMOX[TM] configurations is based on a commercial micron-sized mullite powder and a polysiloxane precursor.

Continuous Nextel[TM] 610 (3M[TM], USA) alumina fibres are used as reinforcement. These fibres are equipped with an organic fibre coating which is removed after composite manufacture (fugitive interface) [8]. Note that the characteristics of the resulting gap between fibre and matrix is of elevated significant since on the one hand propagating matrix cracks shall deflected at this interface preventing the fibres from premature failures and on the other hand sintering effects between fibre and matrix after prolonged exposure to high temperature shall to be avoided. By tailoring this fugitive coating and the resulting gap correctly the for CMC known energy dissipating mechanism such as debonding, crack deflection, crack bridging, and fibre pull-out lead to the associated damage tolerant behaviour of CMC.

The oxide CMC is manufactured via the Polymer Infiltration Pyrolysis (PIP) process. As seen in figure 2 the coated oxide fibres are infiltrated with liquid pre-ceramic matrix slurry and subsequently wound by using a constantly rotating drum or mandrel and a fully automated 6+2 axis robot controlled filament winding process allowing high flexibility in fibre architecture and geometrical shapes. After drying the impregnated fibres are vacuum packed and consolidated in an autoclave process under pressure ($p>10$ bar) at elevated temperature ($T>150°C$). The matrix polymer forms a cross-linked network at this process step bonding the laminate together which leads to a green body that is solid enough for handling and further processing. Subsequent the pre-ceramic matrix is converted to a ceramic by pyrolysis in an inert atmosphere at temperatures above $T=1000$ °C. Due to shrinkage of the pre-ceramic matrix during pyrolysis re-infiltration and pyrolysis cycles can be performed with a polymeric precursor to reduce the open porosity. Obviously the number of re-infiltration cycles depends on the desired material porosity and properties [10]. A typical fibre volume content of the described CMC material is roughly 48-50 Vol.-% by porosity 10-12 Vol.-% and a density of 2.4-2.5 g/cm³.

More information on the material and its manufacturing as well as on UMOX[TM] thermo-mechanical performance can be found in [1,2,11].

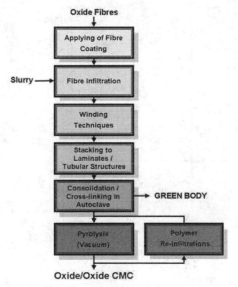

Figure 2. Schematic oxide CMC PIP manufacturing process.

DEVELOPMENT APPROACH

The typical development approach, as schematically shown in figure 3, clearly reveals the interactive and iterative design process of composite structures in aerospace applications between various technical disciplines. Beginning from the top, the general application specification (system requirements) are broken down and narrowed to the specific component or composite structure requirements in terms of performance, basic geometric conditions, and specific environment. Prior to establishing a component design based on input requirements, general design rules as well as specific experiences in composite design including standard solutions for e.g. sealing or sensor implementation are taken into account. The concept design solution has obviously to be carefully reviewed for its manufacturing feasibility, which are again influenced by the envisaged materials and associated manufacturing processes as well as corresponding costs and schedule impacts. The iterative and interactive process results in a concept design which is feasible from manufacturing perspective and which shall comply with the applicable requirements.

In the case of "non-feasibility" due to e. g. inevitable physical boundary conditions or contradicting requirements a feedback with higher level system requirements has to be considered. This may either lead to an adaptation on requirements by reducing for instance margins or the reductions of overambitious performance requirements.

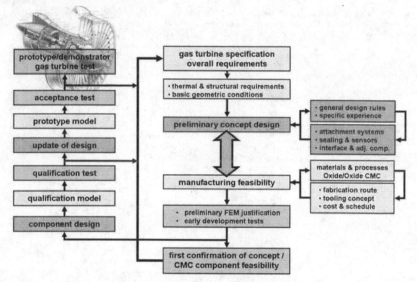

Figure 3. Exemplary development logic for CMC components in aerospace applications indicating the interactive and iterative design process.

Although FEM structural analysis can verify the feasibility of the design comparably fast, technology samples and manufacturing trials (within early development tests) on critical or yet not known characteristics are essential in order to enable an early and profound consolidation of the design (cf. section Early Development Test).

COMBUSTION CHAMBER DESIGN

Within the HiPOC program the combustion chamber is designed for a given test rig which provides similar conditions to gas turbines. Thus the verification of design features such as CMC composition, lay-up, and attachments systems can most suitably be performed.

As already mentioned, ASTRIUM ST focuses on the design of the CMC combustion chamber including its attachment system. Thereby the design has to meet various requirements which mainly can be broken down to geometrical boundary condition of the test rig, the lifetime performance under severe thermo-mechanical loading conditions as well as displacement limitation and stiffness requirements. The design is further influenced by manufacturing capabilities and experiences in advanced CMC fabrication (cf. figure 3) which, in case of the tubular design, is tailored to the UMOX^{TM} manufacturing process. Additional design drivers can emerge from the development outcomes of the other disciplines such as within material and EBC developments as well as optimised cooling configurations. Beside challenging requirements, in particular the latter interactions request for an iterative designing process, according to figure 3, accounting for e.g. manufacturing experiences, CMC adequate design know-how, and FEM analysis capabilities in order to verify the feasibility of the intended design. Similar approaches and challenges as for gas turbine combustion chambers have already been encountered by ASTRIUM Space Transportation, however, during developments of Hot Structures (HS) and Thermal Protection Systems (TPS) for (reusable) re-entry vehicles.

Several design concepts have been established within the HiPOC project and are constantly refined with further findings. Currently two tubular combustion chamber concepts are analysed in detail which are presented in figure 4 and figure 5.

Figure 4. ASTRIUM ST combustion chamber design C 3-1 including attachment systems in assembled state within the test rig (left) and depicted separately (right) showing the upstream integral CMC/metallic attachment system and the rear metallic spring element.

The left image of figure 4 shows the tubular combustion chamber (approximately 140 mm in diameter and 250 mm in length) in an assembled state within the pressurised barrel of the test rig, whereas the right image shows the oxide CMC combustion chamber with its attachment system located outside the severe thermally loaded combustion chamber. Special attention has to be drawn to the attachment system since the right balance between flexibility (decreasing the thermally induced stresses) and stiffness (accounting for allowable displacements and frequency requirements) has to be adjusted. As a consequence the design concept C 3-1 uses an integral manufactured CMC brackets which via metallic brackets are attached to the upstream flange of the test rig. In doing so the attachment system is flexible in radial direction by having a relatively high stiffness in lateral direction leading to a defined condition of the integrated combustion chamber. In contrast to this the downstream metallic spring element is flexible in both directions (radial and lateral) enabling the CMC combustion chamber to expand due to the emerging temperatures. The experiences on CMC adequate design including metallic attachment systems reveals that for the bolted connection of the upstream attachment system (at the interface CMC bracket - metallic bracket) special Thermal Stress-Free Fasteners should be used compensating the mismatch of thermal expansion coefficients effectively [12].

The difference to the second design concept C 3-2, depicted in figure 5, can mainly be attributed to the upstream attachment system. Thereby the combustion chamber is characterised by a "clean" tubular CMC design, i.e. no integral CMC brackets are needed, having the advantage of an eased manufacturing and elevated robustness. The outer bypass of the metallic bracket of the upstream attachment system further leads to an elevated radial flexibility. The rear spring element is in accordance to the C 3-1 design.

Figure 5. ASTRIUM ST combustion chamber design C 3-2 including attachment systems in assembled state within the test rig (left) and depicted separately (right) showing the upstream metallic attachment system and the rear metallic spring element.

Due to the improved robustness of the design without integral CMC stand-off at the upstream attachment system the design concept C3-2 was pre-selected for the detailed FEM analysis.

FEM ANALYSIS AND PRELIMINARY RESULTS

Description of the FE-Model

Based on the current design the FE-Model is modelled with ANSA for the use within the FE-Solver MSC.NASTRAN. All thin components like the combustion chamber or the attachment system are represented by shell elements instead of solids to model the layer set-up in detail gaining for instance stress results for each layer and enabling the application of different failure criteria. For illustration the FE-Model is depicted in figure 6. Mechanical and thermo-mechanical characteristics of UMOX[TM] are implemented from former internal and recent testing campaigns performed within HiPOC [11]. The combustion chamber assembly is set hard mounted at all bolted interfaces to the test rig. Additionally the combustion chamber edge at the inlet (left) is constrained in axial direction due to the support by an upstream sealing.

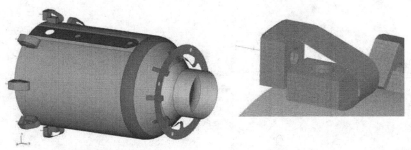

Figure 6. FE-Modelling of the complete combustion chamber (left) and close-up of the FE-Modelling of the upstream attachment system (right).

Preliminary FE Results

The stress and displacement analysis consider the critical thermal and pressure load cases. At present state, however, the implementation of the emerging temperature distribution calculated by computational fluid dynamics (CFD) analysis is still running and thus only preliminary results can be provided.

Figure 7 shows the deformation of the CMC combustion chamber due to the pressure load case. As clearly seen at the top of the upstream attachment brackets the combustion chamber compresses due to the pressure gradient evolving during operation (elevated pressures at the outer cavity and reduced pressure levels within the combustion chamber). However, as seen in the resulting stresses (exemplary provided by figure 8 for stresses in fibre direction at the inner and outer layer of the composite) the evolving stress levels are very low (-10 MPa $< \sigma <$ 10 MPa). Hence the pressure load case can not be considered design driving for this application.

Figure 7. Displacement plot of the CMC combustion chamber resulting from the pressure load case, (black grid: undeformed and green grid: deformed), displacement are scaled for better observation.

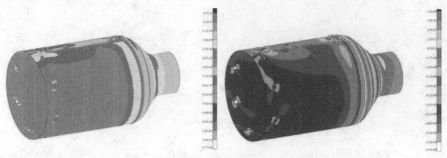

Figure 8. Stress plot of the inner and outer layer resulting from the pressure load case, stress are depicted in fibre direction of the inner layer (left) and the outer layer (right)

In contrast to this, the temperature load case has to be carefully investigated. This is because the combination of both, a harsh gradient within the CMC combustion chamber wall and a relatively high Coefficients of Thermal Expansion (*CTE*), may lead to severe thermally induced stresses that might be difficult to handle. The evolving temperature which has to be refined in the next step by precise CFD analysis results are exemplary depicted in figure 9 and reveal that temperature gradients of approximately 350 K through the thickness of the CMC chamber have to be dealt with.

Figure 10 shows the deformation of the CMC combustion chamber resulting from the preliminary assessed temperature load case. As clearly seen at top of the upstream attachment brackets as well as at the rear of the chamber the temperature field leads to a pronounced thermal expansion of the combustion chamber. The attachment system thereby enables the CMC combustion chamber to expand in order to minimise corresponding mechanical influence on the CMC structure.

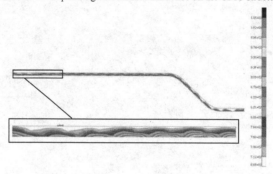

Figure 9. Exemplary temperature distribution within the oxide CMC combustion chamber during operation.

Figure 10. Displacement plot of the CMC combustion chamber resulting from the preliminary assessed temperature load case, (black grid: undeformed and green grid: deformed), displacement are scaled for better observation.

Besides these first results further FEM analysis in particular including the design driving temperature load cases have to be completed and evaluated. In parallel (and in line with the presented development logic of figure 3) early development tests are envisaged within the HiPOC program.

EARLY DEVELOPMENT TESTS

These early development tests include manufacturing trials and technology samples. Manufacturing trials can comprise the fabrication of adequate components for investigations on spring-back and shrinkage effects or can relate to manufacturing parameters such as for the introduction of cooling channels.

Besides the general characterisation of the oxide/oxide CMC materials, as input for FEM and CFD analysis, also technology samples are planed such as to validate degradation effects of cooling channels on the mechanical performance of oxide/oxide CMC. For this reason samples according to figure 11 (left illustration) are manufactured for tensile testing among other mechanical tests. Furthermore technology tests for the attachment system are envisaged (cf. figure 11, right illustration) in order to confirm the stiffness characteristics used in the FE-Model. Additionally, specific thermal property data at the attachment system (e.g. conductance values) can be measured. The test results will then lead to the verification of the FEM tools as well as to the refinement of FEM assumptions and boundary conditions.

Note that, although the combustion chamber is designed for the given test rig corresponding conditions are similar or true to gas turbine environments and the verification of design features such as CMC composition, lay-up, and attachments systems can most suitably be performed. Throughout these developments also a significant spin-off related to the use of oxide CMC for TPS and HS applications can be achieved.

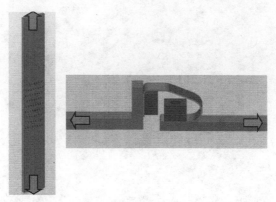

Figure 11. Exemplary technology samples to be tested within early development tests, left: tensile sample with cooling channels for the evaluations of mechanical degradation effects by cooling channels, right: attachments system / bracket stiffness test for evaluation of the FE-Model input parameters.

3. SUMMARY AND OUTLOOK

The paper provides an overview of the German BMBF HiPOC program for the development of oxide/oxide CMC materials and components for the use in gas turbine applications [2]. Thereby primarily ASTRIUM activities and the related development approach are presented including two design concepts for a CMC combustion chamber. First FEM analysis results are available and were presented. However, FEM analyses have to be continued in particular considering precise temperature distributions derived from CFD computations. In parallel early developments test will be performed in order to enable (together with FEM results) an early verification on the feasibility of the current combustion chamber design.

After successful confirmation an oxide/oxide CMC combustor for testing in the high pressure single sector test facility of DLR will be manufactured and qualified. Finally, the CMC combustor will be tested with pressure, temperature, and air-fuel ratio being representative for an aero engine on a regional airplane in conjunction with an advanced fuel injector in order to achieve reduced NO_x emissions compared to conventional TBC-coated metallic liner.

At the end of the program in 2012, the results, including the once for the turbine liners (also referred to as seal segments or blade tracks) developed in parallel in another work stream, will be used to update a performance deck of a regional engine to quantify the improvements in terms of specific fuel consumption and CO_2 emission reduction [2].

ACKNOWLEDGMENTS

This work is part of the HiPOC project (contract number 03X3528) within the WING framework. It is funded in part by the German Federal Ministry of Education and Research (BMBF) and administered by Projektträger Jülich (PTJ). Their support is gratefully acknowledged.

REFERENCES

[1] C. Wilhelmi, T. Machry, R. Knoche, and D. Koch, Processing of Advanced Oxide/Oxide Composites for Gas Turbine Applications Based on Tailored Textile Technologies. *Proceedings of the 34th International Conference on Advanced Ceramics and Composites (ICACC)*, Florida, USA (2011).

[2] M. Gerendás, Y. Cadoret, C. Wilhelmi, T. Machry, R. Knoche, T. Behrendt, T. Aumeier, S. Denis, J. Göring, D. Koch, and K. Tushtev, Improvements of Oxide/Oxide CMC and Development of Combustor and Turbine Components in the HiPOC Program, in press: *Proceedings of the ASME Turbo Expo 2011: Power for Land, Sea Air*, GT2011-45460, Canada (2011).

[3] U. Trabandt, H. Ritter, H. Reinkober, T. Schmid, and H. G. Wulz, CMC Large Panel TPS Applied on X-38 Nose Skirt, *Proceedings of the 30th International Conference on Environmental Systems*, SAE Technical Paper Series 2000-01-2234, France (2000).

[4] U. Trabandt, and K. Handrik, CMC Nose Skirt Panels for X-38 - Successfully Qualification Tested and Fit-Checked, *Proceedings of the 2nd World Space Congress*, IAF-No. p.I-7-09IAF, USA (2002).

[5] R. Knoche, W.-D. Ebeling, and U. Trabandt, Thermo-mechanical Design and Plasma Test Verification on PARES CMC Nose Cap, *Proceedings of the 38th International Conference on Environmental Systems ICES2008*, No. 2008-01-2171, CA, USA (2008).

[6] R. Knoche, E. Werth, W. Rotärmel, H. Hald, E. Brach Prever, A. Denaro, and S. Langlois, Thermo-mechanical Design of the IXV Nose Assembly, *Proceedings of the 6th European Workshop on Thermal Protection Systems and Hot Structures*, ESA, Germany (2009).

[7] R. Knoche, E. Werth, S. Goëk, and M. Weth, Thermo-mechanical Design of the ASTRIUM C/SiC Experiment on SHEFEX II, *Proceeding of the 7th International Conference on High Temperature Ceramic Matrix Composites (HT-CMC-7)*, Germany (2010).

[8] R. Knoche, The Response of C/SiC Ceramic Matrix Composites Exposed to Severe Thermo-mechanical Load Conditions for Hot Structures and Thermal protection Systems in Reusable Launch Vehicles, *Doctoral Thesis*, University of Bremen (2010).

[9] B. Newman and W. Schäfer, Processing and Properties of Oxide/Oxide Composites for Industrial Applications, In: *High Temperature Ceramic Matrix Composites* (Ed. W. Krenkel, R. Naslain, H. Schneider), Wiley-VCH, 600-609 (2001).

[10] W. D. Vogel and U. Spelz, Cost Effective Production Techniques for Continuous Fibre Reinforced Ceramic Matrix Composites, In: *Ceramic Processing Science and Technology*, **51**, 225-259 (1995).

[11] D. Koch, K. Tushtev, K. Rezwan, C. Wilhelmi, S. Denis, and J. Göring, Effect of Microstructure on Room and High Temperature Properties of Oxide/Oxide Composites Developed for Gas Turbine Applications, *Proceedings of the 34th International Conference on Advanced Ceramics and Composites (ICACC)*, Florida, USA (2011).

[12] H. G. Wulz, H. Stark, and U. Trabandt, Joining and Load Transfer Technology for Hot Structures, *Proceedings of the 34th AIAA Thermophysics Conference*, AIAA-2000-2435, USA (2000).

EFFECTS OF PRELOADING ON FOREIGN OBJECT DAMAGE IN AN N720/ALUMINA OXIDE/OXIDE CERAMIC MATRIX COMPOSITE

D. Calvin Faucett and Sung R. Choi[†]
Naval Air Systems Command, Patuxent River, MD 20670

ABSTRACT

Foreign object damage (FOD) of an N720/alumina oxide/oxide ceramic matrix composite (CMC) was assessed with 1.59 mm-diameter hardened steel ball projectiles using an impact velocity range of 150 to 350 m/s at a normal incidence angle. Targets were ballistically impacted under tensile preloading with three different levels of load factors of 0, 30, and 50 %. Frontal impact damage was in the form of craters and increased in size with increasing impact velocity. Subsurface damage beneath the impact sites was associated with fiber/matrix breakage, collapse of pores (compaction), tows' bend/breakage/protrusion, delamination, and formation of cone cracks. Difference in FOD between preloading and non-preloading was noticeable particularly at 350 m/s with a load factor 30 %, where the targets were on the brink of penetration by projectiles, resulting in significant strength degradation as well.

INTRODUCTION

The brittle nature of either monolithic ceramics or ceramic matrix composites (CMCs) or ceramic environmental barrier coatings (EBCs) or ceramic thermal barrier coatings (TBCs) has raised concerns on structural damage when subjected to impact by foreign objects. This has prompted the propulsion communities to take into account for foreign object damage (FOD) as an important design parameter when those materials are intended to be used for aeroengine applications. A significant amount of work on impact damage of brittle monolithic materials has been conducted during the past decades experimentally or analytically [e.g. 1-14], including gas-turbine grade toughened silicon nitrides [15-17].

CMCs have been used and considered as enabling propulsion materials for advanced civilian and/or military aeroengines components. A span of FOD work has been carried out in the past to characterize FOD behavior of some gas-turbine grade CMCs such as state-of-the-art melt-infiltrated (MI) SiC/SiC [18], N720 /aluminosilicate (N720/AS) oxide/oxide [19], N720/alumina (N720/A) oxide/oxide [20], and 3D woven SiC/SiC [21]. Responses to Hertzian indentation were also explored using a SiC/SiC CMC to simulate a quasi-static impact phenomenon [22]. Unlike their monolithic counterparts, all the CMCs investigated have not exhibited catastrophic failure for impact velocities up to 400 m/s, resulting in much increased FOD resistance. However, the degree of damage that resulted in strength degradation was still substantial particularly at higher impact velocities 350 m/s when impacted by hardened steel ball projectiles (1.59 mm diameter).

Most of the FOD work in CMCs has been conducted under a condition that target samples were free-standing during impact, i.e., without any external- or pre-loading applied onto

---

[†] Corresponding author. Email address: sung.choi1@navy.mil

89

the targets. Recently, FOD tests of an N720/alumina oxide/oxide CMC with target samples being tensile-loaded during impact have been performed to determine the aspects of damage morphology as well as of strength degradation [23]. The test was for a simulation of a rotating aeroengine component such as blades where tensile stresses are generated due to centrifugal force. The current paper will present more in-depth assessments of damage morphologies with respect to the degree of preloading. From the authors' best knowledge, this type of FOD work was considered to be unique, as applied to CMC material systems.

EXPERIMENTAL PROCEDURES

Materials

The CMC utilized in this work as target samples has been described elsewhere [20] and is repeated here briefly. The composite was a commercial, 2-D woven, N720™ fiber-reinforced alumina matrix CMC, fabricate by ATK/COIC (San Diego, CA; Vintage 2008). N720™ oxide fibers, produced in tows by 3M Corp. (Minneapolis, MN), were woven into 2-D 8 harness-satin fabrics. The fabrics were cut into a proper size, slurry-infiltrated with the matrix, and 12 ply-stacked followed by consolidation and sintering. The fiber volume fraction of the final composite panels was approximately 0.45. Typical microstructure of the composite is shown in Fig. 1. Substantial porosity and microcracks in the matrix were observed, a characteristic of this class of oxide/oxide CMCs for increased damage tolerance [24,25]. No interface fiber coating was employed. Porosity was about 25%. Basic physical and mechanical properties of the oxide/oxide CMC are shown in Table 1. Note that both interlaminar tensile and interlaminar shear strengths of the composite were added during the course of this work. Target specimens measuring 10-12 mm in width, 50-150 mm in length, and about 2.7 mm in as-furnished thickness were machined from the composite panels.

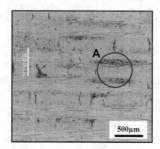   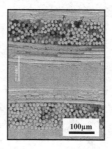

(a) Overall   (b) Region of 'A'

Figure 1. Typical microstructure of an N720/alumino oxide/oxide CMC used in this work: (a) Overall, and (b) Enlarged in 'A' region.

Table 1. Basic physical and mechanical properties of target CMC and projectile materials at ambient temperature

Material		Fiber/matrix	Bulk density (g/cm^3)	Fiber volume fraction	Elastic modulus[1] E (GPa)	Tensile or flexure strength[2] (MPa)	ILT Strength[3] (MPa)	ILS strength[4] (MPa)
Target	Oxide/oxide	Nextel 720 /alumina (2D woven)	2.74	0.45	81	140-145	3.8 (1.6)#	8.6(0.6)
Projectile	Hardened chrome steel (SAE52100)	-	7.78	-	200*	>2000*	-	-

Notes: 1. By the impulse excitation technique, ASTM C 1259 [26]; 2. By flexure specimens in four-point flexure testing (with 20/40 mm spans) and by dog-boned tensile specimens, ASTM C1275 [27]; 3. By ASTM C1468 [28]; 4. By ASTM C1292 [29]. The numbers in parentheses indicate 1.0 standard deviation. * from the literature. ILT: interlaminar tension; ILS: interlaminar shear

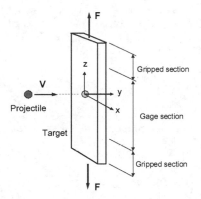

Figure 2. Experimental setup of impact testing under tensile loading (F) for an N720/alumina oxide/oxide CMC

Foreign Object Damage Testing

A ballistic impact gun, as described elsewhere [15,16], was used to conduct foreign-object-damage testing. Hardened (HRC 60) chrome steel-ball projectiles with a diameter of 1.59 mm were inserted into a 300mm-long gun barrel; See the very basic properties of the projectiles in Table 1. A helium-gas cylinder and relief valves were utilized to pressurize and regulate a reservoir to a specific level, depending on prescribed impact velocity. Upon reaching a specific level of pressure, a solenoid valve was instantaneously opened accelerating a steel-ball projectile through the gun barrel to impact the as-furnished 10-12-mm wide side of a target. The target specimens were tensile pre-loaded via two wedge grips, as illustrated in Fig. 2. Each target specimen was aligned such that the projectile impacted at the center of the specimen (along the y-direction) at a normal incidence angle. Three different impact velocities of 150, 250, and 350 m/s were employed in this work. Applied tensile preloading (or stress) during ballistic impact was normalized with respect to the composite strength to get a load factor (α) defined as

$$\alpha = c_{FOD}/c_f \tag{1}$$

where σ_{FOD} is the applied stress during impact and σ_f is the as-received composite strength (=140 MPa). Three levels of load factor, α = 0, 30, and 50 % were used. At least two to three target specimens were used in impact-and-strength tests at each load factor for a given impact velocity. Three targets were also additionally used for impact only with which impact morphologies as well as cross-section's features were to be characterized in more detail.

Post-Impact Tensile Strength Testing
       Post-impact tensile testing was performed to determine post-impact strength of impacted target specimens from which the degree of impact damage was better assessed. Strength testing was carried out via tensile wedge grips with an MTS servohydraulic test frame (Model 312) at a crosshead speed of 0.25 mm/min. The as-received tensile strength of the composite was also determined for a base-line reference using a total of three dog-boned tensile test specimens in accordance with ASTM C 1275 [27].

RESULTS AND DISCUSSION

Projectiles
       It is noted that the steel ball projectiles exhibited by no means any visible damage or plastic deformation upon impact on the oxide/oxide targets even at the highest impact velocity of 350 m/s. This trend was similar to the case observed in the N720/aluminosilicate composite [19]. However, the steel balls impacting harder monolithic silicon nitrides [15-17] or MI SiC/SiC composite [18] were severely flattened or fragmented at impact velocities 350 m/s. Hence, the oxide/oxide composites are relatively 'soft' in response to their ballistic impact, as compared to the hardened steel ball projectiles. This may be derived from difference in elastic modulus: The N720/alumina CMC exhibited a significantly low elastic modulus of E = 81GPa while the hardened steel ball projectiles exhibited a high value of E = 200 GPa. A similar situation was for N720/aluminosilicate CMC (with E = 67 GPa) impacted by the same, hardened steel projectiles [19]. The silicon nitrides [15] and the MI SiC/SiC [18] showed E = 300 and 220 GPa, respectively. The effects of projectile materials such as hardened steel, annealed steel, brass, and silicon nitride were also assessed using silicon nitrides as target materials [17]. The hardness of a projectile material was found to be a key material parameter affecting degree of impact damage onto a given ceramic target material.

CMC Targets
       Figure 3 presents the frontal impact damage of target specimens with respect to impact velocity and load factor. As observed in many CMCs, the ballistic impact generated surface and subsurface damage, including craters with fiber/matrix breakage and some material removal with

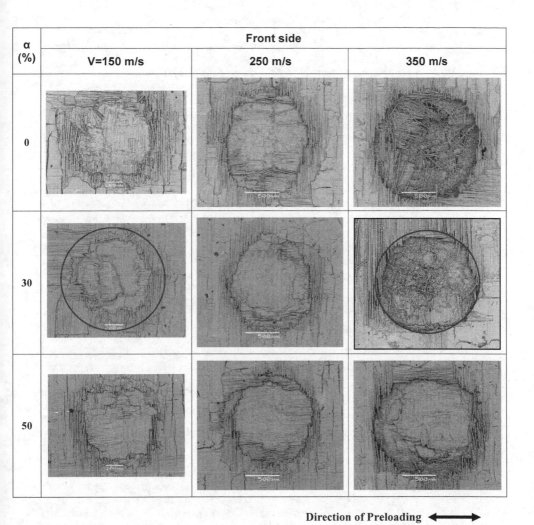

Direction of Preloading ⬌

Figure 3. Frontal impact damages with respect to impact velocity (V) and load factor (α) in an N720/alumina oxide/oxide CMC impacted by 1.59-mm hardened chrome steel ball projectiles. The circle represents the size of a steel ball projectile utilized. All the images are under the same magnification.

their degree of severity being dependent on impact velocity. As seen in Fig. 3 (the overall top-surface), the physical configuration of the craters showed conformity to that of the projectiles dimensions, particularly at impact velocities $\geq$250 m/s, which is an aspect of 'softer' targets impacted by much 'harder' projectiles. It is noted that the effect of load factor (α) on frontal impact damage is seemingly insignificant when viewed as the size of craters. However, although not easily discernable from the figure, it was observed that at α=50 % and V$\geq$250 m/s, fiber/matrix breakage in the vicinity of the impact site was more predominant in the direction normal to applied preloading than in the direction parallel to preloading.

The overall damage on the backside of targets is shown in Fig. 4. The backside damage was negligible at V=150 m/s, intermediate at 250 m/s, and severe at 350 m/s. The backside damage generated at V$\geq$250 m/s was much greater (about 150-250 %) than the respective frontal damage, as depicted in the figure. The damage was associated with fiber/matrix breakage/removal, fiber tows' fracture, bending and their protrusion toward the impact direction. The degree of backside damage was also dependent on load factor. The load-factor effect was observed to be significant when V$\geq$250 m/s and α 30%. Under this condition, the targets were actually on the verge of penetration by the projectiles (This can be seen more clearly from the cross-sections of impact sites, which will be discussed in a later section). Although not presented here, the side view of a target impacted at 350 m/s with α = 50 % showed significant breakage and pushed-out of fibers (or tows), which was also related to a sign of complete penetration. Greater backside damage, as compared to the frontal counterpart, was due to the existence of tensile stresses associated with flexure of a target upon impact and has been observed regardless of the types of CMCs when targets were under a 'partial' support [18-20]. Stress wave interactions could be a secondary effect, if any.

Figure 5 presents the cross-sectional views of target specimens in response to different impact velocities and load factors. The cross-sectional views provide information on many features such as the shape and size of craters, the damage and material compaction beneath the impact sites, the formation of cone-cracks and delaminations, the mode of penetration, and the backside damage, etc.. It is also observed how the progress of overall damage occurred regarding impact velocity and load factor. A complete penetration took place at 350 m/s with α $\geq$30%. Nonetheless, the target specimens as a whole survived without any catastrophic failure upon impact. The material compaction observed at lower impact velocities $\leq$250 m/s was attributed to the nature of target material's 'softness' and 'open' (porous) structure, which was also seen from the N720/aluminosilcate CMC [19]. The formation of cone cracks in brittle materials is common in ballistic impact by spherical projectiles, regardless of material's types such as CMCs, monolithic ceramics or glasses [18-20]. A more detailed cross-sectional view of a target impacted at 350 m/s with α = 30 % is shown in Fig. 6. The aforementioned damage features including tows fracturing, bending, protruding, and then cone cracking, delaminations, and material removal, etc. are well presented within the illustration.

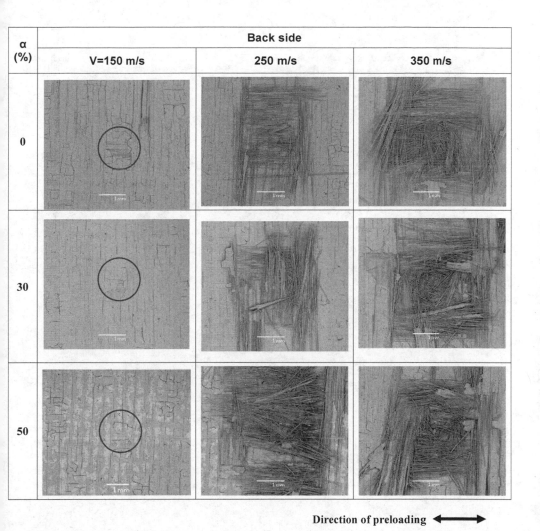

Direction of preloading ←→

Figure 4. Backside impact damages with respect to impact velocity (V) and load factor (α) in an N720/alumina oxide/oxide CMC impacted by 1.59-mm hardened chrome steel ball projectiles. The size (=1.59 mm) of a steel ball projectile (marked with a circle) was included for comparison. Bars=1000 μm.

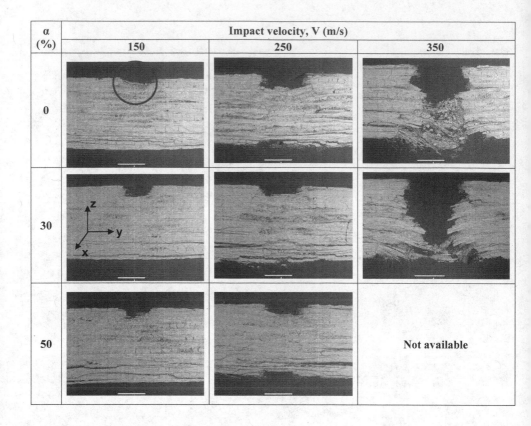

Figure 5. A typical example of the cross-sectional view of a target specimen of N720/alumina oxide/oxide CMC impacted at 340 m/s with a load factor of 30 % by 1.59-mm steel ball projectiles. The size of the projectile (marked with a circle in the top left) was included for comparison. The direction of tensile loading during impact is in the 'x' direction (see an inset). Bar = 1000 μm.

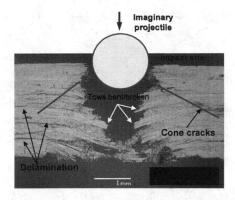

Figure 6. A typical example showing several impact damage features from the cross-section of a target specimen of N720/alumina oxide/oxide CMC impacted at 350 m/s with a load factor of 30 % by 1.59-mm steel ball projectiles. An imaginary projectile was included for comparison.

Post-Impact Strength

The results of post-impact tensile strength testing are shown in Fig. 7, where post-impact tensile strength was plotted as a function of impact velocity and load factor. Also included is the as-received tensile strength determined, $\sigma_f = 140$ MPa. Despite some inherent data scatter, post-impact strength decreased with increasing impact velocity for a given load factor. Similarly, for a given impact velocity, the post-impact strength decreased with increasing load factor, starting from $\alpha > 30$ %. This behavior of post-impact strength degradation is consistent with the results of impact morphologies as already shown. Hence, the combination of damage morphology and post-impact strength can provide much better insight and understanding regarding impact damage analysis and modeling as well.

Figure 8 compares with the post-impact strength determined previously from flexure targets that were impacted under a partial support and then post-strength-tested in a four-point flexure [23]. As it can be seen from the figure, strength degradation in terms of target specimen's configuration was greater in tension than in flexure. Note that the targets impacted in the partial support were strength-tested such that frontal impact sites were placed in tension side of a four-point flexure. Hence, the mode of stresses applied during strength testing with respect to damage orientation is responsible for such a difference in strength degradation between the two different test schemes.

The effect of load factor on strength degradation may be modeled at least conceptually using a fracture mechanics concept when a residual stress field due to impact is introduced. Crack propagation during impact while a target is under tensile preloading may be postulated similar to that of the indentation fracture mechanics. The post-impact strength may be then estimated based on the arrested crack size together with other associated parameters. Fracture toughness or crack growth resistance of the oxide/oxide CMC may also play a central role in governing mechanics. Modeling together with more experimental work will be a task of further study. Also, more post-impact strength data are needed for their improved reproducibility.

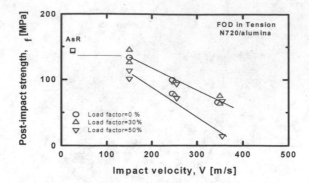

Figure 7. Post-impact strength as a function of impact velocity of N720/alumina oxide/oxide CMC under different levels of tensile preloading, impacted by 1.59 steel ball projectiles.

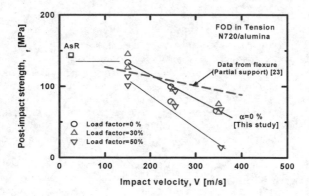

Figure 8. Comparison of the current post-impact strength data at α=0 % with the previous result on flexure targets in a partial support (post-impact- strength tested in a four point flexure) [23].

CONCLUSIONS

The overall impact damage of the N720/alumina oxide/oxide composite was dependent not only on impact velocity but also on load factor. The oxide/oxide composite shows significant impact damage occurring at 350 m/s, particularly for a load factor 30 %. The result of post-impact strength was consistent with the trend in impact damage in terms of impact velocity and load factor. Strength degradation was greater in pure tension than in flexure, due to difference in orientations of applied stresses (during strength tests) with respect to the geometry of impact damages.

Acknowledgements
   The authors acknowledge the support by the Office of Naval Research and Dr. David Shifler.

REFERENCES
1. Wiederhorn, S. M., and Lawn, B.R., 1977, "Strength Degradation of Glass Resulting from Impact with Spheres," J. Am. Ceram. Soc., **60**[9-10], pp. 451-458.
2. Wiederhorn, S. M., and Lawn B. T., 1979, "Strength Degradation of Glass Impact with Sharp Particles: I, Annealed Surfaces," J. Am. Ceram. Soc., **62**[1-2], pp. 66-70.
3. Ritter, J. E., Choi, S. R., Jakus, K, Whalen, P. J., and Rateick, R. G., 1991, "Effect of Microstructure on the Erosion and Impact Damage of Sintered Silicon Nitride," J. Mater. Sci., **26**, pp. 5543-5546.
4. Akimune, Y, Katano, Y, and Matoba, K, 1989, "Spherical-Impact Damage and Strength Degradation in Silicon Nitrides for Automobile Turbocharger Rotors," J. Am. Ceram. Soc., **72**[8], pp. 1422-1428.
5. Knight, C. G., Swain, M. V., and Chaudhri, M. M., 1977, "Impact of Small Steel Spheres on Glass Surfaces," J. Mater. Sci., **12**, pp.1573-1586.
6. Rajendran, A. M., and Kroupa, J. L., 1989, "Impact Design Model for Ceramic Materials," J. Appl. Phys, **66**[8], pp. 3560-3565.
7. Taylor, L. N., Chen, E. P., and Kuszmaul, J. S., 1986 "Microcrack-Induced Damage Accumulation in Brittle Rock under Dynamic Loading," Comp. Meth. Appl. Mech. Eng., **55**, pp. 301-320.
8. Mouginot, R., and Maugis, D., 1985, "Fracture Indentation beneath Flat and Spherical Punches," J. Mater. Sci., **20**, pp. 4354-4376.
9. Evans, A. G., and Wilshaw, T. R., 1977, "Dynamic Solid Particle Damage in Brittle Materials: An Appraisal," J. Mater. Sci., **12**, pp. 97-116.
10. Liaw, B. M., Kobayashi, A. S., and Emery, A. G., 1984, "Theoretical Model of Impact Damage in Structural Ceramics," J. Am. Ceram. Soc., **67**, pp. 544-548.
11. van Roode, M., et al., 2002, "Ceramic Gas Turbine Materials Impact Evaluation," ASME Paper No. GT2002-30505.
12. Richerson, D. W., and Johansen, K. M., 1982, "Ceramic Gas Turbine Engine Demonstration Program," Final Report, DARPA/Navy Contract N00024-76-C-5352, Garrett Report 21-4410.
13. Boyd, G. L., and Kreiner, D. M., 1987, "AGT101/ATTAP Ceramic Technology Development," Proceeding of the Twenty-Fifth Automotive Technology Development Contractors' Coordination Meeting, p.101.
14. van Roode, M., Brentnall, W. D., Smith, K. O., Edwards, B., McClain, J., and Price, J. R., 1997, "Ceramic Stationary Gas Turbine Development – Fourth Annual Summary," ASME Paper No. 97-GT-317.
15. (a) Choi, S. R., Pereira, J. M., Janosik, L. A., and Bhatt, R. T., 2002, "Foreign Object Damage of Two Gas-Turbine Grade Silicon Nitrides at Ambient Temperature," Ceram. Eng. Sci. Proc., **23**[3], pp. 193-202; (b) Choi, S. R., et al., 2004, "Foreign Object Damage in

Flexure Bars of Two Gas-Turbine Grade Silicon Nitrides," Mater. Sci. Eng. **A 379**, pp. 411-419.

16. Choi, S. R., Pereira, J. M., Janosik, L. A., and Bhatt, R. T., 2003, "Foreign Object Damage of Two Gas-Turbine Grade Silicon Nitrides in a Thin Disk Configuration," ASME Paper No. GT2003-38544; (b) Choi, S. R., et al., 2004, "Foreign Object Damage in Disks of Gas-Turbine-Grade Silicon Nitrides by Steel Ball Projectiles at Ambient Temperature," J. Mater. Sci., **39**, pp. 6173-6182.

17. Choi, S. R., 2008, "Foreign Object Damage Behavior in a Silicon Nitride Ceramic by Spherical Projectiles of Steels and Brass," Mat. Sci. Eng. **A497**, pp. 160-167.

18. Choi, S. R., 2008, "Foreign Object Damage Phenomenon by Steel Ball Projectiles in a SiC/SiC Ceramic Matrix Composite at Ambient and Elevated Temperatures," J. Am. Ceram. Soc., **91**[9], pp. 2963-2968.

19. (a) Choi, S. R., Alexander, D. J., and Kowalik, R. W., 2009, "Foreign Object Damage in an Oxide/Oxide Composite at Ambient Temperature," J. Eng. Gas Turbines & Power, Transactions of the ASME, Vol. **131**, 021301. (b) Choi, S. R., Alexander, D. J., and Faucett, D. C., 2009, "Comparison in Foreign Object Damage between SiC/SiC and Oxide/Oxide Ceramic Matrix Composites," Ceram. Eng. Sci. Proc., **30**[2], pp. 177-188.

20. Choi, S. R., Faucett, D. C., and Alexander, D. J., 2010, "Foreign Object Damage in An N720/Alumina Oxide/Oxide Ceramic Matrix Composite," Ceram. Eng. Sci. Proc. **31**[2] pp. 221-232.

21. Ogi, K., et al., 2010, "Experimental Characterization of High-Speed Impact Damage Behavior in A Three-Dimensionally Woven SiC/SiC Composite," Composites Part A, **41**[4] pp. 489-498.

22. Herb, V., Couegnat, G., Martin, E., 2010, "Damage Assessment of Thin SiC/SiC Composite Plates Subjected to Quasi-Static Indentation Loading," Composites Part A, **41**[11] pp. 1677-1685 (2010).

23. Faucett, D.C., Choi, S. R., "Foreign Object Damage in An N720/Alumina Oxide/Oxide Ceramic Matrix Composite under Tensile Loading," presented at MS&T2010, in print *Ceramic Transactions* (2011).

24. Mattoni, M. A., et al., 2005, "Effects of Combustor Rig Exposure on a Porous-Matrix Oxide Composite," J. App. Ceram. Tech., **2**[2], pp.133-140.

25. Simon, R. A., 2005, "Progress in Processing and Performance of Porous-Matrix Oxide/Oxide Composites," *ibid*, **2**[2], pp. 141-149.

26. ASTM C 1259, "Test Method for Dynamic Young's Modulus, Shear Modulus, and Poisson's Ratio for Advanced Ceramics by Impulse Excitation of Vibration," *Annual Book of ASTM Standards*, *Vol. 15.01*, ASTM, West Conshohocken, PA (2010).

27. ASTM C 1275, "Test Method for Monotonic Tensile Behavior of Continuous Fiber-Reinforced Advanced Ceramics with Solid Rectangular Cross-Section Test Specimens at Ambient Temperature," *Annual Book of ASTM Standards*, *Vol. 15.01*, ASTM, West Conshohocken, PA (2010).

28. ASTM C 1468, "Test Method Transthickness Tensile Strength of Continuous Fiber-Reinforced Advanced Ceramics at Ambient Temperature," *Annual Book of ASTM Standards*, *Vol. 15.01*, ASTM, West Conshohocken, PA (2010).

29. ASTM C 1292, "Test Method for Shear Strength of Continuous Fiber-Reinforced Advanced Ceramics at Ambient Temperature," *Annual Book of ASTM Standards*, *Vol. 15.01*, ASTM, West Conshohocken, PA (2010).

FREQUENCY AND HOLD-TIME EFFECTS ON DURABILITY OF MELT-INFILTRATED SIC/SIC

Ojard, G.[2], Gowayed, Y.[3], Morscher, G.[4], Santhosh, U.[5,6], Ahmad, J.[5], Miller, R.[2] and John, R.[1]

[1] Air Force Research Laboratory, AFRL/RXLM, Wright-Patterson AFB, OH
[2] Pratt & Whitney, East Hartford, CT
[3] Auburn University, Auburn, AL
[4] University of Akron, Akron, OH
[5] Research Applications, Inc., San Diego, CA
[6] Structural Analytics, Inc., Carlsbad, CA

ABSTRACT

With the growing interest in ceramic matrix composites for multiple applications, the response of the material to service conditions needs to be understood. A range of durability assessment was undertaken on Melt Infiltrated (MI) SiC/SiC Ceramic Matrix Composite. MI SiC/SiC was tested under 30 Hz fatigue, 1 Hz fatigue, dwell fatigue (2 hour hold cycle) and creep loading. The applied stresses ranged from the micro-cracking initiation point to well above the saturation stress of the material. Test temperatures included room, 815 C and 1204 C. The effects of fatigue loading frequency and dwell (hold) time on the durability of the composite are discussed.

INTRODUCTION

For any material to be considered for insertion into real applications, the material needs to be fully characterized. This is especially true for Ceramic Matrix Composites (CMC) that are being considered for applications in the extreme environments of gas turbine engines. The full extent of the material performance throughout a range of temperatures and durability should be explored. This has been shown by the authors by looking at long term "creep" type testing and residual properties and fatigue testing [1-4]. This is usually considered to be important for long term applications such as industrial gas turbines [5-6]. This work expands the past assessment of this material where durability and the time the material spent at temperature was the focus of the investigation [4].

PROCEDURE

Material Description

The Melt Infiltrated SiC/SiC CMC system chosen for this study was initially developed under the Enabling Propulsion Materials Program (EPM) and is still under further refinement at NASA-Glenn Research Center (GRC) [7-8]. The material description and microstructure has been documented earlier by the authors [1-4].

Mechanical and Durability Testing

Testing for this effort ranged from standard tensile testing per ASTM C1275 (room temperature) and ASTM C1349 (elevated temperature) to durability testing. The creep and dwell fatigue testing was performed per ASTM C1337. The dwell fatigue testing conditions included 2 hour hold period followed by unloading of the sample to a load ratio (R) of 0.05. The fatigue testing was done per ASTM C1360 under load controlled conditions at 1 Hz or 30 Hz using a sine wave with R of 0.05, 0.5 and 0.8. During the creep testing, either total strain or creep strain was recorded. During the dwell fatigue and fatigue testing, total strain was recorded.

RESULTS

Tensile Tests

The tensile behavior of MI SiC/SiC has been reported previously [1,4]. The results are summarized in Figure 1 showing the initiation of micro-cracking stress, proportional limit and crack saturation stress (load fully taken up by the fibers) [4]. As can be seen in Figure 1, the micro-cracking, proportional limit and crack saturation stress are not sensitive to temperature. In contrast, the ultimate tensile strength does decrease for the testing done at 1204°C. (For the room temperature testing, there were 42 tests, at 815°C and 1204°C there were 2 tests each.)

Creep/Dwell Fatigue Testing

Creep and dwell fatigue testing (2 hour hold, R = 0.05 where R is the stress ratio ($\sigma_{min}/\sigma_{max}$)) were performed over a range of times, temperature and stresses. For the testing done, there were a varying range of run-out times set for the testing. When the time for the testing was complete, the test was stopped and the sample removed. The results of this testing is summarized in Figure 2. For this series of testing, the only failures observed were for the creep tests performed at 1204°C. These failures corresponded to either a high stress level or long test duration.

Fatigue Testing – 1 Hz and 30 Hz

In addition to the above work, a series of 1 Hz and 30 Hz fatigue tests were done at 1204°C at R of 0.05. The results of this testing are shown in Figure 3. The run-out (or discontinued) condition for the 1 Hz testing was set at 400,000 cycles while the 30 Hz testing run-out was achieved at a value greater than 40,000,000 cycles. The effect of R on fatigue behavior at 30 Hz and at 1204°C is shown in Figure 4.

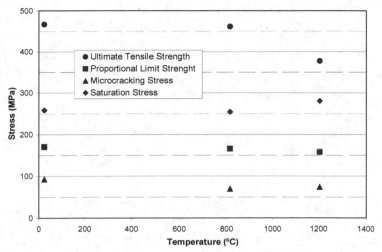

Figure 1. Key Stress Points from Stress-Strain Curves for MI SiC/SiC versus Temperature

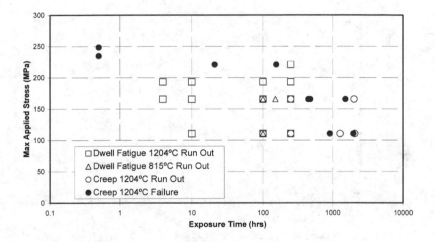

Figure 2. Summary of Creep and Dwell Fatigue Testing at 1204°C

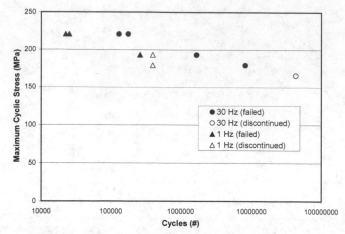

Figure 3. Summary of Fatigue Testing at 1204°C (R = 0.05)

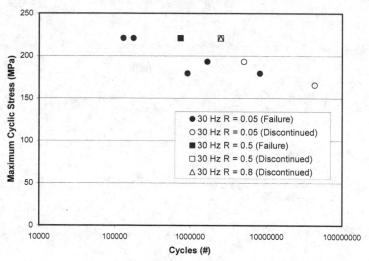

Figure 4. Summary of Fatigue Testing at 1204°C with Differing Load Ratios (R).

Residual Tensile Tests

From the series of creep and dwell fatigue (2 hour hold) testing, there were multiple samples where the testing was stopped at several different exposure temperatures and times. From these samples, a series of residual tensile tests were conducted. Since the evolved strain was recorded during testing as well as the strain from the residual tensile test was known, the data was plotted as the evolved strain during exposure versus the residual strain seen in the tensile test. This is shown in Figure 5. This work has been presented elsewhere by the authors [9]. The data falls near a constant 0.5% strain line as shown in Figure 5. Past published work by the authors has shown that the as-received strain capability is 0.5% (tensile) and failed creep tests achieve near 0.5% strain [9]. Figure 5 shows that the total strain accumulated either by creep, tensile or combination of both tests sums to 0.5% giving guidance on how to use the material.

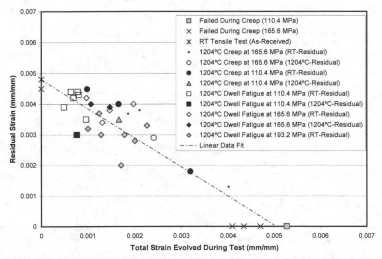

Figure 5. Residual Tensile Results plotted as Evolved Strain vs. Residual Strain (capability)

DISCUSSION & CONCLUSIONS

As noted above, a wide range of durability testing was performed across a variety of tests. This raises the question about which is the best way to compare the data. Some researchers looked at the cycles to failure when looking at frequency effects [10,11]. Other researches looked at the loading rate between tests [12]. Since the range of frequencies tested only varied by an order of magnitude, it was decided to look at time to failure for comparison and not cycles to failure. This type of analysis excludes samples that were discontinued early in the analysis as there is no way to project where they would fail in a direct comparison. This is done for all the creep and fatigue tests (dwell fatigue, 1 Hz and 30 Hz) where R was 0.05 as shown in Figure 6. Note that there were no dwell fatigue failures in this testing series. Figure 6 also includes the key points from the stress-strain curve showing that most of the testing was done between the proportional limit and the saturation stress for the tests run at 1204°C.

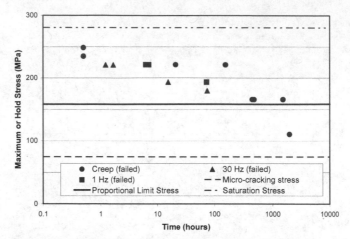

Figure 6. Failed Creep and Fatigue Data Points (1204°C)

A quick review of Figure 6 can lead to the conclusion that the failure is dependent on the time at temperature [4]. This is easily the case when looking at creep tests closer to the micro-cracking stress (long term) and saturation stress (short term). With the current work on R of 0.5 and 0.8, this conclusion can be challenged. This can be seen in Figure 4 where the 30 Hz data points are plotted with differing R ratios. As the R ratio is increased (meaning less cycling between stress levels), longer lifetimes are achieved. For the 220.8 MPa stress level experiments, there were no failures for the R = 0.8 tests (as shown in Figure 4). As R is being increased, the test is becoming more and more like a hold test (creep or dwell fatigue (2 hour hold)).

With the limited 30 Hz testing that was done with different R ratios and no R ratio testing done at 1 Hz, it would appear that analysis would be limited and no additional insight could be achieved. As an alternative, residual tensile testing done on these samples can be used. While the curve in Figure 5 shows that the overall breadth of the data falls on an almost constant strain line, the data does segregate when comparing the dwell fatigue (2 hour hold) versus the creep data. As can be seen in Figure 7, the dwell fatigue data is mostly below the constant strain line while the creep data falls above. This is reinforced when looking at a curve fit of the data (See Figure 8) where the creep data has the appearance of a constant line parallel to the previous lines shown in Figures 5 and 7. The dwell fatigue (2 hour hold) data has a much lower slope and appears to be greatly affected by the load un-load during the test.

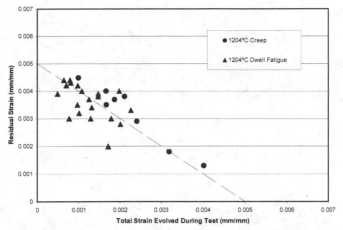

Figure 7. 1204°C Creep and Dwell Fatigue Data from Figure 5

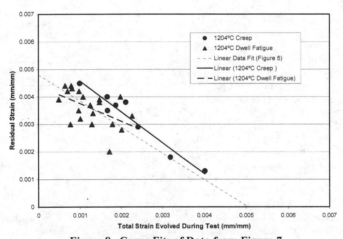

Figure 8. Curve Fits of Data from Figure 7

Figures 7 and 8 show that periodic load un-load cycles degrades the material performance, as indicated by the reduction in failure strain of the dwell fatigue specimens compared against the creep samples (with no unloading occurring). This is consistent with the 30 Hz data shown in Figure 4 where having a higher R is not as detrimental as the lower R on the material performance. The higher R ratio results in less stress cycling meaning that the cracks formed during the test are not being cycled to as great of an extent in the low R ratio baseline tests. This allows any oxide formed in the crack to remain allowing some protection. This is then consistent with the R ratio of 0.8 tests that did not fail during the test.

This study has shown that there is a load ratio (R) effect on the fatigue behavior and no clear evidence of a frequency effect in MI SiC/SiC composites. We suggest that this work be followed with additional testing looking further into possible frequency effects in greater detail for this class of material.

ACKNOWLEDGMENTS

The Air Force Research Laboratory, Materials and Manufacturing Directorate (AFRL/RX) sponsored this work under contracts F33615-01-C-5234 and F33615-03-D-2354-D04.

REFERENCES

1. Y. Gowayed, G. Ojard, R. Miller, U. Santhosh, J. Ahmad, and R. John, "Correlation of Elastic Properties of Melt Infiltrated SiC/SiC Composites with In-Situ Properties of Constituent Phases", Composites Science and Technology 70 (2010) 435–441.
2. Morscher, G.N., Ojard, G., Miller, R., Gowayed, Y., Santhosh, U., Ahmad, J., and John, R., "Tensile Creep and Fatigue of Sylramic-iBN Melt-Infiltrated SiC Matrix Composites: Retained Properties, Damage Development, and Failure Mechanisms", Composites Science and Technology 68 (2008) 3305–3313.
3. Y. Gowayed, G. Ojard, J. Chen, G. Morscher, R. Miller, U. Santhosh, J. Ahmad and R. John, "A Study of the Time-Dependent Response of Sylramic-iBN Melt-Infiltrated SiC Matrix Composites Using Dwell-Fatigue Experiments", Submitted to the Journal of Composites Science and Technology, 2010.
4. Ojard, G., Gowayed, Y., Morscher, G., Santhosh, U., Ahmad, J., Miller, R. and John, R., "Creep And Fatigue Behavior Of MI SiC/SiC Composites At Temperature", published in Ceramic Engineering and Science Proceedings, 2009.
5. Brewer, D., Ojard, G. and Gibler, M., "Ceramic Matrix Composite Combustor Liner Rig Test:, ASME Turbo Expo 2000, Munich, Germany, May 8-11, 2000, ASME Paper 2000-GT-670.
6. Wedell, James K. and Ahluwalia, K.S., "Development of CVI SiC/SiC CFCCs for Industrial Applications" 39th International SAMPE Symposium April 11- 14, 1994, Anaheim California, Volume 2, pg. 2326.
7. Calomino, A., NASA-Glenn Research Center, personal communication.
8. J.A. DiCarlo, H-M. Yun, G.N. Morscher, and R.T. Bhatt, "SiC/SiC Composites for 1200°C and Above" Handbook of Ceramic Composites, Chapter 4; pp. 77-98 (Kluwer Academic; NY, NY: 2005)
9. Ojard, G., Calomino, A., Morscher, G., Gowayed, Y., Santhosh, U., Ahmad J., Miller, R. and John, R., "Post Creep/Dwell Fatigue Testing of MI SiC/SiC Composites", American Ceramic Society, Ceramic Engineering and Science Proceedings, pp. 135-143, 2008.
10. Mall, S. and Engesser, J.M., "Effects of Frequency on Fatigue Behavior of CVI C/SiC at Elevated Temperature", Composites Science and Technology 66 (2006) 863–874.

11. Staehler, J.M., Mall, S. and Zawada, L.P., "Frequency Dependence of High-Cycle Fatigue Behavior of CVI C/SiC at Room Temperature", Composites Science and Technology 63 (2003) 2121–2131.
12. Choi, S.R. and Gyekenyesi, J.P., "Load-rate Dependency of Ultimate Tensile Strength in Ceramic Matrix Composites at Elevated Temperatures", International Journal of Fatigue 27 (2005) 503–510.

MECHANICAL AND MICROSTRUCTURAL CHARACTERIZATION OF REACTION BONDED SILICON CARBIDE PROCESSED WITH PETROLEUM COKE

Rodrigo P. Silva
Universidade Federal do Rio de Janeiro – PEMM/COPPE/UFRJ
Rio de Janeiro, RJ, BRAZIL

Celio A. Costa
Universidade Federal do Rio de Janeiro – POLI/PEMM/COPPE/UFRJ
Rio de Janeiro, RJ, BRAZIL

ABSTRACT

RBSiC has some advantages over conventional silicon carbide sintering processes, such as reduced time and processing cost. However, to achieve a successful processing by this technique is not a simple task, since appropriate raw materials and processing parameters must be well adjusted. This study evaluates the mechanical properties and microstructure of the material obtained from ultrafine SiC particles, with an average particle size of 0.51µm, and petroleum coke. The green body was infiltrated between 1450 and 1600 °C. The processed material had a 2.85 g/cm^3 density, an average four-point bending strength of 193 MPa and modulus elasticity of 270 GPa.

Keywords: silicon carbide, processing, thermal shock, mechanical properties.

INTRODUCTION

Reaction bonded silicon carbide (RBSiC) is a very important advanced ceramic material. Its processing technique consists in infiltration by capillary action of molten silicon into a porous compact, called green body, which is composed by silicon carbide and carbon. The reaction between silicon and carbon nucleates a new SiC, and the final microstructure is composed of old SiC, new SiC, residual Si, C and porosity. RBSiC presents particular advantages in relation to other densification processes, such as low cost, reduced time and lower processing temperatures (1400 - 1600 °C), besides the very low volumetric shrinkage, which allows the manufacture of parts in their nearly net shape[1].

Despite of the above advantages, a high quality material is not a trivial issue to obtain. There are relevant factors that determine, or not, the success of the processing, that is, to obtain a complete or incomplete infiltration, in addition to the respective resulting microstructure. Due attention should be given to the porous preform characteristics (volume fraction of porosity, particles size and type of carbon used as raw material precursor), to the silicon infiltration process and its chemical reaction to form SiC, which have direct influence on the mechanical properties, particularly thermal shock[2,3].

A wide range of raw materials and processing routes may be used in the RBSiC manufacture, resulting in materials with similar microstructures, but with different mechanical properties[2-6]. The objective of this study is to process RBSiC, using sub-micrometer SiC particles and recalcined petroleum coke, and then characterize the material through mechanical tests and microstructure analysis.

MATERIALS AND METHODS

The raw materials used here were the silicon carbide powder, commercially called PREMIX (H.C. Starck) and a commercial petroleum coke, calcined at 1400 °C (Petrocoque). PREMIX is a mixture of SiC UF-15 with carbon, boron and binders, which are sintering additives and green body forming, respectively. The SiC and coke used had an average particles size of 0.51µm and 0.1µm, respectively. Both in submicron size range. To infiltrate the green bodies, silicon metal (Rima S.A.) was used; its reported purity was of 99%.

The coke needed to be recalcined in the laboratory at temperatures at 2100 °C, in order to remove volatiles substances still contained in the coke, which may be released during the infiltration process and, thus, affecting the densification of the material[7].

The green body was prepared with following mass fraction: 80% PREMIX and 20% coke. The mixture was homogenized via dry ball mill, containing alumina balls, for 24 hours. Dry homogenization was carried out so as to keep the PREMIX characteristics, which also brought the advantage of simplifying this process stage, in comparison with wet milling. After homogenizing, the mixture was sieved. The green-bodies were formed by uniaxial pressing with 1.5 MPa, in disks with 70 mm diameter and 8 mm thickness.

The ready to press mixture of SiC (PREMIX) was chosen since it is a powder that easily compact (flowability) and gives good green strength. Furthermore, it was observed that around 1200 °C the first stage of the solid state sintering takes place, forming a skeleton that has good strength to sustain the infiltration process.

The infiltration process was carried out with molten silicon, under argon atmosphere, following the thermal cycle shown in Figure 1. Binder release occurs up to 600 °C. At 1200 °C there is pre-consolidation of the body (the initial stage of solid state sintering), while the infiltration and reactive process occur at 1450 and 1600 °C, respectively.

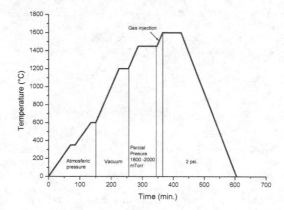

Figure 1 – Heating cicle used to process the present RBSiC, in argon atmosphere.

The infiltrated disks had their larger surfaces grinded for modulus of elasticity measurement by ultrasonic testing. Then, the disks were cut and twenty-two (22) specimens were prepared and tested following the ASTM C 1161, type B. After the four-point flexural tests, three samples were selected for fractographic analysis, namely: the highest, the lowest and the medium strength. Weibull´s statistical analysis was also preformed.

Another test conducted was the thermal shock resistance (ASTM C 1525-04), and thirty-five (35) more flexure samples were similarly prepared. A set of five (5) specimens was submitted to temperatures of 25, 300, 400, 500, 600 and 700 °C. The procedure used was as follows: they were heated at a 5 °C/min, kept for 20 minutes at each temperature and, then, fastly placed in large water tank kept at 25 °C; the specimens did not touch each other and there was a soft sponge at the bottom. The samples were then tested in four-point bending, as ASTM C 1161.

The microstructure of the material was observed by optical microscopy and density was measured by Archimedes method on grinded samples.

RESULTS AND DISCUSSION

The physical characteristics of the powders used as precursor materials are crucial to the infiltration process and sintering, because the properties of the infiltrated material are influenced by the green body properties, which, in turn, has its properties determined by the precursor material used in its composition [2,3].

The processing with ultrafine SiC particles (0.51 μm), similar to the one used in this study, are unusual in the literature, since they generally present infiltration problems of the molten silicon [4]. The mixture used here resulted in a dense infiltrated material with good mechanical properties; however, infiltration problems were verified in some tiny little areas of the disk. The non-infiltrated tiny areas might be originated by the combination of inhomogeneous composition, which showed an individual agglomeration of both carbon and SiC due to the dry milling process used, and pressing. The pressing is able to compact the regions that contains Premix much more and easier (local density increase) than the areas which contains both Premix and carbon. When the silicon melts, the regions with higher local density becomes a barrier to the infiltration process and, consequently, porosity is left behind.

The average density of the infiltrated disks, measured by Archimedes, was 2.85 g/cm³, a value within the range of densities reported in the literature, which are in between 2.7 and 3.1 g /cm³ [2, 3, 4, 6, 8, 9]. The modulus of elasticity measured was 270 GPa. The individual results of the 22 MOR specimens are shown in Figure 2, where it can be observed that the highest MOR was 247 MPa, the lowest 142 MPa and the average 193 MPa.

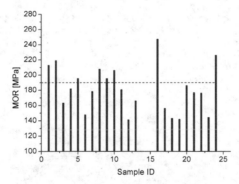

Figure 2 – Four-point bend flexural strength, according to ASTM C 1161.

Scafe et al.[8] performed a study on the mechanical behavior of various RBSiC. In their study, a three-point flexural strength and modulus of elasticity (measured by ultrasound) for various types of RBSiC were carried out. The MOR values ranged from 180 to 667 MPa and the modulus of elasticity varied from 164 to 375 GPa. Comparing to the Scafe's et al.[8] work, it can be verified that the modulus of elasticity of the material processed here (270 GPa) is compatible with the values measured by them for a 51% volume fraction of SiC in the composition, which is similar to the material here processed, as seen in Figure 3. This microstructure is quite refined, with very small SiC grains (dark phase) interconnected by thin Si veins (bright phase), some residual Si lakes and porosity. Carbon was not seen nor detected by x-ray.

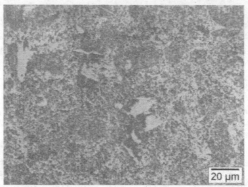

20 µm

Figure 3 – Optical micrograph of the RBSiC microstructure.

Scafe *et al.*[8] obtained, for a 51% vol SiC, an average MOR value of 402 MPa. Such a value was measured with 3-point flexural test, using specimens with 4 x 4 mm cross section and 35 mm outer load span. In order to compare the Scafe *et al.*[8] results with ours, it is necessary to adjust the effective volume[10], since the effective volume tested here is 4 times larger than that used by them. So, assuming the same Weibull modulus for both materials (m = 7), a valid approach, and using the correction of MOR data from our test condition to Scafe condition[11], our average MOR in 3-point bending would be 348 MPa, which could be in the scatter range of a ceramic material. Even though the results a close to each other, improvements must be made in the processing and/or to increase the SiC particle size, in order to improve the infiltration process, as already reported[4].

Ungyu Paik *et al.*[3] processed the same type of material and tested it using four-point flexural test, using specimens with 3 x 4 mm cross section and inner and outer load span of 10 and 20 mm, respectively. They obtained MOR values in between 260 and 310 MPa, which could indicate higher levels of strength. Nevertheless, using the same concept of effective volume used above (now our effective volume is two times larger than the one used by Ungyu Paik) and the same Weibull´s modulus (m = 7), the average MOR obtained here would be 317 MPa, a value even higher than the ones obtained by them. Despite of the differences in methodologies used in the literature to evaluate the flexural strength, the values obtained in this study are within the ranges reported [2,3,4,8,9].

The fracture surface analysis was performed on three specimens, as seen in Figures 4, 5 and 6. Figure 4 shows micrographs of the fractured surface of the specimen that fractured with the highest MOR (247 MPa). The surface of this specimen was almost defect free, having only one significant defect, which was near the tension surface of the specimen. This defect is presented in detail in Figure 4(b).

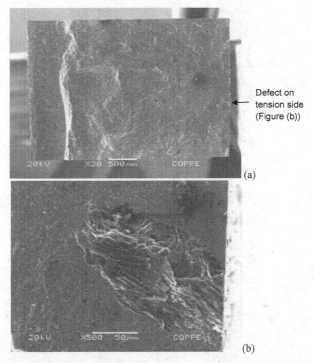

Defect on
tension side
(Figure (b))

(a)

(b)

Figure 4 – Fractography of the highest MOR specimen (a) macrograph of the surface fracture; (b) defect observed on the tension side of the specimen.

Figure 5(a) shows a macrograph of the fracture surface of the specimen that fractured with MOR of 190 MPa, a value near the average flexure strength of the samples. In this Figure, a few white spots along the fracture surface can be observed, whose incidence is higher on the tension side and decreases as it approaches the compression side. This white spots are defects, possibly originated during processing, which decreases the strength of the material, especially if they are located near the tensile surface. The Figure 5(b) shows a typical defect found on the edge of the tension side of the specimen.

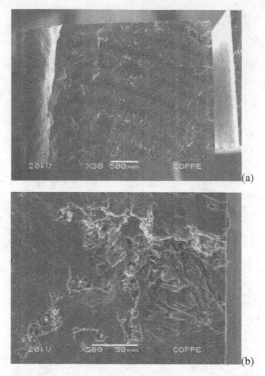

Figure 5 – Fractography of the specimen that fracture with an average MOR. (a) Fracture surface where white spots can be observed; (b) Example of the defects present on the tension side and representative of the white spots in Figure (a).

Figure 6 shows the fracture surface of the specimen that fracture with the lowest MOR (142 MPa). A greater number of whitely colored areas on the tension side of the specimen (Figure 6(a)) can be observed. This colored areas decrease in number as they approach the neutral axis and virtually disappear when the stress turn into compression. These colored areas can either be the porosities left behind due to poor infiltration and/or regions of low intergranular strength (SiC-SiC and/or SiC-Si), but both originated from the processing – a higher magnification of this regions is illustrated in detail in Figure 6(b). Figure 5 and 6 show the presence of the same type of defect, varying only in quantity, but not in location. Specimens with higher amounts of defective regions present lower strength, as the specimen shown in Figure 6. Thus, the elimination of the defective areas may result in specimens with flexural strength above 190 MPa, as the specimen shown in Figure 4.

It was observed that the whitely areas has a higher volume fraction in the tension side of the MOR bars, and almost disappear as the compression side is approached. It suggests that this kind of defect is influenced by the state of stress applied, namely, such defects did not exists in the as processed material. Consequently, the regions of low intergranular strength, which might even be surrounding the porosity, may cause more damage than the pores themselves, since that may act as much larger defects than the observed porosity. Again, the infiltration process must be improved in order to eliminate these areas of low intergranular strength.

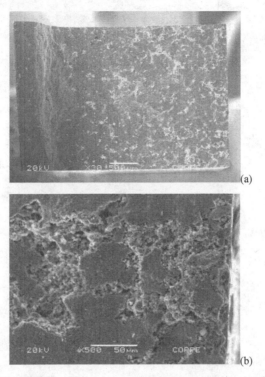

(a)

(b)

Figure 6 – Fractography of the specimen that fracture with the lowest MOR. (a) Fracture surface where large amount of white spots can be observed; (b) Example of the defects present on the tension side and, also, representative of the white spots in Figure (a).

Thermal shock resistance is a very important characterization for applications in which the material/component is subjected to abrupt temperature changes and it cannot fracture. To evaluate the thermal shock resistance, the specimens are heated up to a temperature and, then, immersed in water at 25 °C, which imposes a severe thermal gradient and possible loss of mechanical strength. A parameter used to assess the thermal shock strength is the critical temperature difference (ΔTc), in other words, a temperature difference that produces a 30% reduction in the flexural strength of the material when compared with the average flexural strength of the specimens as received[12]. The curve of residual four-point flexural strength variation versus temperature of the current material is shown in Figure 7, which shows that the material is practically insensitive to 400 °C (375 ° C = ΔTc) temperature. The 30% flexural strength loss is associated with 440 °C (ΔTc = 415 °C). It can also be noticed that strength decrease did not occur in a discontinuous way, evidencing that the material would be able to withstand some level of stress for small increments ΔT above 415 °C.

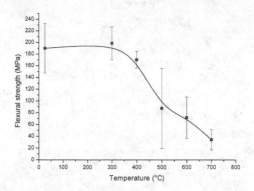

Figure 7 – Residual four-point flexural strength after thermal shock.

CONCLUSION

RBSiC was processed with ultrafine SiC particles and recalcined petroleum coke; a very fine microstructure was obtained, consisting of about 50 vol% SiC and 50% residual silicon metal.

The values of density, modulus of elasticity and four-point flexural strength were 2.85 g/cm^3, 270 GPa and 193 MPa, respectively. These values are in good agreement with values presented in the literature for the same type of material.

The presence of low intergranular strength areas were only observed in specimens that showed flexural strength below 190 MPa. It is proposed that such defects are weak links between SiC-SiC or SiC-C interfaces and they are only activated by tensile stress. Improvements in the processing shall lead to higher mechanical strength by elimination of such areas.

The thermal shock strength of the material processed was of 415 °C, that is, for ΔTC = 415 °C the loss of mechanical strength was of 30% in relation to the material as processed.

To increase the material's strength there are two possible ways: improvements in the processing in order to eliminate the defects and/or to increase the SiC particle size used as a precursor powder to facilitate the infiltration process.

ACKNOWLEDGEMENT

The authors wish to thanks the Brazilian Petroleum Agency (ANP-PRH 35), FINEP, PETROBRAS and DEPROCER supporting this research.

REFERENCES
[1]Matthias Wilhelm, Werner Wruss, Influence of anneling on the mechanical properties of SiC-Si composites with sub-micron SiC microstructures, *Journal of the European Ceramic Society,* **20**, 1205-13 (2000).
[2]Yanxiang Wang, Shouhong Tan, Dongliang Jiang, The effect of porous carbon preform and the infiltration process on the properties of reaction-formed SiC, *Carbon,* **42**, 1833-39 (2004).
[3]Ungyu Paik, Heyon-Cheol Park, Sung-Churl Choi, Chang-Gi Ha, Jae-Won Kim, YEON-GIL JUNG, Effect of particle dispersion on microstructure and strength of reaction-bonded silicon carbide, *Materials Science Engineering,* **A334**, 267-274 (2002).
[4]Matthias Wilhelm, Martin Kornfeld, Werner Wruss, Development of SiC-Si Composites with fine-grained SiC Microstructures, *Journal of the European Ceramic Society,* **19**, 2155-63 (1999).
[5]Yet-Ming Chiang, Robert P. Messner, Chrysanthe D. Terwilliger, Reaction-formed silicon carbide, *Materials Science Engineering,* **A144**, 63-74 (1991).
[6]Leszek Hozer, Jonq-Ren Lee, Yet-Ming Chiang, Reaction-infiltrated, net-shape SiC Composites, *Materials Science Engineering,* **A195**, 131-143 (1995).
[7]Rodrigo P. Silva, Claudio V. Rocha, Marysilvia F. Costa, Celio A. Costa, Effect of coke calcination temperature on the processing of reaction bonded silicon carbide, *Proceedings of the 34th International Conference on Advanced Ceramics and Composites,* **CESP 31**, Issue 2, 9-14 (2009).
[8]E. Scafe, G. Giunta, L. Fabbri, L. Di Rese, G. De Portu, S. Guicciardi, Mechanical Behaviour of Silicon-Silicon Carbide Composites, *Journal of the European Ceramic Society,* **16**, 703-713 (1996).
[9]Matthias Wilhelm, Silvia Werdenich, Werner Wruss, Influence of resin content and compaction pressure on the mechanical properties of SiC-Si composites with sub-micron SiC microstructures, *Journal of the European Ceramic Society,* **21**, 981-990 (2001).
[10]D. R. Bush. Desing ceramic components for structural applications", *Journal of Materials Engineering and Performance,* **2**, 851-862 (1993).
[11]John B. Wachtman Jr., Structural Ceramics, *Treatise on Materials Science and Technology,* **29**, 35, (1989), Academic Press, Inc – USA
[12]ASTM C 1525 – 04, Standard Test Method for Determination of Thermal Shock Resistance for Advanced Ceramics by Water Quenching

Nondestructive Evaluation

IDENTIFICATION OF DAMAGE MODES IN CERAMIC MATRIX COMPOSITES BY ACOUSTIC EMISSION SIGNAL PATTERN RECOGNITION

N. Godin, M. R'Mili, P. Reynaud, G. Fantozzi, J. Lamon

Université de Lyon, MATEIS, INSA-Lyon,
7 Avenue Jean Capelle,
69621 Villeurbanne, France

ABSTRACT

The control of damage is a key factor for the durability and reliability of parts in service. Sound identification of damage mechanisms and kinetics is an important issue with a view to predicting remaining lifetime. For that purpose, analysis of acoustic emission (AE) is an interesting and powerful tool. Damage of fiber reinforced ceramic matrix composites involves several phenomena at various length scales including matrix cracking, fiber debonding, fiber failures. The objective of the present paper is to propose a quantitative approach to damage identification based on signal analysis by pattern recognition. An unsupervised classification method was used to differentiate the signals generated during reference fatigue tests performed on C/SiC samples at high temperature. Each class of signals was then associated to relevant damage mode. A library of signals was then used to identify the damage modes generated during fatigue tests performed at various temperatures (supervised classification method).

INTRODUCTION

Ceramic matrix composites (CMCs) are interesting structural materials for high temperature applications, because of high mechanical properties at elevated temperatures. SiC matrix composites reinforced by carbon fibers (C/SiC) are candidate materials for aeronautical and aerospace applications. Mechanical behaviour and damage modes under static or cyclic loading [1-4] have been investigated by several authors. Today, the challenge is predicting components lifetime in service. To achieve this goal, quantification of damage as well as identification of the various damage modes are required.

An acoustic emission (AE) [5] is a transient ultrasonic wave resulting from the sudden release of stored energy during a damage phenomenon such as fibre breakage, interface debonding, matrix cracking in composite materials. The signals recorded depend on the source, material microstructure and sensor features. However, in identical conditions, similarities exist among the AE signals originating from similar sources. Consequently, it can be considered that a signal represents a source and that the acoustic signature of a damage mode can be determined. The following assumptions are made:

(1) Sources with different amounts of energy released (in terms of intensity and speed) generate different waveforms;

(2) Signal energy is related to the amount of energy released by the source. This parameter is thus very important for the discrimination of sources. It depends in particular on source size and amount of elastic energy stored;

(3) Signals are affected by wave propagation. The discrimination of the AE signatures of different damage mechanisms is thus a major issue.

Common analysis of AE is based on diagrams of cumulative hits or counts or histograms of amplitude. Nevertheless, in the case of composite material, this approach is inadequate. It was improved by grouping signals of similar shapes into clusters using classifier parameters [6]. Many works [7-15] have shown that AE techniques and multivariable classification techniques are the basis of pattern recognition tool. Kostosopoulos [9], Godin [12-13], Moevus [14-15 have identified different classes of AE signals which were attributed to damage modes in oxide/oxide, glass/polyester and SiCf/[Si-B-C] composites. In [15], the AE data were analyzed using an unsupervised multi-variable clustering technique. It was shown that several types of signals can be distinguished, and that the obtained clusters were consistent with the expected damage modes in 3D SiC/SiC composites.

AE based identification of damage modes is not straightforward for two reasons. On the one hand, the AE signals have complex shapes that must be characterised by multiple pertinent descriptors. On the other hand, the acoustic signatures of damage modes are not known *a priori* and they are scattered. A major issue in the use of AE technique is thus to associate AE signatures to pertinent damage mechanisms.

The main objective of the present paper is to propose an analysis of acoustic emission for recognition of the damage phenomena which operate during fatigue tests at high temperatures (> 700°C) on C/SiC. Both an unsupervised and a supervised pattern recognition technique were used to group the AE signals [17-19]. Thus a combination of an unsupervised pattern recognition scheme (k-means method [11]) with the supervised K-Nearest Neighbours method (KNN method [6]) is proposed to take into consideration several AE descriptors. The approach involves two main steps:

(1) AE signals that have similar characteristics are grouped using a clustering method in order to identify acoustic signatures of the damage mechanisms;

(2) Then a supervised classification is conducted, using a library of labelled pre-classified signals that can be used in real time.

EXPERIMENTAL PROCEDURE: AE EXPERIMENTS AND DESCRIPTION OF THE AE SIGNAL

A 3D multi-layered composite made of self-healing [Si-B-C] matrix reinforced ex-PAN (High Resistance) carbon fibres (40% volume fraction) was investigated. Test specimens with 4 mm thickness, 16 mm width, and 30 mm gauge length were tested.

Static and cyclic fatigue tests were conducted in air at various temperatures (700°C, 1000°C and 1200°C) under uni-axial on axis tensile loading. The cyclic fatigue tests at 700°C and 1000°C were conducted under tensile/tensile sinusoidal loading with constant amplitude and 0.25 Hz frequency.

Acoustic emission was monitored using a MISTRAS 2001 data recording device (Euro Physical Acoustics) and two piezo-electric resonant sensors (micro80). Two heat-resistant steel wave-guides (100 mm distant) with 140 mm length and 8 mm diameter were attached to the specimen (figure 1). Silicon high vacuum grease was used for sensor / waveguide and waveguide / specimen coupling. For locating the AE sources, AE wave velocity was determined before tests using pencil lead break procedure. Several breaks were performed on the specimen at various locations between the two sensors. The difference in time of arrival was calculated using the first peak of the signal received at both sensors. In the present work, AE wave velocity was found to be equal to 3200 m/s. Only those AE events located within the gauge length were considered

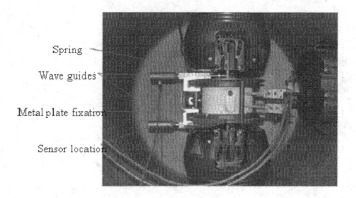

Figure 1. Experimental set-up with AE monitoring for fatigue tests at high temperature.

PATTERN RECOGNITION TECHNIQUES

Figure 2 summarises the procedure which was used for the analysis of the AE signals.

Relevant AE descriptors

AE signals were initially described using 18 descriptors (amplitude, duration, energy, number of counts, rise time, counts at peak, etc.). The descriptors were then normalized in order to obtain data in the interval [-1; 1]. The correlation matrix of the 18 descriptors was calculated and subjected to complete link hierarchical clustering [14]. The resulting dendrogram enabled to eliminate redundant descriptors. To keep significant information about each signal, a maximum correlation coefficient $R = 0.75$ [19] was chosen, which corresponds to a non correlation coefficient $1-R = 0.25$ (R is the Pearson coefficient). For this level of non correlation the 8 less correlated descriptors were kept: rise time, average frequency, reverberation frequency, energy, initiation frequency, rise time/duration, amplitude/rise time, amplitude/average frequency. Then a Principal Component Analysis (PCA) was performed in order to define new uncorrelated descriptors obtained from linear combination of the eight ones mentioned above and to reduce the data set size. At the end, the data were expressed in a 4 dimensional orthogonal basis. The first Principal Component (PC) is a linear combination of the 8 descriptors which accounts for most of the variability. Then, depending on data variability, additional components were introduced. In the present study, four principal components accounted for more than 95% of the initial data set variance.

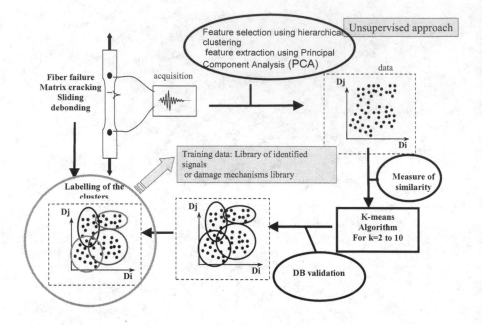

Figure 2. Flow chart representation of the pattern recognition method proposed for the analysis of the AE data showing the main steps.

Similarity measure: a weighted Euclidean distance

In order to take into account that the principal components do not contain the same quantity of information, the distance between two points in this 4D-space was defined as

$$d_w^2(X,Y) = \sum_{i=1}^{4} \lambda_i . (X_i - Y_i)^2$$ where λ_i is the i^{st} eigenvalue, X_i and Y_i are the i^{st} coordinates of the

vectors X and Y. The distance d_w^2 is a squared weighted Euclidean distance.

Clustering of AE events by k-means method (unsupervised method)

The k-means method consists in minimizing the sum of squared distances from all the vectors of a cluster to its centre. This method assumes that the cluster number k is known and specified in advance. The k-means algorithm can then be described as follows:
1) Initialize the coordinates X_i of cluster centre for each class C_i
2) Assign each descriptor vector X to a class C_i such as the weighted Euclidean distance between X and the cluster centre X_i is the lowest among all the cluster centres
3) Compute the coordinates of the new cluster centres
4) If there is no change in the co-ordinates of the cluster centres, then the algorithm has converged and the procedure is done; otherwise go to step 2.

One major problem in performing cluster analysis is to determine the optimal number of clusters for a given data set. To evaluate the quality of cluster partitions, the Davies-Bouldin validity index DB [20] was used. The best clustering result corresponds to minimum of the DB index. So, several clustering steps were performed with k varying from 2 to 10. Then, the Silhouette index [21] was used to evaluate the quality of clustering.

K-Nearest-Neighbour classifier (K-NN) [12; 18] (Supervised method)

The K-NN method is a supervised classification technique that uses a training set. An unknown signal is classified according to its K-Nearest Neighbours in the training set. In the present study, neighbourhood was evaluated using the lowest Euclidean distance. The matrix of distances from each signal to the points of the training set was computed. The class of this signal is that one of the largest number of objects among the K-neighbours. Optimum number K is determined using a cross-validation procedure (method of the "leave one out"). Each signal in the training set is considered as an unknown. For each value of K, the method is:
1) Split the training set into two parts A and B, where B contains only one vector
2) Apply the K-NN method to the single vector B and classify B by its K-nearest neighbours in A
3) Determine if B is classified correctly
4) Repeat steps1 to 3 for every possible choice of B
5) Calculate the percentage of correctly classified vectors (average classification rate).
The optimum number K of the nearest neighbours is that one that minimises the classification error.

RESULTS AND DISCUSSION

Unsupervised Classification

Optimal clustering was obtained for 4 classes at 700°C and 1000°C and at 1200°C only when $\sigma > 200$ MPa. Feature correlation is illustrated in Figure 3a, on amplitude vs. rise time plots for a four-class solution. Class overlapping was limited, indicating satisfactory clustering. The A, B, C, D clusters were obtained for all the specimens and they displayed the same mean values. Clusters A and B were composed of high-intensity events. Cluster A was composed of AE events with high amplitude greater than 70 dB. A-type signals were characterized by: the highest energy, duration and amplitude. B-type signals could be distinguished by lower energy, highest rise time. Clusters C and D contained events of lower intensity. They exhibited lower energy and apparent frequency but C-type signals had the shortest rise time and the corresponding waveforms were thus really different. Cluster D could be distinguished by the lowest average frequency and the lowest energy.

At 1200°C, a five-class solution (Fig. 3b) was obtained when $\sigma \leq 200$ MPa. The additional class denoted A' at 1200°C under low stresses (≤ 200 MPa) exhibits similarities with class A (high amplitude, high energy and high average frequency), but rise time is shorter. Feature correlation is shown on Figure 3b. The amplitude vs. rise time plot for the five cluster solution exhibits scatter. This algorithm gave reproducible clustering of AE signals (4 or 5 classes depending on the testing conditions), without requiring preliminary knowledge of the damage sources.

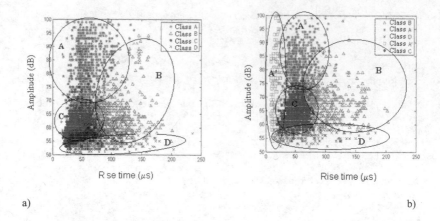

R se time (μs)   Rise time (μs)

a)   b)

Figure 3. Amplitude vs. rise time correlation diagram for a static fatigue test realized to a) (1200°C-230 MPa) b) (1200°C-150 MPa)

In cyclic fatigue at 700 °C and 1000 °C at 0.25 Hz, the optimal number of classes was also four. Segmentation into four classes (A, B, C', D) is illustrated on the amplitude vs. rise time correlation diagram (Fig. 4). 3 classes contained signals similar to those obtained previously for the static fatigue tests (A, B and D). Class C 'is similar to class C recorded in static fatigue in terms of very short rise time (about 20μs) and amplitude (65 dB to 70 dB). But unlike class C, class C' corresponds to higher energy when compared to class B and energy of the same order of magnitude when compared to class A. For class C', the source mechanism was assumed to be similar to that of Class C, but operating at larger scale.

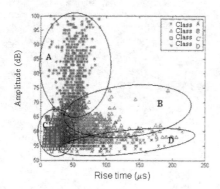

Rise time (μs)

Figure 4. Amplitude vs. rise time correlation diagram for a cyclic fatigue test realized at 700°C with an amplitude of 0/130 MPa and a frequency of 0,25Hz

Labelling of the clusters.

The clusters had to be assigned appropriate damage modes. Energy was the highest for A and B signals. It is usually assumed that AE energy is commensurate with the energy dissipated by the source, whatever the source. So, they were attributed to the most significant damage mechanisms observed in this type of composite during the first loading cycle: matrix cracking and fibre failures with maybe multiple fibre failures. Prior to composite ultimate fracture, some fibre breaks may be recorded. When resulting overloading on the surviving fibres is too large, yarns can fail leading to ultimate fracture. The signals corresponding to this phenomenon may have higher energy than those due to individual fibre breaks. However, there is no *a priori* information about the energy of fibre break signals in comparison with the energy of matrix cracking. For this reason, the discrimination between the energy released by matrix cracking and by fibre failure is not easy. Moreover, classes C and D were the most active during unloading/reloading cycles carried out during the static fatigue tests contrary to classes A and A' (figure 5). Classes C and D were thus associated to phenomena at interfaces during the unloading/reloading cycles such as growth of debond cracks between fibres and matrix, interfacial slip. In addition, classes A and C were more active in the rupture zone near composite ultimate failure

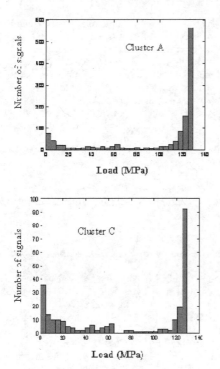

Figure 5. Histogram in load of the acoustic activity of classes A and C obtained during cyclic fatigue (700°C, 0/130 MPa),

From detailed analysis of the activity of each class, it is possible to propose a label for the different classes of signals:

- Classes A and A' were composed of signals with high energy and high amplitude but with different rise times (the shorter ones for Class A'). Classes A and A' may be associated to the same mechanism. Class A' only appeared during the tests conducted at 1200 ° C under σ < 200 MPa. Note that for σ > 200 MPa, the lifetime of composite was very short (<10h) with collective failure of fibres in longitudinal yarns. Class A rose more steeply in the fracture zone before ultimate fracture than it did elsewhere. Class A can be attributed primarily to collective fibre breaks and class A' to individual fibre breaks. Indeed, on SEM images of the fracture surfaces at 700°C, 1000°C and 1200°C (for an applied stress below 200 MPa) multiple failures of fibres were observed. That is the reason why class A can be assigned to failure modes of major importance e.g. multiple fibre breaks and certainly seal-coat cracking.

- Class B was characterized by a long rise time and relatively high amplitude that was lower than that of Classes A and A'. The activity of class B increased with stress during initial loading. The B-type signals (long rise time) may correspond to matrix cracking in transverse and in longitudinal yarns. The activity of class B in terms of number of signals or energy was lower than that of class A. This is due to the fact that the matrix contained pre-existing cracks created during cooling down from the processing temperature.

- Class C contained events of lower intensity and shorter time rise. Acoustic activity of this class was more sensitive to unloading/reloading cycles and it was very significant during initial loading, irrespective of temperatures. Recovery of activity of this class was observed at the end of the test, for the static fatigue test performed at (1000°C-215 MPa) (16 signals of type C among 57 detected) for example. So, this class may be mainly attributed to the phenomena associated to fibre matrix debonding in the longitudinal yarns. Under cyclic fatigue, the quantity of C'-type signals was lower than that of C-type signals obtained in static fatigue. In cyclic fatigue, yarn/yarn debonding was observed unlike in static fatigue. This labelling of class C' is also consistent with the fact that the yarn/yarn debonding phenomenon is more energetic than the fibre/matrix one. Thus, class C' may correspond to interply debonding.

- Class D corresponds to low energy signals. This class was the less active during the first loading cycle and was also the most sensitive to unloading-reloading cycles. This activity can be related to interfacial phenomena, such as fibre/matrix interfacial sliding and cracks closure during the unloading steps of cycles. It can also be noticed that class D activity under cyclic fatigue was more significant than that under static fatigue, which supports assignment of this class to interfacial sliding that was promoted during cycles.

Supervised Classification: Classification of the data by the K-Nearest-Neighbours classifier.

The supervised classification technique requires a data base of signals that have been labelled: the training set. This training set was created by merging AE data collected during reference static fatigue tests (1200 ° C-150 MPa) and cyclic fatigue tests (700 ° C-0/130MPa). As described in the previous section, the analysis of AE signals, based on microscopy and mechanical behaviour of composite led to 6 types of AE signals and to the following labelling of classes: Class A: Collective fibre breaks, Class A': individual fibre breaks, Class B: matrix cracking, Class C: fibre/matrix debonding, class C': yarn/yarn debonding, Class D: sliding at fibre/matrix interfaces and closure of matrix cracks after unloading. In order to establish the training set of labelled signals for the supervised analysis, the same amount of signals (350 signals) in each class (A, A', B, C, C' and D) was used. This training set included all the damage modes that may operate in this composite. The supervised method consists in comparing each detected signal to those of the library and to assign it to the class of the K nearest neighbours in the library.

First, it is necessary to determine the number K of nearest neighbours required to obtain the best classification. For this purpose, a self-validating procedure was used which consists in comparing each signal of the training set to all the others. The classification error rate was estimated for all the values of K using the "leave-one-out" method. The optimal value of K corresponded to the lowest value of error, and in the present study K = 13, and error = 2.5%. Then, for each AE signal of a given test, the Euclidean distance to all signals of the training set was calculated and the signal was labelled according to the class the most frequent class among its 13 nearest neighbours.

This analysis was applied to AE signals recorded during fatigue tests at high temperatures (Fig. 6). It was observed that in addition to classes A, B, C and D, a small amount of type A' and type C' signals was identified. This effect can be explained by the fact that at 700°C a few fibre breaks and inter tow debonding occured in static fatigue. The results obtained using the supervised classification technique indicated that the fraction of type A' signals increased with temperature in static fatigue (2% to 5% at 700°C and 1000°C and about 15 % at 1200°C) and the fraction of type C' signals increased under cyclic fatigue (1% to 6% in static fatigue and about 17% in cyclic fatigue). At 700°C, the amount of type A signals was significantly greater than that of type A' ones. Under static fatigue, the amount of type-C' signals was very low, whereas in cyclic fatigue it became predominant compared to that one of signals of Class C. These results are therefore more satisfactory than those obtained using the unsupervised classification, since individual fibre failures, collective fibre failures, fibre debonding, bundles debonding can be distinguished whatever temperature, load level and loading mode (static or cyclic fatigue) are.

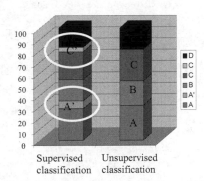

a) Static fatigue (700°C-150 MPa)

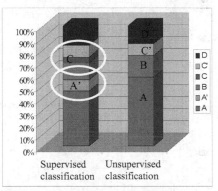

b) Cyclic fatigue *(700°C-0/130 MPa)*

Figure 6 . Comparison of the proportion of signals obtained for each class with the two classification techniques a) for the test carried out in static fatigue at (700°C-150 MPa), b) for the test carried out in cyclic fatigue at *(700°C-0/130 MPa)*,
(A : collective fiber breaks, A' : individual fiber breaks, B : matrix cracking, C : debonding between fiber and matrix, C' : wire / thread decohesion , D : fibre/matrix interfacial sliding and cracks closing)

CONCLUSION

Unsupervised and supervised pattern recognition methods have been developed for classification of AE signals. Two methods were implemented: PCA + k-means (unsupervised classification) and K nearest neighbours (supervised classification). The unsupervised method gave reproducible classification of AE signals in 6 different classes (A, A', B, C, C' and D), from their acoustic signature, without preliminary knowledge of the damage mechanisms. Each class of signals could be well assigned to one or more damage modes (fibre break, matrix cracking, etc.). In a second step, the supervised classification method allowed the signals to be classified with a better accuracy regardless of temperature, applied stress and loading mode (static or cyclic). This latter method is very interesting because it can be used more easily and it can identify signals recorded in small quantity. Nevertheless, a preliminary classification using the unsupervised method is necessary to establish a significant training set.

ACKNOWLEDGMENTS

The authors thank Snecma Propulsion Solide – SAFRAN Group, CNRS and the DGA for supporting this work under the research program: 'Modeling, extrapolation, validation of the lifetime of CMC'.

REFERENCES

[1]Dalmaz, A., Reynaud, P., Rouby, D., Fantozzi, G. & Abbe, F. Mechanical behaviour and damage evolution during cyclic fatigue at high temperature of a 2,5D Cf/SiC composite. Composite Science and Technology,58, 693-699 (1998).

[2] Lamon J. Micromechanics-based approach to the mechanical behavior of brittle matrix composites. Composites Science and Technology, 58, 2259-2272 (2002).

[3]Shuler, S.F., Holmes, J.W. & Wu, X. Influence of loading frequency on the room temperature fatigue of a Cf/SiC composite. Journal of the American Ceramic Society, 76, 2327-2336 (1993).

[4]Lee, S.S., Nicholas, T. & Zawada, L.P. Fatigue damage mechanisms and environmental effects on the long-term performance of matrix composites. Mechanical Testing of Ceramics and Ceramic composites, ASME, 117-155 (1994).

[5] Beattie AG. Acoustic emission, principles and instrumentation. J Acoustic Emission. Vol. 2. p. 95-128 (1983).

[6]Jain AK, Murty MN, Flynn PJ. Data clustering: a review. ACM Computing Surveys, 31(3), 264-323 (1999).

[7] Ono, K., Huang, Q. Pattern recognition analysis of acoustic emission signals. Progress. in Acoustic Emission VII, The Japanese Society for NDI. 69-78 (1994).

[8]Yan T., Holford K., Carter D. and Brandon J. Classification of acoustic emission signatures using a self-organization neural network. Journal of Acoustic Emission; 17, Number1/2, 49-59 (1999).

[9] Kostosopoulos, V., Loutas, T.H.,'Kontsos, A., Sotiriadis, G., Pappas, Y.Z. On the identification of the failure mechanisms in oxide/oxide composites using acoustic emission. NDT & E International, 36: 571-580 (2003).

[10]Pappas, Y.Z., Kontsos, A., Loutas, T.H., Kostosopoulos, V. Failure mechanisms analysis of 2D carbon/carbon using acoustic emission monitoring NDT & E International; 31, 571-580 (1998).

[11]Likas A., Vlassis N., Verbeek J. The global k-means clustering algorithm., Pattern Recognition; 36(2), 451-461 (2003).

[12]Godin, N., Huguet, S., Gaertner, R. Clustering of acoustic emission signals collected during tensile tests on unidirectional glass/polyester composite using supervised and unsupervised classifiers NDT & E International, 37, 253-264 (2004).

[13]Godin N, Huguet S, Gaertner R. Integration of the Kohonen's self-organising map and k-means algorithm for the segmentation of the AE data collected during tensile tests on cross-ply composites. NDT&E Int., 38(4), 299-309 (2005).

[14]Moevus M, Rouby D, Godin N, R'Mili M, Reynaud P, Fantozzi G, Fayolle G. Analyse of damage mechanisms and associated acoustic emission in two SiC/[Si-B-C] composites exhibiting different tensile curves. Part I: Damage patterns and acoustic emission activity. Composite Science and Technology, 68 (6), 1250-1257 (2008).

[15] Moevus M, N. Godin, D. Rouby, R'Mili M, Reynaud P, Fantozzi G, Farizy G. Analyse of damage mechanisms and associated acoustic emission in two SiC/[Si-B-C] composites exhibiting different tensile curves. Part II: Unsupervised acoustic emission data clustering. Comp Sci Technol; 68 (6),1258-1265 (2008).

[16] Davies D. L. and Bouldin D. W. A cluster separation measure, IEEE Trans. Patt. Anal. Machine Intell., 1, 224-227 (1979).

[17] Momon S. Durée de vie des composites C/SiC à matrice auto-cicatrisante à hautes températures sous air. *Thesis*. 2008. Lyon: Institut National des Sciences Appliquées de Lyon. France

[18] Momon S., M. Moevus, N. Godin, M. R'Mili, P. Reynaud, G. Fantozzi, G. Fayolle, Acoustic emission and lifetime prediction during static fatigue tests on ceramic-matrix-composite at high temperature under air, Composites Part A: Applied Science and Manufacturing, Volume 41, Issue 7, Pages 913-918 (2010).

[19] S. Momon, N. Godin , P. Reynaud, M. R'Mili, G. Fantozzi, Unsupervised and supervised classification of AE data collected during fatigue test on CMC at high temperature, submitted to Composites Part A: Applied Science and Manufacturing.

[20] P. Rousseeuw. Silhouettes: a Graphical Aid to the Interpretation and Validation of Cluster Analysis, J. Comput. Appl. Math. 20, 53-65 (1987).

[21] K. Hattori, M. Takahashi, A new nearest-neighbor rule in the pattern classification problem. Pattern Recognition;32, 425-432 (1999).

AN INDENTATION BASED NON-DESTRUCTIVE EVALUATION TECHNIQUE FOR THERMAL BARRIER COATING SPALLATION PREDICTION

J. M. Tannenbaum, K. Lee and B. S.-J. Kang
Mechanical and Aerospace Engineering, West Virginia University
Morgantown, West Virginia, United States

M. A. Alvin
National Energy Technology Laboratory, Department of Energy
Pittsburgh, Pennsylvania, United States

ABSTRACT

A load-based micro-indentation method for NDE of TBCs exposed to thermal loads has been developed. TBC thermal cyclic exposure tests were performed where after a defined thermal cycling period, surface stiffness response was measured to assess damage accumulation and identify macroscopic debonding failure sites. Microstructural analyses were conducted correlating YSZ adherence to TGO/BC and evidence of internal crack formations with the measured surface stiffness responses. Finite element analyses indicate that high YSZ/BC interfacial rumpling leads to the development of both in-plane and out-of-plane residual stresses upon cooling. Additional rumpling of this interface as a result of non-uniform oxide growth leads to enhanced residual stresses. Average stress levels within the TBC have been found to decrease with accumulated thermal exposure, yet the variance of these values was found to increase primarily due to enhanced residual stress variation across the coupon which can be identified by our micro-indentation technique. As a result, areas producing relative increases in surface stiffness response enable early detection of initial TBC spallation locations. In addition, finite element analyses of YSZ/TGO/BC interfacial stresses generated upon cooling provide an explanation for the experimentally observed micro-cracking and failure patterns.

INTRODUCTION

Under elevated temperatures, the presence of high thermal stresses within TBC systems leads to the development of creep cavities and wedge cracks which ultimately form microcracks and cause interfacial debonding in the ceramic top coat/bond coat interface. The TGO, a major contributor to this increase in thermal stress forms between the top coat and bond coat, and increases in thickness as thermal loads are applied. Furthermore, due to thermal expansion mismatches, as the TBC cools to ambient temperature large residual compressive stresses (2-6 GPa)[1,2] are generated within this layer and contribute to ratcheting/rumpling of the interfacial morphology. The building of these residual stresses is ultimately responsible for TBC spallation failure.[1-5] Laser scattering and various other NDE methods have shown promise in the detection of sub-surface TBC defect detection. However, these techniques are not able to detect interfacial microcracks, and the prediction of TBC spallation prior to its occurrence remains a significant hurdle to overcome by any NDE technique due to the TBC's complex multi-layer structure and rough surface morphology.[6-9]

In this research, a load-based micro-indentation method for NDE of TBC spallation and damage assessment is presented. A non-destructive loading/partial unloading testing methodology has[10,11] been developed where in stiffness responses (unloading slope) of TBC coupons subjected to cumulative thermal cyclic loadings were analyzed to predict spallation location and assess overall TBC degradation state. Specifically, the measured stiffness responses at various thermal loading cycles were used to generate time-series color maps for correlation with the accumulation of TBC residual stress states. Through this procedure, regions having higher stiffness responses can be linked to increases in out-of-plane residual stress at the YSZ/TGO interface that initiate TBC spallation failure. A TBC thermal cyclic loading plan was conducted where, after each thermal loading, cross-sectional microstructural analyses were carried out to identify damage accumulation and evolution of spallation failure (i.e., TGO thickness, ratcheting, localized oxidation, etc.). The microstructural analyses were then correlated to measured TBC surface stiffness responses at various thermal cyclic loadings. The results indicate that this load-based micro-indentation test methodology is capable of predicting spallation sites on TBCs before its actual occurrence.

LOAD BASED MICRO-INDENTATION EXPERIMENTS

The TBC coupons used in this study consisted of a 1/8" thick by 1" square nickel base single crystal super alloy RenéN5 substrate followed by a 300 μm MCrAlY bond coat layer and subsequent yttria stabilized zirconia layer applied by an air plasma spray (APS) or electron beam-physical vapor deposition (EB-PVD) process. Seven APS coupons were evaluated, three isothermally and three cyclically exposed to temperatures of 1100°C in ambient air environments. Where as only one EB-PVD coupon was evaluated throughout this study and was cyclically exposed. Isothermal high temperature exposures (coupons B, C and D) were performed in 40 hour increments where in the coupon was brought to a temperature of 1100°C within 15 minutes followed by a 40 hour dwell period after which the coupon was removed from the furnace within 15 minutes as well. Although Coupon A followed a similar thermal ramping rate as the other isothermally exposed TBC coupons, it was not exposed in 40 hours increments but rather various dwell times (10, 15, 20, 50, etc.).

Opposing that of the isothermally exposed coupons, all four cyclically exposed coupons (E, F, G and H) were heat treated using a similar ramping and dwell procedure. Each cyclic exposure consisted of the coupon being brought to 1100°C within 15 minutes followed by a 45 minute dwell period at this temperature. Subsequently the coupons were removed from the furnace within 15 minutes. The cyclically exposed coupons were exposed to 40 cycles per heat treatment. Throughout these cycles the samples were exposed to, but may have never reached room temperature conditions (24°C). Both the cyclic and isothermal heat treatments were thermally exposed in a horizontal tube furnace able to maintain a set temperature of 3°C along with a programmable linear actuator on which the coupons were placed and inserted into the furnace.

Following high temperature exposure the samples were weighed, imaged, and evaluated using a micro instrumented indentation technique for TBC degradation. A multiple partial unloading indentation procedure was utilized throughout this study, capable of allowing unloading slopes at multiple loads to be analyzed. A spherical tungsten carbide (WC) indenter having a radius of 793.5 μm, combined with a piezoelectric actuator (3.6 nm Resolution, Physik Instrumente, P-239.9S, 180 μm) and a high accuracy (±0.15% Accuracy, Honeywell, Model 31, 100 lb.) load cell completed the indentation chain, Figure 1. The large radius indenter and high precision actuator minimized coating damage during unloading surface stiffness response evaluation. All indentation testing was performed on the same system enabling maximum consistency over the course of the entire study.

Performed on each coupon's surface were 20 evenly spaced micro-indentations 4 mm from the sample's edges. Unloading slopes obtained on each TBC coupon are presented using a contour color mapping scheme. The procedure used to generate each color map is as follows: a) acquire unloading slopes at a given load for each location upon which indentation was performed, b) assign each location a value according to the average and standard deviations obtained on the entire coupon, c) employ the use of a finite element interpolation algorithm to generate each contour color map using these values. Although slope data is obtained at multiple loads (20~160 N), the stiffness responses recorded at 90±3 N were shown to have manufactured the clearest degradation trends and were ultimately enlisted to predict spallation failure location.

Demonstrating the non-destructive nature of this micro-indentation testing technique when using a large radius spherical (793.5 μm) indenter with relatively low maximum applied loads (~160 N), an APS/MCrAlY/RenéN5 coupon was subjected to high temperature thermal exposure as well as micro-indentation testing and a second coupon was subjected to only high temperature thermal exposure. Observed to fail at comparable cumulative thermal cycles, the micro-indentation was found to have no significant effect on ultimate TBC system life. Furthermore, Figure 2 displays an optical micrograph of a failed APS/MDL/RenéN5 coupon with surface spallation cracking paths not linking to the indented areas. Additionally, cross-sectional Scanning Electron Microscopy (SEM) analysis of areas beneath previously indented regions displayed no additional indentation induced YSZ or TGO cracking. Moreover, YSZ plan section SEM analysis of the indented regions, similar to that of the cross-sectional images displayed no indentation induced cracking when compared to that of non-indented areas on the coupon for both APS/MCrAlY/RenéN5 and EB-PVD/MDL/RenéN5 TBCs, Figures 3 and 4. Through this process it was determined that the proposed micro-indentation degradation evaluation technique does not promote TBC system micro-cracking or debonding.[12]

APS/MCrAlY/RenéN5 Coupon A

An APS/MCrAlY/RenéN5 coupon was exposed to isothermal heating at 1100°C under ambient air condition. The coupon was periodically removed from the furnace (i.e., after 85, 185, 300, 400 and 450 cumulative hours of exposure at 1100°C) for surface stiffness response measurement as well as visual inspection for TBC spallation. After approximately 400 isothermal loading hours, a small visible corner spallation between Side 1 and Side 2 was observed. Following micro-indentation evaluation, the coupon was subjected to an additional 50 hours of isothermal heating (to a cumulative total of 450 hours), and upon cooling to room temperature, a corner spallation between Side 1 and Side 4 was observed. Again, following micro-indentation evaluation, the coupon was placed in the furnace for an additional 50-hour isothermal heating (for a cumulative total of 500 hours). Upon cooling to room temperature, edge spallation along Side 1 was observed as shown in Figure 5. These observations correlated well with sites of measured higher surface stiffness response. Also shown in Figure 5, the time-series measured surface stiffness responses are plotted using surface color maps, which show that at this location (prior to TBC spallation) a relatively higher surface stiffness response was identified in comparison to adjacent regions (i.e., the red regions shown in Figure 6). The time-series color maps correlate well with the progression of corner spallation (initiated at the corner of Side 1 and Side 2 (400 hours)).

After two weeks in storage, the coupon was placed in the furnace for an additional 100-hour isothermal heating at 1100°C. Following this heat treatment the coupon's YSZ was still intact and was stored in a sealed plastic bag. After approximately 10 days of storage, the YSZ layer was found to have completely separated from the substrate; this phenomenon is generally referred to as "desk-top failure".[13]

APS/MCrAlY/RenéN5 Coupons B, C and D

Similar to the test methodology for Coupon A, Coupons B, C, and D were subjected to isothermal heating at 1100°C with an average of 40-hours of heating per cycle. As shown in Figure 6, edge spallation of Side 1 was observed immediately after 400 cumulated isothermal hours on Coupon B. The experimental result correlated well with the color map generated after 360 isothermal hours, which predicted edge spallation to occur along Side 1. As for Coupons C and D, no edge or corner spallation was observed immediately following 400 cumulated isothermal hours. After routine micro-indentation tests, the specimens were stored in a sealed plastic bag. Subsequently, about one and a half weeks later, coupon C was observed to have partial corner spallation between Side 2 and Side 3 as shown in Figure 7. For Coupon D, after about three weeks in storage, edge spallation along Side 1 was observed, as shown in Figure 8. Again, for both TBC coupons, the spallation sites correlated well with measured higher stiffness response sites, as shown in Figures 7 and 8.

As shown in Figures 5 to 8, the evolution of the color maps correlated well with the progression of spallation site(s). Furthermore, we propose that the measured stiffness response can be correlated to the residual stresses (in-plane and out-of-plane) developing at the top coat/TGO/bond coat interfacial region. It is argued that the locations experiencing higher in-plane residual compressive stress (as compared to other regions subjected to the same thermal loads) will have more ratcheting/rumpling at the TBC/TGO/bond coat interface and lead to the enhancement of the out-of-plane residual tensile stress occurring at the convex site of the TGO wave and eventually initiate micro-spallation. Past research by Padture[3], Schlichting[4] and others[14-17] has suggested that prior to TBC spallation (whether it be an APS or EB-PVD TBC), there is an increase in out-of-plane residual stress normal to the metal/ceramic interface and the severity of these residual out-of-plane stresses increases with thermal cyclic loadings. Coupled with the interfacial imperfections (e.g. TGO thickness, ratcheting, localized oxidations, etc.), this build-up of residual out-of-plane stresses will eventually cause TBC spallation failure.[3]

APS/MCrAlY/RenéN5 Coupons E, F, G and EB-PVD/MDL/RenéN5 Coupon H

Coupons E, F, G and H were cyclically exposed to ambient air high temperature environments for a cumulative total of 400 (E and H) and 440 (F and G) hours at 1100°C. The accumulated cycles at which all four cyclically exposed coupons failed varied from 400 to 440 total cycles. Spalling first, coupon E was exposed to a total of 400 cycles.

Failure of this coupon did not occur immediately upon cooling to room temperature but rather two days later, following routine sample imaging and weighing along edge 1, as shown in Figure 9. For coupon F, complete spallation (from the substrate) was observed after exposure to 440 cycles, as shown in Figure 10. Due to this, no direct correlation between indentation results and failure pattern could be made, yet some relationship between the contour color map and TBC spallation failure pattern can be seen at the corner of edges 1 and 4. Future SEM analysis will be able to determine the location of initial spallation for experimental validation. For coupon G, corner spallation at edges 1 and 4 was observed after 440 total cycles, as shown in Figure 11. Coupon H, the only EB-PVD/MDL/RenéN5 TBC in this study, having been cyclically exposed at temperatures of 1100°C in ambient air for a cumulative total of 640 hours experienced large scale spallation primarily along edges 1 and 4, Figure 12.

Demonstrating, as did the isothermally exposed coupons, contour color maps generated after 360 cycles for coupon E, 400 cycles for coupon G and 600 cycles for coupon H display direct correlations with the locations of initial spallation failure. In summary, regions having higher measured surface stiffness responses, as shown in Figures 9, 11 and 12, for coupons E, G and H respectively, coincide with edge spallation failure locations.

DISCUSSION

The test results show that this newly developed micro-indentation test methodology is capable of detecting increases in out-of-plane residual tensile stress (in terms of measured increases in stiffness response) at locations prior to TBC spallation. Furthermore, cross-sectional scanning electron microscope (SEM) imaging of the failed EB-PVD/MDL/RenéN5 coupon displayed those regions having a higher measured surface stiffness response prior to failure experience additional rumbling/roughening at the YSZ/TGO/BC interface, Figure 13. A direct comparison between these two images reveals a substantially greater amount of oxidation in Figure 13(a) than that of Figure 13(b). Moreover, Figure 13(b)'s YSZ/TGO interface appears to be more uniform than that of the nonlinear geometry found in Figure 13(a). This less uniform interface geometry found in Figure 13(a) is believed to be the source of the high surface stiffness responses obtained in this region and ultimately led to the prediction of this location as that where failure would initiate. Additionally, the unloading slope averages obtained following each heat treatment did not continuously increase along cumulative thermal exposure, but rather steadily decrease. This reduction in unloading stiffness is the result of an overall coating degradation. Yet, displaying a trend opposite to that of the average unloading stiffness response, standard deviation within these data sets was found to increase with cumulative high temperature exposure. This increasing standard deviation within these data sets exists due to an increasing division between low and high residually stressed regions within the TBC system.

Finally, it is important to note that this TBC spallation criterion is specimen-independent and thus a portable micro-indentation unit may be developed for spallation detection/prediction on industrial-size TBC turbine components. Development of a prototype portable micro-indentation unit for measurement on flat and curved surfaces has been demonstrated.[12] With further research development, the potential to achieve this goal is reasonably high. Furthermore, without prior knowledge of thermal cyclic history, this micro-indentation test methodology is capable of detecting the potential spallation site (s), i.e. one does not need to start with a "fresh" TBC component and conduct micro-indentation tests at various thermal cycles in order to predict TBC spallation. Current NDE/NDT techniques can only detect TBC spallation after its occurrence, thus spallation site prediction prior to its actual occurrence by this micro-indentation test method is unique and has the potential to further the development of remaining life prediction models for TBC turbine components subjected to various thermal cyclic loading conditions.

FINITE ELEMENT ANALYSIS

A parametric finite element analysis of a TBC system has been conducted in which the TGO is modeled using a sinusoidal wave accounting for any rumpling/ratcheting that may occur during its growth. In this manner, the residual stresses produced upon cooling may be evaluated along oxide wave amplitude. The resulting residual stresses may then be applied to the free edge of a full TBC model for elastic modulus measurement using a multiple loading/partial unloading micro-indentation test methodology. Through this procedure residual stress variance along oxide growth within the TBC will be reviewed. Additionally, correlation between unloading surface stiffness response and residual stress will be reviewed as TGO amplitude is varied as well. Throughout this study, TGO thickness will remain the same and only the rumpling aspect ratio is varied where λ is defined as the oxide wave length, A amplitude and t thickness, as shown in Figure 21.[18-20] Although this model does not recreate the exact geometrical parameters along the YSZ/TGO/BC interface, it allows stress magnitudes occurring at imperfections to be estimated.

Material properties imposed on this model include the elastic modulus, Poisson's ratio, yield stress, tangent modulus and coefficient of thermal expansion for all layers, as shown in Table 1.[4] Simulating a TBC cooling effect, reductions in temperature are imposed on this wave and entire TBC system. In order to increase resolution, a full scale section of TBC system is modeled to investigate stress translation.[19,20] Having a width of 10000 μm and a total height of 3260μm, this model contains all layers; TBC: 353μm, TGO: 1μm, BC: 209μm and Substrate: 3000μm, as shown in Figure 15. For FEA residual stress calculation, a reduction in thermal load from 1100°C to 23°C is applied. Isotropic bilinear elastic-plastic behavior is employed with a total of 31720 elements. Radial stresses are applied to the outer cylindrical surface of the axisymmetric model, the magnitude of which were found from the aforementioned full scale TBC model. A 500 μm total model width and 353 μm total height is assumed, as shown in Figure 16. Standard nano-indentation data reduction methods were also used to obtain elastic modulus measurements from the TBC system.[21]

FEA RESULTS AND DISCUSSION

Although local YSZ/TGO/BC interfacial residual stresses have been evaluated along increasing wave amplitude elsewhere[22,23], whether these variations alter the overall TBCs residual stress state is undetermined. Due to this, an expanded model is created to evaluate residual stress effects across the entire system domain. Avoiding singularity, all residual stress measurements obtained from this model was evaluated at sufficient distances from the model's boundaries. Upon applying a temperature difference of -

1077°C, both the in-plane and out-of-plane residual stresses in the aforementioned model were found to vary in all layers along increasing wave amplitude. The in-plane compressive residual stresses occurring within the YSZ steadily increase along wave amplitude and were found to be maximum at the TBC surface in all cases, as shown in Figure 17. Now, by applying only the in-plane YSZ residual compressive stresses to the outer surface of the YSZ only model, it is shown that indentation unloading slope increases with TGO wave amplitude as well, Figure 18. Thus, these findings are consistent with the experimental expectations that localized unloading surface stiffness response will increase as compressive stress strengthens due to TGO oxide growth with the associated interfacial ratcheting/rumpling morphology, making these regions more prone to spallation failure.

CONCLUSION

The unloading surface stiffness responses obtained from TBC coupons along thermal loadings has proven to be a viable analysis tool for the determination of spallation failure location. Using a multiple partial unloading procedure, slopes from various loads were able to be evaluated for abnormalities or trends along ever increasing thermal exposures at 1100°C. Failure locations for both isothermal and cyclically exposed APS/MCrAlY/RenéN5 as well as one EB-PVD/MDL/RenéN5 TBC coupons were correctly predicted with a 90±3 N evaluation load. Furthermore, assuming a sinusoidal TGO wave pattern, large scale TBC models were developed to investigate the relationship between residual stress and wave amplitude. Results indicate that these two parameters act proportionally, thus providing conformation to the previously discussed experimental results where indentation unloading surface stiffness response is found to increase along wave amplitude. This result indicates that regions having higher unloading surface stiffness response contain greater residual stresses and are more prone to spallation failure. Ultimately, residual stresses within the TGO and YSZ due to thermal expansion mismatches create subtle changes in measured stiffness response. These increases in unloading slope produce a numerical value from which the initial spallation failure location is predicted. The development of this non-destructive evaluation technique is exclusive in that it is not only able to detect spallation site location but do so prior to its occurrence.

ACKNOWLEDGMENTS

This research is supported by U.S. Department of Energy, National Energy Technology Laboratory under RES Contract DE-FE0004000. The support of Richard Dennis, NETL Turbine Technology Manager, is much appreciated.

REFERENCES
[1] D. M. Lipkin and D. R. Clarke, Measurement of the Stress in Oxide Scales Formed by Oxidation of Aluminum-Containing Alloys, *Oxidation of Metals*, **45**, 3/4 (1996).
[2] K. W. Schlichting, K. Vaidyanathan, Y. H. Sohn, E. H. Jordan, M. Gell and N. P. Padture, Application of Cr^{3+} Photoluminescence Piezo-Spectroscopy to Plasma-Sprayed Thermal Barrier Coatings for Residual Stress Measurement, *Material Science and Engineering*, **191**, 1/2 (2000).

[3] Nithin P. Padture, Maurice Gell, Eric H. Jordan, Thermal Barrier Coatings for Gas-Turbine Engine Applications, *Science*, **296**, 5566 (2002).

[4] K. W. Schlichting, N. P. Padture, E. H. Jordan and M. Gell, Failure Modes in Plasma-Sprayed Thermal Barrier Coatings, *Materials Science and Engineering*, **A342**, 1/2 (2003).

[5] M. A. Alvin, F. Pettit, G. Meier, N. Yanar, M. Chyu, D. Mazzotta, W. Slaughter, V. Karaivanov, B. Kang, C. Feng, R. Chen and T-C. Fu, Materials and Component Development for Advanced Turbine Systems, *5th International Conference on Advances in Materials Technology for Fossil Power Plants*, Marco Island, FL, October 3-5 (2007).

[6] E. Ellingson, R. Visher, R. Lipanovich and C. Deemer, Optical NDT Techniques for Ceramic Thermal Barrier Coatings, *Materials Evaluation*, **64**, 1 (2006).

[7] W. A. Ellingson, J. A. Todd and J. Sun, Optical Method and Apparatus for Detection of Defects and Microstructural Changes in Ceramics and Ceramic Coatings, *United States Patent 6,285,449*, September 4 (2001).

[8] K. Ogawa, D. Minkov, T. Shoji, M. Sato and H. Hashimoto, NDE of Degradation of Thermal Barrier Coating by Means of Impedance Spectroscopy, *NDT and E International*, **32**, 3 (1999).

[9] N. Goldfine, D. Schlicker, Y. Sheiretov, A. Washabaugh, V. Zilberstein, T. Lovett, Conformable Eddy-Current Sensors and Arrays for Fleetwide Gas Turbine Component Quality Assessment, *Journal of Engineering for Gas Turbines and Power*, **124**, 4 (2002).

[10] C. Feng and B. S.-J. Kang, *A Simple Indentation Measurement Technique for Surface Mechanical Property Evaluation*, Materials Science & Technology 2007 Conference and Exhibition, Detroit, Michigan, September 16-20 (2007).

[11] C. Feng, J. M. Tannenbaum, B. Kang and M. A. Alvin, *A Load-Based Multiple-Partial Unloading Micro-Indentation Technique for Mechanical Property Evaluation*, to be published in Experimental Mechanics (2009).

[12] B. S.-J. Kang, C. Feng, J. M. Tannenbaum and M.A. Alvin, A Load-based Depth-sensing Indentation Technique for Damage Assessment of Thermal Barrier Coatings, *Proceeding of ASME Turbo Expo 2009*, Orlando, FL, June 8-9 (2009).

[13] J. Smialek, D. Zhu, M. Cuy, Moisture-Induced Delamination Video of an Oxidized Thermal Barrier Coating, *Scripta Materialia*, **59**, 1 (2008).

[14] X. Y. Gong and D. R. Clarke, On the Measurement of Strain in Coatings Formed on a Wrinkled Elastic Substrate, *Oxidation of Metals*, **50**, 5/6 (1998).

[15] D. R. Clarke and W. Pompe, Critical Radius for Interface Separation of a Compressively Stressed Film from a Rough Surface, *Acta Materialia*, **47**, 1749 (1999).

[16] C. H. Hsueh, P. F. Becher, E. R. Fuller, S. A. Langer and W. C. Carter, Surface-Roughness Induced Residual Stresses in Thermal Barrier Coatings: Computer Simulations, *Materials Science Forum*, **308-311** (1999).

[17] C. H. Hsueh and E. R. Fuller, Jr., Analytical Modeling of Oxide Thickness Effects on Residual Stresses in Thermal Barrier Coatings, *Scripta Materialia*, **42**, 8 (2000).

[18] M. Y. He, J. W. Hutchinson and A. G. Evans, Large Deformation Simulations of Cyclic Displacement Instabilities in Thermal Barrier Systems, *Acta Materialia*, **50**, 5 (2002).

[19] J. Rosler, M. Baker and K. Aufzug, A Parametric Study of the Stress State of Thermal Barrier Coatings, Part I: Creep Relaxation, *Acta Materialia*, **52**, 16 (2004).

[20] J. Rosler, M. Baker and M. Volgmann, Stress State and Failure Mechanisms of Thermal Barrier Coatings: Role of Creep in Thermally Grown Oxide, *Acta Materialia*, **49**, 18 (2001).

[21] A. Bolshakov, W. C. Oliver and G. M. Pharr, Influences of Stress on the Measurement of Mechanical Properties Using Nanoindentation: Part II. Finite Element Simulations, *Journal of Materials Research*, **11**, 3 (1996).

[22] D. Naumenko, V. Shemet, L. Singheiser, W. Quadakkers, Failure Mechanisms of Thermal Barrier Coatings on MCrAlY-Type Bondcoats Associated with the Formation of the Thermally Grown Oxide, *Journal of Materials Science*, **44**, 7 (2009).

[23] J. M. Tannenbaum, K. Lee, B. S.-J. Kang and M. A. Alvin, Non-Destructive Thermal Barrier Coating Spallation Prediction by a Load-Based Micro-Indentation Technique, *ASME International Mechanical Engineering Congress & Exposition*, November 12-18, Vancouver, British Columbia (2010).

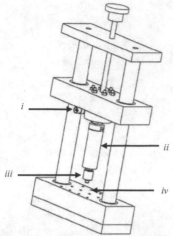

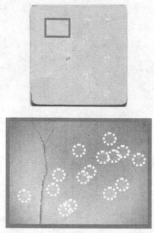

Figure 1. Experimental Setup, i) ±0.15% accuracy Honeywell Model 31 100 lb. load cell, ii) 3.6 nm resolution Physik Instrumente P-239.9S 180 μm piezoelectric actuator, iii) spherical tungsten carbide (WC) 750μm radius indenter, and iv) coupon stage.

Figure 2. Topcoat optical micrograph for an EB-PVD/MDL/René N5 obtained after a cumulative 640 cycles of heat treatment at 1100°C in an ambient air environment (note, indentation locations denoted by circles).

(a)   (b)

Figure 3. Micrographs of an APS/MCrAlY/René N5 obtained after a cumulative 375 isothermal heat treatment hours at 1100°C in an ambient air environment, a) non-indented region and b) indented region.

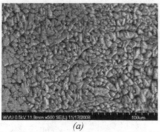

(a)   (b)

Figure 4. Micrographs of an EB-PVD/MDL/René N5 obtained after a cumulative 60 cycles of heat treatment at 1100°C in an ambient air environment, a) non-indented region and b) indented region.

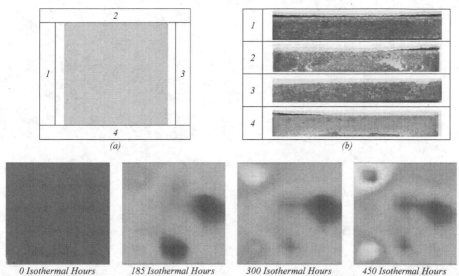

0 Isothermal Hours   185 Isothermal Hours   300 Isothermal Hours   450 Isothermal Hours

Figure 5. APS/MCrAlY/RenéN5 coupon A after a cumulative 500 isothermal heat treatment hours at 1100°C in an ambient air environment and 2 weeks storage, a) surface and b) edge and surface stiffness response color maps.

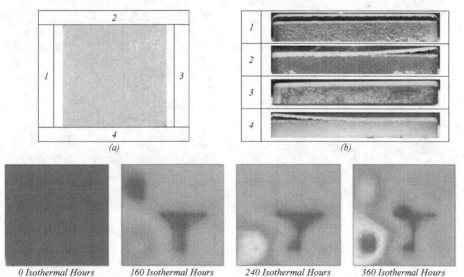

0 Isothermal Hours   160 Isothermal Hours   240 Isothermal Hours   360 Isothermal Hours

Figure 6. APS/MCrAlY/RenéN5 Coupon B after a cumulative 400 isothermal heat treatment hours at 1100°C in an ambient air environment, a) surface and b) edge and surface stiffness response color maps.

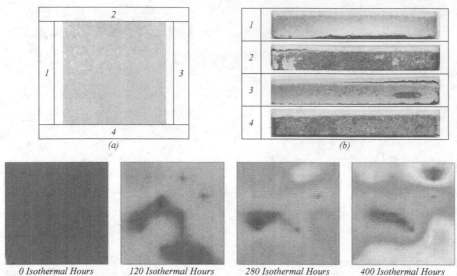

Figure 7. APS/MCrAlY/RenéN5 Coupon C after a cumulative 400 isothermal heat treatment hours at 1100°C in an ambient air environment and 1.5 weeks storage, a) surface and b) edge and surface stiffness response color maps.

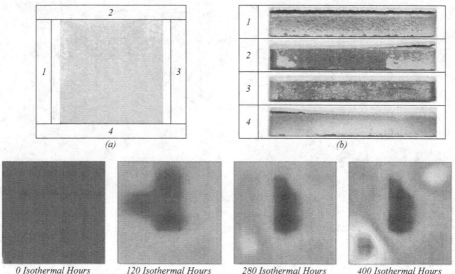

Figure 8. APS/MCrAlY/RenéN5 Coupon D after cumulative 400 isothermal heat treatment hours at 1100°C in an ambient air environment and 1.5 weeks storage, a) surface and b) edge and surface stiffness response color maps.

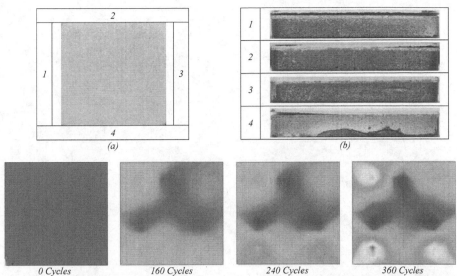

Figure 9. APS/MCrAlY/RenéN5 Coupon E after a cumulative 400 cycles of heat treatment at 1100°C in an ambient air environment and 2 days in storage, a) surface and b) edge and surface stiffness response color maps.

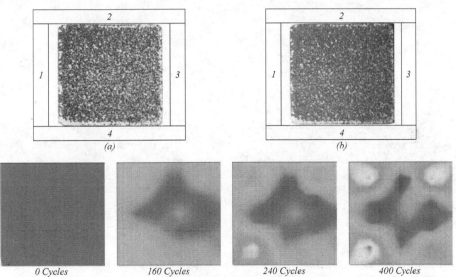

Figure 10. APS/MCrAlY/RenéN5 Coupon F after a cumulative 440 cycles of heat treatment at 1100°C in an ambient air environment, a) YSZ Top Coat, b) substrate and surface stiffness response color maps.

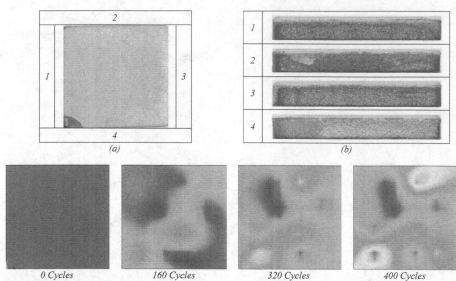

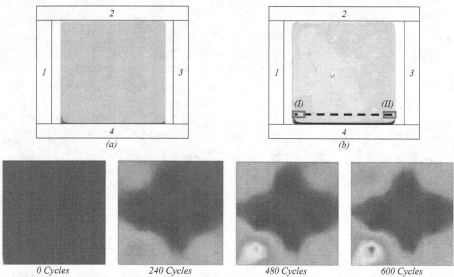

Figure 11. APS/MCrAlY/RenéN5 Coupon G after cumulative 440 cycles of heat treatment at 1100°C in an ambient air environment, a) surface and b) edge and surface stiffness response color maps.

Figure 12. EB-PVD/MDL/René N5 Coupon H after cumulative 640 cycles of heat treatment at 1100°C in an ambient air environment, a) intact YSZ, b) failed YSZ and surface stiffness response color maps.

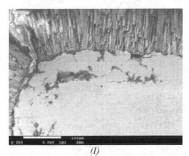

(I)   (II)

Figure 13. EB-PVD/MDL/RenéN5 coupon edge cross sectional scanning electron micrographs after 640 isothermal cycles, I) high and II) low residually stressed region.

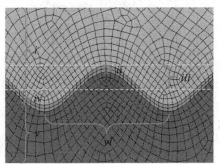

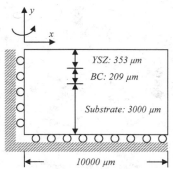

Figure 14. Thermal barrier coating finite element design, i) Yttria stabilized zirconia, ii) Thermally grown oxide thickness, t, iii) wave amplitude, A, iv) Thermally grown oxide, v) Bond coat, vi) Wavelength, λ.

Figure 15. Thermal barrier coating finite element schematic, including: yttria stabilized zirconia, thermally grown oxide, bond coat and substrate.

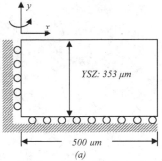

(a)

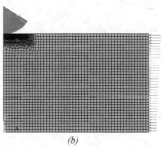

(b)

Figure 16. Stressed thermal barrier coating elastic modulus evaluation by use of an indentation method, a) schematic and b) mesh.

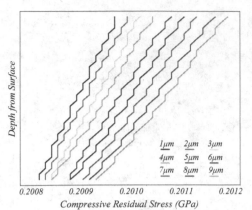

Figure 17. Yttria stabilized zirconia in-plane residual compressive stress generated upon cooling.

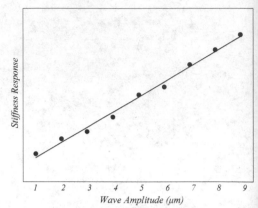

Figure 18. TBC surface stiffness response with increase thermally grown oxide wave amplitude.

Table 1. TBC System Material Properties

Material/Property	YSZ	TGO	BC	Substrate
Elastic Modulus (GPa)	50	375	211	150
Poisson's Ratio	0.1	0.25	0.3	0.25
Yield Stress (MPa)	250	1000	422	841
Tangent Modulus (GPa)	1	3.75	0.889	1.05
Thermal Expansion Coefficient ($10^{-6}/°C$)	11	8	14	14.5

DETERMINATION OF APPARENT POROSITY LEVEL OF REFRACTORY CONCRETE USING ULTRASONIC PULSE VELOCITY TECHNIQUE AND IMAGE ANALYSIS

Anja Terzić[1], Ljubica Pavlović[1], Vojislav Mitić[2]
Institute for Technology of Nuclear and other Raw Mineral Materials, Franchet d'Esperey 76, 11000 Belgrade, Serbia
Institute of Technical Sciences, Serbian Academy of Science, Knez Mihailova 35, 11000 Belgrade, Serbia

ABSTRACT
Aim of this paper is to establish the correlation between sintering process, porosity and important thermo-mechanical property of refractory concrete, i.e. creep. Creep deformation was investigated according to standard laboratory procedure applied at three temperatures: 1200, 1300 and 1400°C. Corundum and bauxite based refractory concretes were investigated. The concretes are varying in chemical and mineralogical composition. Both loss of strength and degradation of material occur when refractory concrete is subjected to increased temperature and compressive static load. Measuring of thermo-mechanical properties can indicate and monitor the changes within microstructure. Variation of refractory concrete microstructure, as a consequence of sintering process, during exposure to constant compressive load and constant elevated temperature during certain time-intervals was investigated using scanning electron microscope and Image Pro Plus program for image analysis. Obtained results of the investigation proved that creep can be useful method when type of refractory concrete is to be chosen for an application.

1. INTRODUCTION
Monolithic elements used for linings of metallurgical furnaces and other plants operating at high temperature, such are linings for oil refinery plants, thermal insulation in plants and in objects, linings in nuclear power plants, linings in chemical and petrochemical industries, etc., are made of either shaped or unshaped refractory material. Unshaped refractory materials, i.e. refractory concrete, refractory mortar and shotcrete, have numerous advantages: simplified building of refractory linings, economic aspect i.e. cheaper process of manufacturing, possibility of damaged lining reparation, etc. [1].
Thermo-mechanical properties, including creep deformation, are among most important properties of refractory concrete since they can determine its performance in various applications. Creep deformation is measured in terms of applied compressive load which concrete can withstand at elevated temperature. Microstructure of the material changes when refractory concrete is subjected to compressive load and high temperature: apparent porosity increases, pores become bigger and cracks within structure open. It resuts in loss of strength and material degradation. Therefore, development and change of concrete microstructure can be directly monitored by measuring either of these properties. This asumption is also supported with work by other authors who investigated similar correlation on various refractory materials. For example: *M. Boussuge* investigated thermo-mechanical properties on industrial refractories, *D.L.Y. Kong* investigated metakaolin geopolymers exposed to elevated temperatures, *B.T. Tamtsia* investigated early-age short-term creep of hardening cement paste, *A.A. Wereszczak* investigated creep of CaO/SiO2-containing MgO refractories, etc. [2 - 8].
According to the scientific definition, creep is plastic deformation which is time dependent function of an investigated material at constant temperature and constant compressive load (0.2 MPa). Creep takes

---

[1] corresponding author, e-mail: anja.terzic@gmail.com, phone no: + 381 11 3691 722

place at temperatures above $0.5 \cdot T_m$ (T_m is the melting temperature of an investigated material). Creep test can be carried out in different modes (compressive, tensile and bending). However, creep is generally investigated in compressive mode, because compressive force is the most dominant strain existing in a plant [9-13].

There are three stages within the creep curve: primary, secondary and tertiary creep. Secondary state creep is the dominating region in the creep curve. Therefore, material which is deforming by creep spends the longest period of time in secondary region of creep. During secondary state creep isothermal sintering process occurs. Rate of creep is always determined by the rate of the slowest diffusion particle movement along the fastest diffusion path.

Beside temperature and stress, creep of refractory concrete is affected by the porosity of the material, average grain size, chemical and mineralogical composition of the refractory concrete, firing temperature of samples and the texture and the microstructure of the material [14-17].

The following equation describes assumed relationship between a property of the refractory concrete (x) which varies during the sintering process and duration of the sintering process (t). It is a quantitative description of power law creep which relates to the isothermal sintering process of refractory concrete during secondary state creep:

$$x = k \cdot t^n \qquad (1)$$

Where: k – time constant; and n – constant which describes mechanism of sintering.

If variable x is dimensional change (linear shrinkage), then Pines' equation of sintering can be applied [18]:

$$\Delta l/l_0 = k \cdot t^n \qquad (2)$$

Where: Δl – linear shrinkage of a concrete sample, (mm); and l_0 – initial linear dimension of a concrete sample, (mm).

The activation energy of the sintering process can be calculated using the following equation:

$$E = (R \cdot T_1 \cdot T_2 / (T_1 - T_2)) \cdot \ln (v_1/v_2) \qquad (3)$$

Where: E - activation energy, (kJ/mol); R – gas constant, (J/mol °C); T – temperature, (C°); v - rate of sintering process ($v = \Delta l / \Delta t$), (mm/min); Δl - shrinkage of a sample; and Δt - duration of shrinkage process.

Besides equation (2) suggested by German [19], the other equation which was derived from Frenkel's model of sintering can be used for calculating the activation energy of the sintering process:

$$\Delta l / l_0 = (\sigma \cdot t) / (2 \cdot r \cdot \eta) \qquad (4)$$

Where: σ - specific surfacial energy of boundary grains at the contact; r - diameter of grains; $\eta = A \cdot \exp (-E / R \cdot T)$ - viscosity.

Final form of the equation (4) is:

$$\Delta l / l_0 = c \cdot T^2 \cdot \exp (-E / R \cdot T) \qquad (5)$$

Where: c – constant which merges all constants from equation (4).

If equation (5) is to be transformed into its logarithmic form, energy of activation of sintering process could be calculated as the coefficient of the curve slope from the diagram which describes dependence between $\log (\Delta l/l_0) \cdot (1/T^2)$ and $1/T$.

The goal of this work is to use correlation between results of creep testing and type of sintering process for prediction of behavior and refractory concrete microstructural change during concrete actual service-life.

2. EXPERIMENTAL

The experiments were performed on two different types of concrete containing different volume fractions and different refractory aggregates. First type of concrete (B sample) contained bauxite as aggregate. Other type of concrete (C sample) was prepared with corundum as aggregate. Aggregates had different granulations (bauxite particle sizes ranging: 0-1, 1-4, 4-6 mm and corundum particle size ranging: 0-1, 1-3, 3-5 mm). Both types of concrete were prepared with high aluminate cement SECAR 70 (Lafarge). The chemical compositions of the investigated concretes are as follows: B sample (Al_2O_3 – 62.88 %, SiO_2 – 21.17 %, CaO – 8.26 %) and C sample (Al_2O_3 – 93.62 %, SiO_2 – 0.7 %, CaO – 5.97 %).

Thermo-mechanical properties of C and B samples were experimentally determined according to standard laboratory procedures [20-24]. Twenty cubic samples (10 for each series) whose dimensions were 10 cm x 10 cm x 10 cm were investigated. After 7 days of curing in climate chamber (at temperature 20°C), samples were demoulded and then dried at 110°C for 24 hours. Afterwards samples were transferred into an electric furnace and fired at 1100°C for 4 hours. Such concrete specimens were tested on compressive mechanical strength using conventional laboratory hydraulic pressure device [20].

Creep test was performed on C and B concrete samples (twenty samples, ten samples for each series) which were cylindrical; height 50 mm and diameter 50 mm. Hole (diameter 5 mm) was drilled in the center of each concrete sample. Concrete samples were dried at 110°C for 24 hours and afterwards pre-fired at 1100°C for 4 hours. Surface of samples was polished with diamond paste. Prefired samples were heated at rate of 5°C/min from room temperature (20°C) up to testing temperature (1200, 1300 or 1400°C) in the compressive creep apparatus (Netzsch, Germany) and then submitted to a constant compressive static load (0.2 MPa) at temperatures of 1200, 1300 and 1400°C respectively. Each test on specific temperature lasted for 30 hours. During this period secondary state creep was reached. Investigation was performed according to valid standards [22].

Apparent porosity of the refractory concrete samples was investigated with optical microscope (Olympus, CX31-P) accompanied with PC program for image analysis. The original microscope images were transmitted to the image processor by a color camera. The Image Pro Plus (IPP) program (Materials Pro Analyzer, Version 3.1, Media Cybernetics, Silver Spring, MD, USA) was used in the experiment. Results of image analysis were correlated with results of the creep testing.

Same specimens from the creep testing were used in IPP analysis. The specimens were covered with thin chalk-powder film before surface damage was investigated. The film provided better contrast and differentiation of damaged and non-damaged surfaces. Digital photographs of the samples surface were taken before and after each thermal treatment. Different (damaged and non-damaged) surfaces of the samples were marked with different colors using IPP tools. Thus, higher resolution and sharper difference in damaged and non-damaged surfaces on the specimens could be obtained. When the appropriate color is selected, it is possible to quantitatively measure the ratio and level of damaged and non-damaged areas by means of image analysis using a statistical approach.

The images processed in the analyzer were converted into binary form as white features in front of the black background. The binary images were filtered to reduce as much as possible the other features captured together with the target crack images. Then, a final retouching was performed on the images to eliminate the remaining undesirable features and defects by using painting software. In this stage, enhanced images were ready for quantitative analysis. Program contains a procedure for systematic collection of the image analysis data by dividing the total observation area into squares. Following a similar procedure, a transparent grid was attached on each plane section before the analysis. IPP basically works on comparing colors of different objects and calculating squares in marked area. At least 10 photographs per sample were analyzed in order to obtain a reliable characterization of the microstructure. The ratio between sample surface area and damaged surface area was calculated for each refractory concrete sample and, thus, surfacial apparent porosity was determined.

3. RESULTS AND DISCUSSION
Results of the investigation of thermo-mechanical properties are presented in Table 1 and results of creep deformation testing are presented in Table 2.

Creep deformation curves of B and C refractory concrete samples are given on Fig. 1. The percentage of the compressive creep deformation increases with rising temperature. As it has been noted previously, the creep curve consists of primary, secondary and tertiary creep. Secondary state creep can last for a long period of time and it is of prime importance. The analysis of the creep curves of samples B and C (Fig. 1.) shows that maximal deformation occurs at temperature 1400°C for both samples. They are 4.6 and 4.48 % for B and C, respectively. Higher percentage of deformation for B sample is consequence of its higher apparent porosity. This statement is also supported with results of IPP analysis.

Primary creep lasts for 6 hours approximately for both samples. Transition of primary into secondary state creep is hardly noticeable at 1200°C. However, the transitions at 1300 (after 6 hours for B and C) and 1400°C (after 10 hours for B and after 15 hours for C) are clearly visible. Onset of tertiary state creep was not detected due to the short interval of investigation (30 hours).

The fine matrix situated in inter-aggregate space of concrete sample changes with increasing temperature. Its viscosity significantly diminishes above 1200°C. Therefore, the plastic deformation of concrete samples at 1400°C is higher than deformation at 1200 and 1300°C. The formation of certain amount of liquid phase was noted. Secondary mullite crystallizes from the liquid phase. The amount of mullite depends on Al_2O_3 : SiO_2 ratio. Mullite has influence on the deformation at higher temperatures. The creep deformation of the refractory concrete at 1400°C is higher than deformation at 1200 and 1300°C due to the surplus of the liquid phase. 90 % of the deformation has already taken place at 1400°C. Rate of deformation significantly decreases in following interval. After 15 hours spent at 1400°C very little deformation can be observed. Initial deformation is caused by insufficient amount of mullite in structure, whose formation requires sufficient time period and high temperatures. Mullite formation induces a structural reinforcement that causes a more rapid mechanical hardening of concrete at high temperatures.

As power law creep can be applied on creep curves presented on Fig 1., calculation of sintering constant (n) and activation energy (E) was also performed. Results of dimensional change $(\Delta l/l_0)$ of B and C samples were obtained during creep test at 1200, 1300 and 1400°C. The durations of the dimensional change (shrinkage) of concrete samples were registered by an automatic writer for each time interval. Using equations (2) and (5) approximate sintering constant, i.e. coefficient of sintering reaction mechanism (n) and activation energy (E) have been calculated. The sintering rates and temperature dependences were calculated using coefficient of direction (slope) of the function $\Delta l = f(\Delta t)$. Fig 2. shows numeric results for coefficient of reaction mechanism of sintering process (n) and the rate of sintering (v) for B and C concretes at temperatures 1200, 1300 and 1400°C, and also gives conclusion concerning type of sintering mechanism which occurs during secondary state creep:

The coefficient n describes the mechanism of particle transport during sintering process in secondary state creep. From results exposed on Fig. 2 it can be seen that the most dominant mechanisms are: the surfacial diffusion, diffusion along grain boundary and plastic-viscous flow.
Approximate activation energy (E) was calculated using equation (3) from results obtained at 1300°C. Results are in Table 3.

The microstructure of concretes was examined using scanning electron microscopy. SEM (SEM JEOL JSM-5300) photomicrographs of refractory concrete samples heat treated at 1100°C are shown on Fig. 3. and 4. Aggregate particles sizing few millimeters which are surrounded with fine matrix composed of micronic size particles can be seen on Figures 3 and 4. Also, there can be seen pores of various sizes. Porosity of these samples was calculated using IPP method and (average) results are: 33.2 and 27.1 % for B and C sample respectively, measured at 1100°C. Higher porosity of B sample is the reason why its creep deformation is also higher than in case of C sample, as it was previously assumed (Fig 1.)

By using IPP method, parameters such are maximal, minimal and average pore diameter (D_{max}, D_{min}, D_{av}), pore roundness (R) and number (N) of pores within superficial pores were obtained. Digital photos (as explained in chapter 2.) of concrete samples were used in analysis. Results are presented in Tab. 1. According to IPP analysis: average pore diameter increases untill temperature of approximately 1300°C is reached and afterwards significant shrinkage occurs. Shinkage of pores is a conccquence of sintering process. B sample has smaller average pore diameter although its apparent porosity is higher on all temperatures of investigation. It is consequence of better choice of aggregate granulation. Ideal roundness would be 1.00, and for investigated pores roundness is 1.05 to 1.15. That means that pores are almost spheric. N is smaller for C sample, which also indicates that apparent porosity of C concrete is smaller than in case of B concrete, i.e. most of pores are surfacial.

Microstructure of samples used in creep tests at various temperatures is presented on Fig. 5 (a, b, c) and 6. (a, b, c). Samples fired at 1200°C (Fig. 5a and 6a) exhibit little structural change when compared to the samples heat treated at 1100°C. Structural changes arc noticeable on the samples exposed to 1300°C. There can be noticed a formation of liquid phase and emersion of initially formed crystals. The microstructures of samples investigated at 1400°C are significantly different than the microstructure of the rest of samples. Formation of mullite is noticed in the structure. Mullite provides structural reinforcement and makes rate of the creep slower (rate of deformation). Larger amount of secondary mullite is noticed in the sample B, and it can explain the smaller deformation of these samples during secondary creep, when compared with concrete C. Phases analyses confirmed the presence of mullite in concrete samples. Beside previously written, there is significant difference in porosity. Namely, samples of bauxite concrete have noticeably higher porosity than concrete with corundum.

4. CONCLUSION

Investigation of time dependent viscoplastic deformation at various constant temperatures and constant static load, i.e. creep deformation, lead to the following conclusions: corundum based concrete samples show smaller deformation than bauxite based concrete samples investigated at same temperature (1200, 1300 and 1400°C). Therefore corundum based concrete is more adequate for application in certain proposed extreme conditions in plants operating at high temperatures. At these temperatures material is in the secondary state creep which is a characteristic for most of concrete actual "life" in actual service. Deformation of concrete which occurs at high temperature could damage the plant's lining. However, a structural reinforcement occurs within concrete at temperatures above 1300°C due to the formation of secondary mullite situated in inter-aggregate space, which was noted by SEM technique. This affects creep and decreases rate of plastic deformation as it is proved in case of isothermal sintering at 1400°C for 30 hours. Furthermore, this affects all mechanical properties of refractory concrete (mechanical strength and refractoriness are increased, porosity decreased, etc.) and "life" of refractory concrete lining is elongated.

Microstructure of refractory composites was also investigated in this paper. Apparent porosity, pore distribution and pore size of concrete samples were investigated using Image Pro Plus program. Results presented in this paper contribute to the idea of including other test methods for investigation of mechanical properties and microstructure such are nondestructive methods instead of commonly used standard laboratory procedures. Benefits from using image analysis are numerous: it is nondestructive, simple and fast method; same samples could be used for further tests; there is financial benefit in minimizing number of samples for testing – there is saving in material and in time; entirely new and important information about damages and porosity of surface could be obtained - precise diameters of pores, roundness, number of pores in a section, etc.; as surfacial damage level is measured, results could be useful for prediction of sample behavior during further testing or application.

Investigations presented in this paper confirmed that results of creep deformation test, image analysis and quantitative description of sintering process are interconnected and that they, as well their interconnection, can be useful when type of refractory concrete is to be chosen for application in a metallurgical furnace or some other plant operating at high temperature.

ACKNOWLEDGEMENTS

This work has been supported by Serbian Ministry of Science under project 19012 and 16004.

REFERENCES

[1] Z. Bazant, M.F. Kaplan, Concrete at High Temperatures, Material Properties and Mathematical Models, Concrete Design and Construction Series, Longmann Group, London, 1996.
[2] M. Boussuge, Investigation of the thermomechanical properties of industrial refractories: the French programme PROMETHEREF, Journal of Material Science, 43 [12] (2008) 4069-4078.
[3] DLY Kong, JG Sanjayan, K. Sagoe-Crentsil, Factors affecting the performance of metakaolin geopolymers exposed to elevated temperatures, Journal of Material Science 43 [3] (2008) 824-831.
[4] VWY. Tam, CM. Tam, Assessment of durability of recycled aggregate concrete produced by two-stage mixing approach, Journal of Material Science, 42 [10] (2007) 3592-3602.
[5] BT. Tamtsia, JJ. Beaudoin, J. Marchand, The early-age short-term creep of hardening cement paste: AC impedance modeling, Journal of Material Science, 38 [10] (2003) 2247-2257.
[6] DDL. Chung, Review: Improving cement-based materials by using silica fume, Journal of Material Science, 37 [4] (2002) 673-682.

[7] AA. Wereszczak, TP. Kirkland, WF. Curtis, Creep of CaO/SiO2-containing MgO refractories, Journal of Material Science, 34 [2] (1999) 215-227.

[8] M. Posarac, M. Dimitrijevic, T. Volkov-Husovic, A. Devecerski, B. Matovic, Determination of thermal shock resistance of silicon carbide/cordierite composite material using nondestructive test methods, Journal of European Ceramic Society 28 (2008) 1275–1278.

[9] 11F. Sinonin, C. Olgonon, S. Maximilien, G. Fantozzi, Thermomechanical Behavior of High Alumina Refractory Castables with Syntetic Spinel additions, Journal of American Ceramic Society 83 (2000) 81-92.

[10] I.A. Altun, Effect of temperature on the mechanical properties of self-flowing low cement refractory concrete, Cement and Concrete Research 31 (2001) 1233-1239.

[11] F. Cardoso, M. Innocentini, M. Miranda, F. Valenzuela, V. Pandolfelli, Drying behavior of hydratable alumina bonded refractory castables, Journal of the European Ceramic Society

[12] M. Zawrah, N.M. Khalil, Effect of mullite formation on properties of refractory castables, Ceramics International 27 (2001) 689–694.

[13] N. Hipedinger, A. Scian, E. Aglietti, Magnesia–ammonium phosphate-bonded cordierite refractory castables: Phase evolution on heating and mechanical properties, Cement and C24 (2004) 797–80[20]

[14] E. Karadeniz, C. Gurcan , S. Ozgen, S. Aydin, Properties of alumina based low-cement self flowing castable refractories, Journal of the European Ceramic Society 27 (2007) 1849–1853.

[15] D.N. Boccaccini, M. Canio, T. Volkov Husovic, Service Life Prediction for Refractory Materials, Journal of Material Science 43 [12] (2007) 4079-4090.

[16] M. Bugajski, R. Schwaiger, Self-flowing Castables – a new Type of Unshaped Refractory Products, Veitsch-Radex Rundsch, Wien, 1996.

[17] C. Parr, E. Spreafico, T.A. Bier, A. Mathieu, Calcium Aluminates Cements (CAC) for Monolithic Refractories Technical paper References F 97, LAFARGE Aluminates, 75782 Paris, Cedex 16, 1997

[18] H. Feldborg, B. Myhre, A.M. Hundere, Use of microsilica in a light-weight self-flowing refractory castable, UNITECR '99, Berlin, Verlag Stahleisen: Dusseldorf, Germany, 1999

[19] M. Prassas, J. Phalippou, J. Zarzycki, Science of Ceramic Processing, Wiley-Interscience Publication, New York, 1996.

[20] R.M. German, Sintering Theory and Practice, Wiley-Interscience Publication, New York, 1996.2.oncrete Research 34 (2004) 157–164.

[21] Standard: JUS B. D8. 304

[22] Standard JUS B. D8. 303

[23] Standard JUS B. D8. 301

[24] Standard JUS B. D8. 312

Determination of Apparent Porosity Level of Refractory Concrete

Table1. Thermo-mechanical properties of concrete samples B and C

Parameter:	B	C
Bulk density at 1100°C, (g/cm^3)	2.03	2.46
Water absorption at 1100°C, (%)	15.2	10.1
Refractoriness	20 SK / 1540°C	34 SK / 1755°C
Refractoriness under compressive load (0,2 MPa)		
- Ta, (°C)	1300	1500
- Te, (°C)	1570	>1600
Compressive strength at 1100°C, (MPa)	16.9	25.3
Apparent porosity at 1100°C, (%)	33.2	27.1

Table 2. Linear creep deformation ($\Delta l/l_0$) of B and C concrete samples after 5 and 30 hours of thermal treatment

	B	C	B	C	B	C
	$\Delta l/l_0$ at 1200°C, (mm)		$\Delta l/l_0$ at 1300°C		$\Delta l/l_0$ at 1400°C	
after 5 hours	-1.86	-2.25	-3.10	-3.24	-4.28	-4.17
after 30 hours	-2.57	-2.94	-3.60	-3.72	-4.6	-4.48
$\Delta(\Delta l/l_0)$, %	0.71	0.69	0.5	0.48	0.32	0.31

Table 3. Approximate activation energy (E) for B and C concrete

T, (°C)	E_B, (kJ/mol)	E_C, (kJ/mol)
1200-1300	89.2	84.5
1300-1400	76.9	82.1

Table 4. Results of Image Pro Plus analysis for C and B concrete samples

	C					B				
T	D_{max}	D_{min}	D_{av}	N	R	D_{max}	D_{min}	D_{av}	N	R
(°C)	(mm)	(mm)	(mm)			(mm)	(mm)	(mm)		
110	0.046	0.0042	0.0067	9	1.08	0.056	0.00129	0.003	51	1.07
1100	0.057	0.00448	0.0077	14	1.1	0.073	0.00137	0.0035	74	1.12
1200	0.072	0.0045	0.0084	22	1.13	0.079	0.00138	0.0037	81	1.14
1300	0.089	0.0046	0.0089	26	1.138	0.085	0.00138	0.004	80	1.22
1400	0.084	0.00455	0.0086	24	1.091	0.082	0.00130	0.0038	75	1.17

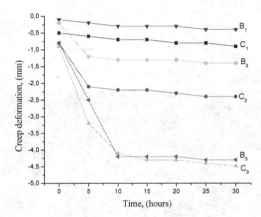

Fig 1. Creep deformation curves obtained during creep tests on B and C sample at three temperatures: B_1 and C_1 – at 1200°C, B_2 and C_2 – at 1300°C and B_3 and C_3 – at 1400°C.

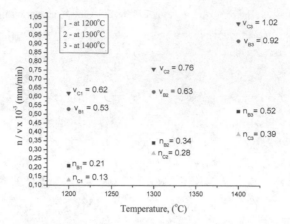

Fig 2. Results for coefficient of sintering (n), rate of sintering (v) and mechanism of reaction for B and C samples: B_1 - surfacial diffusion, B_2 - diffusion along the grain boundary, B_3 - plastic-viscous flow, C_1 - surfacial diffusion, C_2 - surfacial diffusion and C_3 - diffusion along the grain boundary.

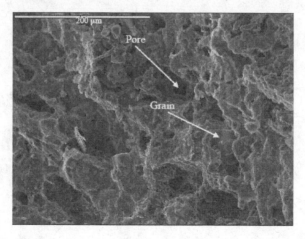

Fig 3. SEM of B refractory concrete sample heat treated at 1100°C: porosity is clearly visible and obviously higher than in case of sample C (Fig.4)

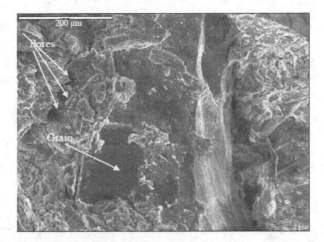

Fig 4. SEM of C refractory concrete heat treated at 1100°C: large corundum grains surrounded with matrix are visible

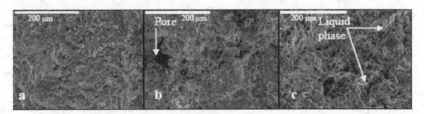

Fig 5. SEM of B sample heat treated at: a)1200°C; b) 1300°C; c) 1400°C: pores within B sample are bigger; and emersion of liquid phase can be seen – white shade on SEM microphotograph

FIG 6. SEM of C Sample Heat Treated at: A) 1200°C; B) 1300°C; C) 1400°C: Pores are Visibly Smaller than in B Sample; Liquid Phase is Noticeable

Processing-Microstructure-Properties Correlations

SINTERING BEHAVIOR OF LITHIUM-TITANATE PEBBLES: MODIFICATIONS OF MICROSTRUCTURE AND PORE MORPHOLOGY

D. Mandal[1*], D. Sen[2], S. Mazumder[2], M R K Shenoi[1], S. Ramnathan[3], D. Sathiyamoorthy[4]

[1]Chemical Engineering Division, Bhabha Atomic Research Centre, Trombay, Mumbai-400085, India; *Corresponding author: dmandal@barc.gov.in, dmandal@iitb.ac.in

[2] Solid State Physics Division, Bhabha Atomic Research Centre Trombay, Mumbai-400085, India,

[3] Advance Ceramic Section, Bhabha Atomic Research Centre Trombay, Mumbai-400085, India,

[4]Powder Metallurgy Division, Bhabha Atomic Research Centre, Vashi, Sector 20, Navi Mumbai 4000705, India;

ABSTRACT

Morphological modifications of mesoscopic structure in lithium titanate pebbles with sintering have been investigated using small angel neutron scattering (SANS), helium pycnometer, mercury porosimeter and electron microscopy. Lithium titanate, used in the study, was synthesized by solid state reaction of a mixture of lithium carbonate and titanium dioxide at 750°C. Green pebbles of lithium titanate were fabricated from synthesized powder by extrusion and spherodization technique. The green pebbles were sintered at 900°C for different duration. From SEM analysis and SANS it was found that the average size of the pores initially increases slightly with sintering time and then decreases although the density of the pebble increases monotonically with sintering time.

1. INTRODUCTION

Li^6 produce tritium when scattered with fast neutron, through (n, α) reaction; $Li^6 + n \rightarrow He + T$. In-situ generation of tritium is necessary as tritium is not available in nature and has half life 12.3 years. Lithium titanate (Li_2TiO_3), is considered as one of the suitable material for the test blanket module (TBM) of fusion reactor [1, 2] for its attractive characteristics, such as the high lithium density, high thermal and chemical stability, tritium recovery at low temperature, thermal conductivity etc [3] compared to other TBM materials viz. lithium- zirconates (Li_2ZrO_3) and lithium-ortho-silicate (Li_4SiO_4).

Alvani *et. al*. [4] reported that both the density and the grain size of Li_2TiO_3 pebbles are the key parameters in determining tritium release rate. If the tritium release is not unduly affected, a higher density is favourable from the viewpoint of the lithium loading and the Li^6 enrichment. The density of lithium-based ceramics also has an effect on its thermo-mechanical performance which is a critical issue for assessing the reliability of solid breeder blanket concepts over the lifetime of the component [5]. The pore size and pore size distribution in the pebble play an important role. For dense pebbles (above 85% of the theoretical density) with a low open porosity, the grain size plays a major

165

role in the tritium release rate, which is increased by decreasing the grain size and consequently by increasing the grain boundaries. It is suggested that the microstructural requirements of the Li_2TiO_3 pebbles in the Helium Cooled Pebble Bed (HCPB) Blanket design for a DEMO Fusion Reactor as 0.3–1.2μm in diameter and 85-90% of the theoretical density [6].

Small angle neutron scattering (SANS) is an important non-destructive technique to study the mesoscopic structures in porous materials [7–12] as SANS probes density fluctuations in mesoscopic length scale. Compared to other complementary techniques like mercury porosimetry, BET, etc., to study the pore structure in a porous material, SANS possess some special advantages, in particular, probing closed pores in addition to open pores. In the present study, the effects of sintering time on pore morphology in Li_2TiO_3 pebbles have been investigated using SANS.

Scattering measurements, using neutrons or x-rays along with electron microscopy and porosimetric techniques can throw light on the evolution of microscopic and mesoscopic structures of such ceramic materials. This paper discusses about the experimental results on the effects of various thermodynamic parameters on the micro and mesoscopic structures of lithium titanate pebbles during sintering.

2. EXPERIMENTAL PROCEDURE

2.1 Synthesis of Li_2TiO_3

Li_2TiO_3 used in this study was synthesized by the solid state reaction of a homogeneous mixture of lithium carbonate and titanium dioxide at 750°C (Reaction-1).

$$Li_2CO_3 + TiO_2 \rightarrow Li_2TiO_3 + CO_2 \qquad (1)$$

2.2 Fabrication of Li_2TiO_3 pebbles

Spherical pebbles of Li_2TiO_3 of 0.8 to 1.0 mm diameter were fabricated from synthesized Li_2CO_3 fine powder (particle size < 48μm) using a binder.

2.2 Sintering of Li_2TiO_3 pebbles

The green pebbles were sintered at 900°C for different duration. The initial raise in temperature was 15°C per minute. After achieving the temperature at 900°C the constant temperature was maintained for different duration, 30 min (Sample No-23), 1 hrs (Sample No-26), 2 hrs (Sample No-24), 3 hrs (Sample No-25), 5 hrs (Sample No-27), and 6 hrs (Sample No-28), in different set of experiment.

The complete process of synthesis and fabrication Li_2TiO_3 pebble is named as solid state reaction process, developed by D. Mandal *et. al.* and described in details in somewhere else [13-14].

2.2 Density and porosity

The density of pebbles was measured by using a helium Pycnometer. The density of the green pebbles was about 60-70% of theoretical density (TD). Open and close porosity of each samples were measured by Hg-porosimeter. The results are shown in Table-1.

Table 1 Variation of density and porosity of pebbles with sintering duration

Sample Number	Sintering Time (hr)	Density (Helium Pycnometer) (gm/cc)	Porosity by Mercury Porosimeter		
			Accessible Porosity (%)	Inaccessible Porosity (%)	Total Porosity (%)
23	0.5	3.2195	6.46	23.77	30.23
26	1.0	3.2183	5.77	22.41	28.18
24	2.0	3.2176	2.74	22.99	25.73
25	3.0	3.2172	1.59	22.11	23.70
27	5.0	3.2142	0.92	20.13	21.05
33	6.0	3.2165	0.73	14.23	14.96

3. RESULTS AND DISCUSSION

3.1 SEM Analysis

Microstructures of the sintered pebbles were studied by Scanning Electron Microscopy (SEM) analysis. Fig. 1(a) to Fig. 1(c) show the SEM micrographs of sintered pebbles. It was observed that the internal porosity gets reduced with increasing duration of sintering time. During first three hours of sintering, the grain size does not change appreciably. However, after five hours of sintering, it started decreasing. It has been observed that the average grain size becomes 2.2μm after six hours of sintering.

Fig. 1 SEM micrographs for (a) 0.5 hr (b) 1hr (c) 6hr sintered samples

3.2 Small angle neutron scattering

Scattering experiments have been performed using a double crystal based medium resolution small-angle neutron scattering facility (MSANS) at the Guide Tube Laboratory of Dhruva reactor, Mumbai, India [15]. The scattered intensities have been recorded as a function of wave vector transfer q [$= 4\pi\sin(\theta)/\lambda$, where 2θ is the scattering angle and λ (= 0.312 nm) is the incident neutron wavelength]. The specimens under SANS investigations were placed on a sample holder with a circular slit of 1.5 cm diameter. Measured SANS profiles have been corrected for background, transmission and instrument resolution [16].

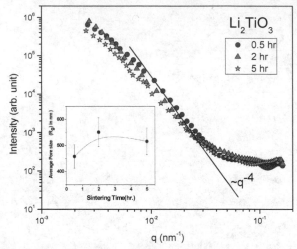

Fig. 2 SANS profiles of the samples sintered at 900 °C for 0.5 hr, 2hr and 5 hr. Average pore size obtained from SANS profile is depicted in the inset.

The radius of gyration (R_g) of the pores was estimated from the scattering data at very low q region. SANS profiles of the samples sintered at 900oC for different duration of sintering and average pore size obtained are shown in Fig. (2). The value of the R_g for the samples sintered for 0.5hr, 2 hrs and 5 hrs was estimated to be 458, 551 and 516 nm. This implies that at the initial/intermediate stage of sintering there has been some growth in pore size by the coalescence of the neighbouring pores. However, pore size decreases afterwards because of the shrinkage of the pores due to modification of pore-neck via mass transport. Pore size distribution for the sintered samples was calculated assuming a polydisperse spherical particle model. The estimated pore size distributions are depicted in Fig. (3).

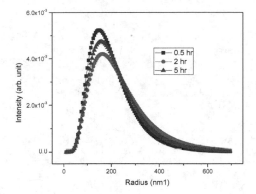

Fig. 3 Estimated pore size distributions in Li₂TiO₃ pebbles sintered at 900 °C for different sintering time.

4. CONCLUSION

The effect of sintering time on the pore size for Li_2TiO_3 pebbles was investigated using SEM and SANS. It has been revealed that the sintering temperature of 900°C is optimum to produce Li_2TiO_3 pebbles fabricated by extrusion and spherodization technique. It was observed from SEM analysis that the pore size for bigger pores decreases with increase in sintering time initially. For relatively smaller pores, as observed from SANS, the size increases slightly and then decrease with sintering time. This may be due to coalescence of some pore network during intermediate stage of sintering.

ACKNOWLEDGMENT

The authors are thankful to Shri S. K. Ghosh, Director, Chemical Engg. Group, Shri Hanmanth Rao, Head, Chemical Engg. Division, BARC for their guidance to carry out the work. The authors thank, Dr. S. Majumder for his supports in SEM analysis. The authors are also thankful to Shri M C Jadeja and Shri B. K. Chogule for their constant assistance in carrying out the work.

REFERENCES

1. J.G. van der Laan, H. Kawamura, N. Roux, D. Yamaki, J. Nucl. Mater, **283-287**, (2000) 99-109.
2. A. Ying, M. Akiba, L.V. Boccaccini, S. Casadio, G. Dell'Orco, M. Enoeda et al., J. Nucl. Mater, **367-370** (2007) 1281-1286.
3. J. M. Miller, H. B. Hamilton, J. D. Sullivian, J. Nucl. Mater, 212-215 (1994), 877-880
4. C. Alvani, P.L. Carconi, S. Casadio, F. Pierdominici, in: J.V. van der Laan (Ed., NRG), Proceedings of the 7th International Workshop on Ceramic Breeder Blanket Interactions, Petten, 1998, 4-69–4-80.
5. A. Ying, M. Akiba, L.V. Boccaccini, S. Casadio, G. Dell'Orco, M. Enoeda et al., J. Nucl. Mater, 367-370 (2007) 1281-1286.
6. A. Deptuła, M. Brykała, W. Ładaa, T. Olczaka, B. Sartowskaa, A.G. Chmielewski, D.Wawszczaka, C. Alvani. Fusion Engineering and Design, 84 (2009), 681–684.

7.   D. Sen, A.K. Patra, S. Mazumder and S. Ramanathan; *J. Alloys and Compounds,* **361,** (2003) 270-275

8.   D. Sen, T. Mahata, A.K. Patra, S. Mazumder and B.P. Sharma; *J. Physics: Condensed Matter* **16**, (2004) 6229-6242

9.   S. Mazumder, D. Sen, A.K. Patra, S.A. Khadilker, R.M. Cursetji, R. Loidl, M. Baron and H. Rauch; *Phys. Rev. Lett.* **93** (2004) 255704.

10. D Sen, J Bahadur, S Mazumder, V Bedekar and A K Tyagi *J. Phys.: Condens. Matter* **20** (2008) 035103

11. I. W. Chen, X.-H. Wang, Nature 404 (2000) 168.

12. P.K. Krishnankutty-Nair, K. Keizer, A.J. Burggraaf, T. Okubo, H. Nagamoto, S. Morooka, Nature 358 (1992) 48.

13. H. Sato, T. Minami, S. Takata, T. Yamada, Thin Solid Films 236 (1993) 27.

14. D. Mandal, D. Sathiyamoorthy, V Govardhana Rao, JAEA-Conf. 2009-006, Proceedings of the Fifteenth International Workshop on Ceramic Breeder Blanket Interactions, 3-4 Sept. (2009) 50-64

15. D. Mandal, M R. K Shenoi, S. K. Ghosh, Fusion Engineering and Design 85 (2010) 819–823.

16. S. Mazumder, D. Sen, T. Saravanan, P.R. Vijayaraghavan, J. Neutron Res. 9, (2001), 39–57.

17. J.A. Lake, *Acta Crystallogr.* 23, (1967), 191–194.

SILICON CARBIDE BASED SANDWICH STRUCTURES: PROCESSING AND
PROPERTIES

Alberto Ortona*, Claudio D'Angelo, Simone Pusterla,
ICIMSI, SUPSI, Galleria 2, 6928, Manno, Switzerland

Sandro Gianella
Erbicol SA
Viale Pereda 22 P.O. Box 321, 6828, Balerna, Switzerland

ABSTRACT
Sandwich structured composites have been widely studied and applied at ambient temperature in
aeronautical, automobile and naval applications. For high temperature applications, an integral ceramic
sandwich structure made of Ceramic Matrix Composites (CMC) skins could be applied to structural
components. As far as cores are concerned, some carbide (e.g. SiC) cellular foams are among the most
appropriate materials, because of their outstanding thermo-mechanical properties,. These foams can
withstand protracted oxidative exposure with low material degradation. This paper presents some
mechanical properties of SiC based sandwich structured CMCs, assembled with an innovative in-situ
joining method, which is performed during sandwich manufacturing and allows the production of
complex shapes at low costs. Flat sandwiches with SiC cellular cores and skins made of C or SiC
fibers and polymer derived SiC matrices showed three point bending strength higher than plain foams
and, most of all, a marked toughening behavior.

INTRODUCTION
A sandwich structured composite is obtained by joining two thin but stiff skins to a lightweight but
thick core [1]. The core is usually made of a lower strength material, but its thickness offers to the
overall structure high flexural stiffness and low density. These structures are widely employed at
ambient temperatures for aeronautical, automobile, naval applications.

For high temperature applications sandwich structured composites become extremely useful because
they can combine weight reduction and stiffness with better heat flow management.
Reticulated SiC foams bear high thermal loads [1] and in comparison with Reticulated Vitreous
Carbon (RVC) foams, they can operate for a long time at higher temperatures (1400°C) in oxidative
environments [1] [3].

Hoefner et al. [4] produced a sandwich in which skins were made of SiC tape casted bulk sheets and
the core of expanded poly (silsesquioxane) co-fired after sandwich assembly. During bending tests,
authors report that the brittle behavior of the skins caused sandwich crushing. Van Voorhees and Green
bonded alumina based ceramic sheets to alumina foams using conventional cements. Conversely to
previous work, they report sandwich which exhibits non catastrophic failures under three point bending
[7].

For thermo-structural applications, CMC skins together with ceramic foams need to be employed. The
first attempt to realize a high temperature structured sandwich made of CMC skins and a Chemical
Vapor Infiltration (CVI) SiC foam core was reported by Fisher in 1985 [5]. Recently, NASA

* Corresponding author : SUPSI-ICIMSI Galleria 2 CH-6928 Manno, tel +41586666611, fax +41586666620, e-
   mail alberto.ortona@supsi.ch

developed a sandwich made with C_f-SiC_f/SiC_m skins and a SiC core, processed through integral densification of the CMC and foam core [6]. In this work, SiC is deposited by CVI both onto fibers preforms and the core. These structures do not present a joining layer. Authors point out that skins and core coefficient of thermal expansion must be similar both for processing and operation.

If bonding between skins and core must be employed, skin to core bonding must be carefully optimized in order to avoid local delaminations under load. An ideal way to accomplish this is trough in situ join CMC preforms with subsequent densification via liquid metal infiltration and reaction bonding was first reported by Krenkel and al. [8].

Based on the same concept, this work presents a procedure to assemble ceramic sandwich structures in any shape with industrially developed manufacturing techniques. Skins and core are assembled in situ during the early stages of sandwich manufacturing (Figure 1). Once CMC preforms and the ceramic core are brought together, they can be further densified with any well known technique employed for CMC manufacturing. Polymer Impregnation and Pyrolysis (PIP) were employed in this work.

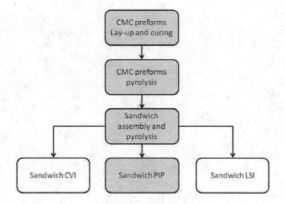

Figure 1 CMC SiC sandwich manufacturing procedure

MATERIALS
Two types of sandwich were produced with different CMC skins made of graphite fibers in one case, and SiC fibers in the other. The carbon fibers employed were Toray T300 (Toraika J), with a pyrolytic Carbon interphase (thickness 0,26 μm) applied by Chemical Vapor Deposition (CVD) and a fiber density of 1.73 g/cm³. The woven fabric was a satin 3HS with an areal weight of 200 g/m².
SiC fibers, without interphase, were Tyranno LOXM (UBE, J), fiber density 2.48 g/cm³ and plain weave with an areal weight of 200 g/m². Both fabrics were pre formed with phenolic novolac resin (Hexion specialty chemicals). For CMC PIP processing, a polysilazane preceramic polymer Ceraset 20 (Kion Specialty Polymers, Clariant, CH) was utilized.

A slurry with two SiC powders of different grain size and Ceraset polymer was made to bond CMC preform skins to the ceramic core. The fine SiC powder is UF-15 α-SiC (H. C. Starck, D), it has a specific surface area from 14 to 16 m²/g with a particle size distribution D90% = 1.2 μm. The coarse α-SiC powder (Washingto/n Mills, USA) has grit size F800.

ERBISIC-R foams, produced with the replica method, (Erbicol SA, CH) were used as porous ceramic core. The foam bulk material is made of α-SiC powder in a β-SiC/Silicon matrix. Bulk average density is 2.83 g/cm³. Reticulated foams are an isotropic three-dimensional network of hollow ceramic ligaments. Pore sizes are circa 5 mm and are equivalent to those of commercial reticulated foams with 10 Pores Per Inch (PPI). Foams characteristics are reported in Table 1.

Table 1 ERBISIC-R foam dimensions and weights.

Sample ID	L [cm]	W [cm]	t [cm]	Volume [cm³]	Mass [g]	Density [g/cm³]	Porosity
Erbisic R0	18.5	13.9	1.5	385.72	150.84	0.39	0.86
Erbisic R1	18.5	13.6	1.5	377.40	134.31	0.36	0.87

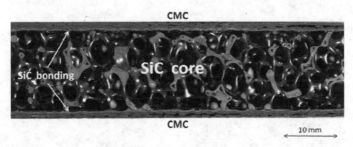

Figure 2 Sandwich assembly

EXPERIMENTAL

Preforms were obtained by laying up fabric plies, pre-impregnated with a small amount of phenolic resin (2% of the fabric weight), and curing (T =150°C. p = 2 bar, t=60 ') in an autoclave. Wider laminates were fabricated from which, panels were cut with the same in-plane dimensions of the SiC foams. To characterize CMC skins, one sample (200x20 mm²) was also cut from these laminates and processed in the same conditions of the sandwiches.

For phenolic resin pyrolysis, samples were slowly (50 °C/h) brought to 1000°C in flowing Argon. After pyrolysis, laminates were measured and weighted; their characteristics are reported in Table 2

Table 2 Pyrolysed composite preform properties. Porosity is calculated upon density of fibers.

Laminate ID	Fiber	N° of layers	Thickness [cm]	Volume [cm^3]	Mass [g]	porosity %
S0-Top	C	4	0.130	33.67	21.60	64
S0-Bottom	C	4	0.135	34.96	21.90	66
S 1-Top	SiC	4	0.100	25.39	27.77	55
S 1-Bottom	SiC	4	0.100	25.39	27.75	56

On the pyrolysed porous preforms, a slurry made of 30%/w fine SiC powders, 30%/w coarse SiC powders and 40% /w preceramic polymer and mixed by wet ball milling for 24h, was deposited onto one side of each laminate. Due to the high slurry viscosity (v=8100 m Pas) it did not fully infiltrate the preforms, acting as a thick (~ 350 µm) bonding layer (Figure 8). Laminates were placed on the top and the bottom of each foam as per Figure 2, cured in the autoclave and pyrolysed in the furnace, as previously described.

The assembled sandwiches were then impregnated with a brush on the outer surfaces of the skins with the polymeric precursor and pyrolysed. The operation was repeated 5 times to ensure a sufficient densification [9], necessary to attain an adequate skins strength for bending tests (Figure 6).

Each sandwich panel was cut in five pieces cut with a diamond saw (139x25x18.6 mm^3) for S0 and (139x25x17 mm^3) for S1.

During S0 cutting, the bottom face of samples 3, 4, 5, 6 debonded from the ceramic core, while samples 1, 2 remained partially bonded.

Sandwich S0   Sandwich S1

Figure 3 Failure modes of the sandwiches S0, with CMC skins(C$_f$/SiC$_m$) and S1, with CMC skins (SiC$_f$/SiC$_m$)

CHARACTERISATION

Ambient temperature three point bending tests (Figure 3) were performed with an universal testing machine (Zwick/Roell Z050 Germany) using a span of 100 mm. Cross head speed was 0.01 mm/s. Set up and dimensions were chosen to compare data with previous works [1][7]. Tensile tests on single CMC were performed with the same apparatus and a cross head speed of 0.008 mm/s.

Samples taken from S0 and S1 were milled and polished for microscopy. Observations were performed using an optical microscope (Leica DMLM Germany). Images were acquired using a digital camera (Leica DFC 280 Germany).

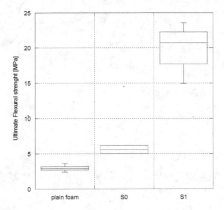

Figure 4 Min, max and average flexural strength values of S1, S2 samples and of plain foams with equivalent (86%) porosity [2].

RESULTS
Figure 4 shows that S1 sandwich structured composites present an average (calculated on the five samples) flexural strength which is higher than those of the experimental plain foams [1] and of S0 foams with partial bonding. It also shows that S0 sandwich with poor bonding has a slightly higher ultimate flexural strength if compared to that of plain foams.

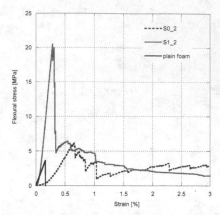

Figure 5 Stress-deflection curve under three point bending of sandwiches S0 and S1 and a plain foam with equivalent porosity (86%) [2].

As already evidenced in previous works [7], despite the fragile behavior of plain foams and SiC_f/SiC_m of the CMC, the whole sandwich structure experienced a marked toughening behavior (***Figure 5***). During the bending tests, the stress-strain behavior on initial loading is almost linear elastic until reaching a maximum corresponding to the formation of a large single crack going through the core. After this event, the load gradually decreases, but without sandwich failure. Essentially every load drop into the chart of ***Figure 5*** corresponded to one or more ligaments tears. After the first load drop, skins were not fractured. Stresses into the skins were thus below the ultimate stresses of both C_f/SiC_m and SiC_f/SiC_m CMC (***Figure 6***).

In sandwich S0 with C_f/SiC_m skins, bonding was very weak and after foam failure, which almost corresponded to that of plain foams, the core completely debonded from the skins (***Figure 7***) which, were subsequently sliding against the foam. In sandwich S1, for higher deflections, skins failure was also observed (***Figure 7***). As a result of the described assembly procedure, the bonding layer wraps the struts on the foam side (***Figure 8*** A) and, on the CMC side, infiltrates the bundles of its external fabric and some fibers within these bundles (***Figure 8*** B). Bonding thickness is uniform along the bonding area and, due to slurry shrinking during pyrolysis, presents an uniform crack pattern subsequently filled during the following PIP cycles.

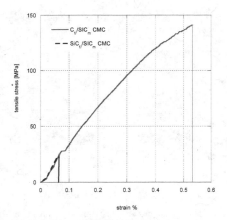

Figure 6 Tensile tests. Stress/strain curves of CMC laminates

DISCUSSION

Joining CMC to the foam before their densification allows a strong grip of the bonding layer to the sandwich skins. That is not a chemical bond, as CMC bundles are entangled into the polymer derived ceramic layer (*Figure 8* B). At the foam to bonding layer interface, due to the several sandwich thermal cycles during its processing, the different coefficients of thermal expansion (α) of the SiC foam and C_f/SiC_m CMC (Table 3) caused struts failure and/or pull out from the bonding layer, depending on the strut thickness (i.e. foam porosity). This was not the case (Table 3) of sandwich S1 with SiC_f/SiC_m CMC skins.

Table 3 Properties of the sandwich constituent materials under study (* from literature)

	Material	α [10^{-6} K^{-1}]
Foam	Si-SiC (foam)	4-5*
S0	C_f/SiC_m (PIP)	1-2*
S1	SiC_f/SiC_m (PIP)	4-5*

Debonded foams failure mode was similar to plain foams, with a central vertical crack (*Figure 7*). Failure was originated mainly because normal stresses exceeded core strength.

In bonded samples (S1 series) an angled crack ($\theta \sim 30°$) was observed (Figure 7*Figure 7*). In this case foam broke because shear stresses were becoming predominant in this sandwich configuration (which depends on skins and core thickness and elastic moduli) [7]. Cracks position was different from sandwich to sandwich in agreement with the fact that, unlike normal stresses, shear is approximately constant throughout the core volume in three-point bending [7]. For a more precise sandwich failure prediction, a more detailed analysis taking into account all principal stresses, should be performed using an appropriate failure criterion.

Debonded, S0 sandwiches still present a toughening behavior under bending. This can be ascribed to foam strut-by-strut failure mode, evidenced by load drops in the stress strain curves of Figure 5 and by acoustic emissions. This failure mode was found to be similar to fiber pull out in CMC fracture mechanic [7].

Figure 7 Sandwiches failure modes of debonded (S0) and bonded (S1) sandwiches.

Bonding layer was able to bear local principal stresses. This is because the slurry, due to its controlled viscosity, infiltrated the CMC preform bundles and few fibers inside the bundles (***Figure 8*** B) and locked foam struts by embedding (***Figure 8*** A).

Figure 8 CMC-foam bonding A: foam embedding into the bonding layer. B: bonding partial infiltration of the outer bundles

CONCLUSIONS

A new integrated assembly method for silicon carbide sandwich structured composites was presented. The assembly was performed during component manufacturing in an integrated fashion. This manufacturing technique takes advantage of well established polymer composites compaction techniques and can be easily adopted to produce samples of complex shapes. Bonding is strictly related to the matching of skins and cores coefficients of thermal expansion. Under bending, bonded sandwich possessed higher ultimate strengths than those of the equivalent plain foams. All the sandwiches, after reaching a maximum stress, all samples did not experience a catastrophic failure. In these sandwich configurations, failure was observed into the ceramic core due to shear stresses

REFERENCES

[1]   Wikipedia contributors. Sandwich structured composite. Wikipedia. The Free Encyclopedia. Wikipedia. The Free Encyclopedia. 29 Oct. 2010. Web. 2 Nov. 2010.

[2]   Ortona A., Pusterla S.. Fino P.. Mach F. R. A., Delgado A., Biamino S., Aging of reticulated Si-SiC foams in porous burners, Advances in Applied Ceramics, 2010,Vol. 109, n°4, pp 246-251

[3]   Mach R. A., v. Issendorff F., Delgado A. Ortona A., Experimental investigation of the oxidation behavior of Si-SiC-foams, Advances in Bioceramics and Porous Ceramics: Ceramic Engineering and Science Proceedings, 2009, 299-311.

[4]   Hoefner T., Zeschky J., Scheffler M., Greil P., Light Weight Ceramic Sandwich Structure from Preceramic Polymers, Ceramic Engineering and Science Proceedings, 2004, 559-564

[5]   Fisher R. Burkland C, Busmante W., Ceramic composite thermal protection systems, Ceram. Eng. Sci. Proc. Vol. 6, no. 7/8, pp. 806, 1985

[6]   Hurwitz F. I., Steffier W., Koenig. Kiser J., Improved Fabrication of Ceramic Matrix Composite/Foam Core Integrated Structures, NASA Tech Briefs, 2009, 36.

[7]   Van Voorhees E.J. Green D.J Failure behavior of cellular-core ceramic sandwich composites, *J. Am. Ceram. Soc.* 1991, 74, 2747–2752.

[8]   Krenkel W., Henke T., Mason. Fuentes N., Martinez Esnaola M. Daniel J.M., In-situ joined CMC components, CMMC 96 – First International Conference on Ceramic and metal Matrix Composites, Key Engineering Materials, 1997, 313-320.

[9]   Sherwood W.J., Composite fabrication and CMCs. in: Colombo P., Riedel R., Sorarù G.D., Kleebe H.-J., editors, Polymer derived Ceramics, Lancaster, DEStech Publications, Inc. 2010.

CHEMICALLY BONDED PHOSPHATE CERAMICS WITH DIFFERENT FIBER REINFORCEMENTS

H. A. Colorado[1,2,*], C. Hiel[3,4] H. T. Hahn [1,5]

[1]Materials Science and Engineering Department, University of California, Los Angeles, CA 90095, USA Email: hcoloradolopera@ucla.edu
[2]Universidad de Antioquia, Mechanical Engineering. Medellin-Colombia
[3]Composite Support and Solutions Inc. San Pedro, California
[4]Mechanics of Materials and Constructions. University of Brussels (VUB)
[5]Mechanical and Aerospace Engineering Department, University of California, Los Angeles
*Corresponding author: Email: hcoloradolopera@ucla.edu

ABSTRACT
The main goal of this paper is to present Wollastonite-based chemically bonded phosphate ceramics (CBPCs) with the different fiber reinforcements. Wollastonite based CBPCs are composites materials itself fabricated by mixing a special formulation phosphoric acid with Wollastonite powder, through the Thinky Mixing technique. These CBPCs have different crystalline (Brushite and Wollastonite) and amorphous (silica, amorphous calcium phosphates) phases. Glass, graphite, Basalt and SiC fibers were introduced through a manual process that optimizes the fiber impregnation by the liquid mixing. Bending strength was obtained with the three point bending test. The microstructure was analyzed with Optical and Scanning Electron Microscopes.

Bending strength results showed that fibers improve the material for more than 14, 20, 16 and 21 times with respect to the CBPC matrix for glass, graphite, basalt and SiC fibers reinforced composites respectively.

INTRODUCTION
Traditional sintered ceramics, even though they have been used for many years, nowadays constitute an important field for industry especially as a high temperature resistant material. However, sintering is energetically harmful for the environment and is expensive when implemented in large scale production. The alternative is chemical bonding, obtained in a new class of ceramics formed by chemical reactions, the chemically bonded ceramics (CBCs) [1]. These ceramics are an inexpensive product that can be produced in high volumes. CBPCs bridge the gap between the attributes of sintered ceramics and traditional hydraulic cements. Often there is a need for materials with properties in between these two. Chemically bonded ceramics can meet this need [2]. Thus, they not only have high compression and bending strength [1] and high stability in acidic and high temperature environment of ceramics, but their processing is inexpensive, castable, and environmentally beneficial.

Chemically bonded ceramics (CBCs) are high-performance cement-based materials with properties approaching those of commercially fired ceramics. The main purpose of the CBCs is to fill the gap between cements and conventional ceramics, by materials with higher compression and bending strength than conventional cements but with an inexpensive and environmental friendly castable processing conducted at low temperatures [3], as opposed to fusion or sintering at elevated temperatures in traditional ceramics and cements. These characteristics enable the CBCs to repair either cements or ceramics parts. The setting is adjustable and can be controlled to be even faster than ultra fast setting cements. Moreover, CBPC is a green material since their phases are biocompatible [4].

The bonding in such CBPCs is a mixture of ionic, covalent, and van der Waals bonding, with the ionic and covalent dominating. In traditional cement hydration products, van der Waals and hydrogen bonding dominate [1]. Besides the presented advantages of CBCs as castable cementitious materials, they are known for their good compressive properties but relatively poor tensile strength. Composite reinforcements (such as fibers or particles) can improve tensile properties enough that the material can be used in structural applications under tension loads. CBC reinforced with particles ([9]) and fibers ([8], [10], [11]) has been produced before, showing the feasibility of CBPC as a building material.

In this research, Chemically Bonded Phosphate Ceramics (CBPCs) were fabricated as the result of mixing calcium silicate (wollastonite, $CaSiO_3$) and phosphoric acid (H_3PO_4), to produce calcium phosphates (i. e. brushite or monetite), silica, sometimes quartz, and depending on the reactivity, some remaining wollastonite powder [4]. The pot life of the resin, the reaction rate and the reaction heat are a function of the chemical composition of raw materials (acidic liquid and powder), the temperature of setting and the wollastonite powder size with its aging grade ([1]-[7]). Next, different fiber (glass, graphite, basalt and SiC fibers) reinforced composites were fabricated. The objective was to fabricate and compare the bending strength of different fiber reinforced CBPCs. The bending strength was improved more than 14, 20, 16 and 21 times with respect to the CBPC matrix for glass, graphite, basalt and SiC fibers reinforced composites respectively.

EXPERIMENTAL PROCEDURE
For CBPC samples, 120g of an special phosphoric acid formulation (from Composites Support and Solutions-CS&S) and 100g of natural Wollastonite powder (M200 from CS&S, see Table 1) were mixed to obtain a 1.2 ratio of liquid/powder. The mixing process of Wollastonite powder and phosphoric acid formulation was conducted in a Planetary Centrifugal Mixer (Thinky Mixer® AR-250, TM). The raw materials (acid formulation and Wollastonite powder) were cooled at 3°C in order to extend the pot life [4], and then mixed at room temperature for 2min.

Table 1 Chemical composition of wollastonite powder.

Sample	CaO	SiO_2	Fe_2O_3	Al_2O_3	MnO	MgO	TiO_2	K_2O
M200	46.25	52.00	0.25	0.40	0.025	0.50	0.025	0.15

Fibers were impregnated manually, in a knead process, before they were introduced to the molds. For all samples, the amount of fibers introduced was 15wt%. Glass fibers Textrand 225 from Fiber Glass Industries, Graphite fibers Tenax(R)-A 511, Basalt fibers BCF13-1200KV12 Int from Kammemy Vek, and SiC fibers Nicalon™ were used in this research.
A Teflon fluoropolymer mold was used to avoid adhesion of the CBPC. Samples were released after 48 hours and then dried at room temperature in open air for 5 days. Then, samples sections were ground (using silicon carbide papers of 500, 1000, 2400 and 4000 grit) and dried in a furnace (at 60°C for 24 hours and then at 100°C for 24 hours) in order to stabilize the weight

loss. Next, they were mounted on an aluminum stub and sputtered in a Hummer 6.2 system at conditions of 15mA AC for 30 sec to obtain a thin film of Au of around 1nm to be observed in the SEM. The SEM used was a JEOL JSM 6700R in a high vacuum mode. For each type of sample, five specimens were tested through the three point bending test. These tests were performed on an Instron 4411 at a crosshead speed of 2.5 mm/min. Samples were cut from pultruded rectangular bars, with dimensions $9.0 \times 12.7 \times 200 mm^3$. The span length was 127mm.

ANALYSIS AND RESULTS
Cross section view images of CBPCs composites reinforced with fibers are presented in Figure 1. Figure 1a is a cross section view image of the CBPC reinforced with glass fibers, showing 1 fiber. The fiber glass appears fully detached. Figure 1b is a lower magnification of Figure 1a, which shows a homogeneous distribution of the fibers, a good matrix impregnation to fibers, and some cracks growing through the matrix and fiber-matrix interfaces. On the other hand, Figure 1c is a cross section view image of the CBPC reinforced with graphite fibers, showing four fibers. Unlike the glass fibers reinforcement, graphite fibers interface is difficult to identify, which proofs a better matrix-fiber bonding. Figure 1d is Figure 1c at lower magnification, which also shows a homogeneous distribution of the fibers and a good matrix impregnation to fibers. Through whole sample, no cracks appear in matrix and matrix-fiber interfaces as well. Figure 1e is a cross section view image of the CBPC reinforced with basalt fibers, showing one fiber. In this case, the basalt fiber interface is difficult to identify. However, unlike graphite fibers, some of them have partial or zero bonding. Figure 1f shows some fibers almost without interfacial cracks, as well as some cracks growing through the matrix.

Finally, Figure 1g is a cross section view image of the CBPC reinforced with SiC fibers, showing one entire and two partial fibers. The interface between the SiC fiber and the CBPC is easy to identify besides interfacial cracks are absent. Figure 1h is a lower magnification of Figure 1g, which shows a homogeneous distribution of the fibers and a good matrix impregnation to fibers.

Table 2 summarizes three point bending results for CBPCs reinforced with different fibers. For all samples, the fiber content was 15wt%. The bending strength was improved more than 14, 20, 16 and 21 times with respect to the CBPC matrix for glass, graphite, basalt and SiC fibers reinforced composites respectively. The standard deviation was higher for graphite and SiC fibers than for glass and basalt fibers, which are more fragile, so it is easy to produce some small damages during the manual impregnation, which increases the variability in the bending strength. A pultrusion process as was reported before for CBPCs reinforced with glass and graphite fibers [12] can reduce significantly the variability since the impregnation process can be controlled and performed repetitively.

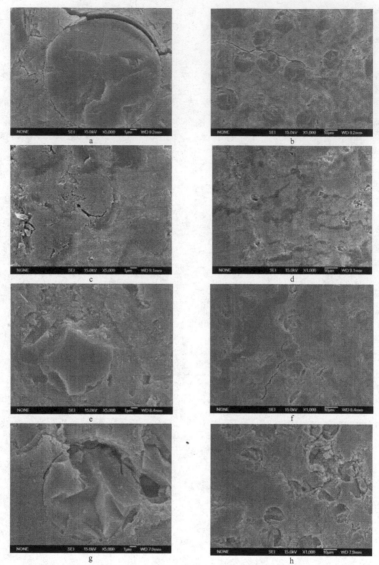

Figure 1 SEM Cross section view images for Wollastonite-based CBPCs reinforced with a) and b) glass fibers; c) and d) graphite fibers; e) and f) basalt fibers; g) and h) SiC fibers.

Table 2 Three point bending tests results for CBPCs with different fiber reinforcement

Sample		Width (mm)	Thickness (mm)	Span (mm)	Max. Bending strength (MPa)
CBPC (reference)	Mean	12.7	9.0	127	9.02
	Standard Dev.	0.01	0.01	0	3.4
CBPC with glass fibers	Mean	12.7	9.0	127	134
	Standard Dev.	0.2	0.2	0	5.6
CBPC with graphite fibers	Mean	12.7	9.0	127	186.5
	Standard Dev.	0.01	0.1	0	16.2
CBPC with basalt fibers	Mean	12.7	9.0	127	146.7
	Standard Dev.	0.2	0.1	0	14.3
CBPC with SiC fibers	Mean	12.3	9.0	127	195.7
	Standard Dev.	0.2	0.3	0	10.1

Figure 2 shows typical three point bending curves of results presented in Table 2 for CBPCs reinforced with different fibers. Basalt and glass fibers composites have larger strains meanwhile graphite and SiC fibers composites showed the larger bending strengths.

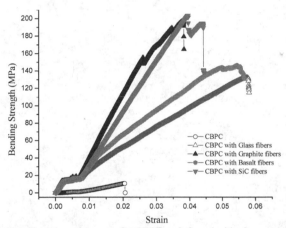

Figure 2 Typical bending strength curves for CBPCs reinforced with glass, graphite, basalt and SiC fibers.

DISCUSSION

Figure 3 shows a representation of the CBPCs reinforced with fibers. As the shrinkage occurs in the matrix, the compression forces lead to an interfacial microcraking [12]. Wollastonite and silica grains usually are observed either partially or with zero bonding to the amorphous calcium phosphate matrix. Another phase, brushite, appears without interfacial cracks. Microcraking is also present in the ceramic matrix-fiber interfaces, depending on the fiber type. Figure 3a shows a representation for non-bonded fibers. All fibers have interfacial cracks, without bonding to the ceramic matrix. This cases is close to the CBPCs reinforced with glass fibers, Figure 1a and b. On the other hand, Figure 3b shows a representation for all bonded fibers to the ceramic matrix. This case is similar to the CBPCs reinforced with graphite fibers (Figure 1c and d) and CBPCs reinforced with SiC fibers (Figure 1g and h).

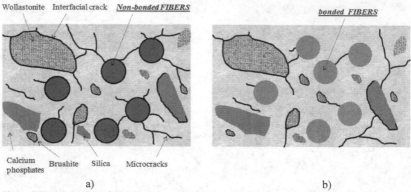

a)   b)

Figure 3 Representation of CBPC reinforced with a) bonded and b) non-bonded fibers to CBPC matrix

The ceramic composites reinforced with basalt and SiC fibers have very few interfacial cracks, so they are very close to the case of fiber interfaces with bonding.

For glass fiber-CBPC and graphite fiber-CBPC interfaces, it has been proposed before [12] a multifactor mechanism involving the bulk and the interface in fiber reinforced CBPCs. For the glass fiber case, fiber detaching is similar to a complex interaction that occurs in the ceramic matrix itself between shrinkable (brushite and amorphous calcium phosphates) and non-shrinkable phases (wollastonite and silica), and also affected by the dissimilar crystalline microstructures with poor coherence among them. This problem is of course worst when the temperature is increased either during the exothermic reaction produced during the setting of the ceramic, or during the external drying process conducted to stabilize the weight loss. For the glass fiber case, similarly, the interaction occurs between the shrinkable amorphous calcium phosphates and the non-shrinkable fiber glass.

For the case of the CBPCs with graphite fibers, in addition to the phenomena associated to the fiber glass, some bonds are possible to bind the matrix-fiber. These are possible when OH molecules are attached to the fiber. For both glass and graphite fibers, not bulk chemical reaction with CBPC seem to occur. However, for the SiC anb Basalt fibers, a bulk chemical reaction with CBPC seem to occur, which suggest they are participating in the setting of the ceramic. More research is being conducted over this topic.

Also, the improvement in bending strength for more than 100 times with respect to the CBPC matrix for all fiber reinforced composites is a promising result for structural applications at high temperature, since some of the fibers work well at those environments. Thermal shock experiments are now under research.

CONCLUSIONS
The improvement in bending strength for more than 14 (134MPa), 20 (186.5MPa), 16 (146.7MPa) and 21 (195.7MPa) times with respect to the CBPC matrix (9.02MPa), for glass, graphite, basalt and SiC fibers reinforced composites respectively; is a promising result for structural applications at high temperature, since some of the fibers work well at those environments. Thermal shock experiments are now under research.
Results show the CBPC as a possible substitute material for high performance cements and ceramics, in applications involving castability, high bending or compression strength and high temperature environments.

Acknowledgements
The authors wish to thank to the NIST-ATP Program through a grant to Composites and Solutions Inc. (Program Monitor Dr. Felix H. Wu) and to Colciencias from Colombia for the grant to Henry A. Colorado.

REFERENCES
[1]. Della M. Roy. (1987). Science, New Strong Cement Materials: Chemically Bonded Ceramics. Vol. 235: 651-58.

[2]. A. S. Wagh. (2004). Chemical bonded phosphate ceramics. Elsevier. Argonne National Laboratory, USA, 283.

[3]. S. Y. Jeong and A. S. Wagh. (2002). Chemical bonding phosphate ceramics: cementing the gap between ceramics, cements, and polymers. Argonne National Laboratory report.

[4]. H. A. Colorado, C. Hiel and H. T. Hahn. (2009). Processing-structure-property relations of chemically bonded phosphate ceramics composites. Bull. Mater. Sc. Ref.: Ms No. BOMS-D-09-00499R1.

[5]. Arun S. Wagh and Seung Y. Jeong. (2003). Chemically bonded phosphate ceramics: I, a dissolution model of formation. J. Am. Ceram. Soc., 86 [11] 1838-44

[6]. Arun S. Wagh and Seung Y. Jeong. (2003). Chemically bonded phosphate ceramics: II, warm-temperature process for alumina ceramics. J. Am. Ceram. Soc., 86 [11] 1845-49.

[7]. Arun S. Wagh and Seung Y. Jeong. (2003). Chemically bonded phosphate ceramics: III, reduction mechanism and its application to iron phosphate ceramics. J. Am. Soc., 86 [11] 1850-55.

[8]. Sugama T. and Taylor M. (1994). Interfacial and Mechanical Behavior of Fiber-Reinforced Calcium Phosphate Cement Composites. Cement & Concrete Composites 16 (1994) 93-106.

[9]. H. A. Colorado, C. Hiel and H. T. Hahn. (2010). Chemically bonded phosphate ceramics composites reinforced with graphite nanoplatelets. In prints, Composites Part A. 2010.

[10].   Sid A. Dimitry. (1991). Characterization of reinforced chemically bonded ceramics, Cement and Concrete Composites, Volume 13, Issue 4, 257-263.

[11].   Sean Wise, Kevan Jones, Claudio Herzfeld and David D. Double. (1998). Cropped steel fiber reinforced chemically bonded ceramic (CBC) composites. Mat. Res. Soc. Symp. Proc. Vol. 114. Materials Research Society.

[12].   H. A. Colorado, H. T. Hahn and C. Hiel. Pultrusion of glass and carbon fibers reinforced
         Chemically Bonded Phosphate Ceramics. To appear at Journal of Composites Materials.
         2010.

PRECERAMIC-POLYMER-BONDED SiC PREFORMS FOR HIGH VOLUME FRACTION SiCp/Al COMPOSITES

Kuljira Sujirote[a], Kannigar Dateraksa[a], Sukunthakan Ngernbamrung[a], Ryan McCuiston[b], Trinmet Sungkapan[c] and Jessada Wannasin[c]
[a]National Metal & Materials Technology Center, Klong Luang, Patumthani 12120, Thailand
[b]King Mongkut Institute of Technology, Tungkru, Bangkok 10140, Thailand
[c]Prince Songkla University, Had Yai, Songkla 90112, Thailand

Preceramic polymer polycarbosilane (PCS) and silicon carbide (SiC) powders were used as the starting materials for the fabrication of porous SiC ceramic preforms. Since the aim of these preforms was to make SiC/Al MMC with high solid loading, maximum packing density of pressed compact was designed using Dinger-Funk distribution with an exponent of 0.37. Additional in-situ SiC from an organic–inorganic transformation during heat treatment process acted as the bonding material between SiC particles. The effect of the SiC powder:PCS ratio on the porosity and flexural strength of the porous SiC ceramics were investigated. Open porosity of the preforms varied from 30 to 45%. Fracture strength of the porous SiC ceramics was in the range of 20-70 MPa. No strength degradation was observed from thermal shock test at ΔT lower than 300°C. An infiltrated SiCp/Al composite exhibited flexural strength of 655±104 MPa.

INTRODUCTION

Application of SiC particle reinforced aluminum metal matrix composites (MMCs) with high SiC loadings (e.g. 50–75 vol%) feature attractive mechanical and thermophysical properties[1]. Owing to a high thermal conductivity coupled with a low thermal expansion coefficient, this type of MMC has now found widespread application in high preformance power-and microelectronics for thermal management and packaging purposes[2]. Additionally, high volume fraction Al/SiC MMCs show a high potential for structural applications in sectors where low weight, high stiffness and strength as well as low thermal distortion and good damage tolerance are required, such as in sporting goods, automotive engineering, aerospace industry, high accuracy mechanical engineering or defense technology[3][4].

In order to improve the packing of the silicon carbide powder, ternary powder blends were created. The blends were designed by an iterative approach. A baseline blend was created and the cumulative percent finer than (CPFT) distribution of the blend determined. The CPFT of the blend was then compared to the CPFT of the Funk-Dinger distribution equation with a distribution modulus selected for maximum packing fraction. The volume percent of the different size fractions in the blend were adjusted to produce a close to the Funk-Dinger distribution. The Funk-Dinger distribution [5][6] equation is

$$\frac{CPFT}{100\%} = \frac{D^n - D_S^n}{D_L^n - D_S^n}$$

Where D is the powder diameter, D_S is the smallest powder size in the blend, D_L is the largest powder size in the blend and n is the distribution modulus. To achieve the maximum packing fraction, n should be equal to 0.37. Ideally, the estimate of 20% porosity or approximately 80% green density could be achieved. However perfect sphere particles with perfect packing and no friction rarely exist. In practice, green density at approximately 60% is realized.

In order to increase SiC weight fraction in aluminum metal matrix composite (MMC), two approaches were applied to the SiC preformed fabrication; cold isostatic press (CIP) and additional in-situ SiC from preceramic polymer transformation. The CIP increases density of uniaxial pressed samples in the range of 10-20%. Although effect on density increase of the in-situ SiC is not as apparent as that of CIP, binding characteristics of the preceramic polymer remarkably assist formability of the SiC mixes. Among many organic precursors, allyl-hydridopoly-carbosilane, called SMP-10 [10], was selected. It is the synthesis of a polycarbosilane that has nominally a 1:1 Si:C ratio, and with H as the only other component. It has the advantage of a high ceramic "yield" (weight of ceramic to that of polymer) of near stoichiometric SiC[7].

High solid loading SiC/Al MMC could be fabricated via squeeze casting technigue[8]. However, SiC preform with high solid loading would lead to high capillary force and high threshold pressure during infiltration. Temperature increase could reduce viscosity of the molten metal but would as well deteriorate structural integrity of the preform and high temperature gradient would damage the composite during cooling down period. Hence, a special technique so-called Gas-Induced Semi-Solid (GISS) process[9] is applied. It was recognized that the viscosity of semi-solid slurries provided an attractive opportunity to incorporate ceramic particulates[10]. In addition, improvements reported in relation to those of the conventionally processed castings were reduced segregation, reduced porosity and ingot cracking, reduced grain size; and increased uniformity in the distribution of the γ-7' phases [11].

Target flexural strength of SiC prefrom in this study is to withstand threshold pressure of molten aluminum which is typically lower than 20 MPa at temperature higher than 660°C[12]. Therefore, objectives of this study is to systematically explore SiC preform fabrication technique that will produce SiC preform with opmimum porosity and high enough mechanical properties to withstand threshold pressure during molten GISSed aluminum infiltration. The specimen will be thermal shock proof test and mechanical strength of the resultant SiCp/Al MMC will be determined.

EXPERIMENTAL PROCEDURE

SiC mix

Figure 1 shows the CPFT distributions of the four powders used in this research. F1000 -W and F1200-FW are black abrasive grade powders with an average particle size 6.3 and 4.4 microns, respectively. The 2.5 and 0.7 micron powders are green SiC with average particles size of 2.2 and 0.9 microns, respectively. While the average particles size of F1000-W, F1200-FW and 2.5 micron are different, the largest and smallest particles sizes are very similar. Two different ternary blends were designed. They were designated as Mix 1 and 2. Their compositions are shown in Table 1

Table 1 Particle size of SiC powders and compositions of Mix 1 and 2.

	F1000-W	F1200-FW	2.5 M	0.7 M
Avg. Particle Size (μm)	6.3	4.4	2.2	0.7
Mix 1 (vol%)	23	64	0	13
Mix 2 (vol%)	66	0	20	14

The CPFT distributions for Mix 1 and 2 are shown in Figure 2. It is readily seen that both Mix 1 and 2, when compared to the Funk-Dinger distribution, have a large volume of excess particles that are large in size and are missing a volume of particles small in size. This is a result of using only three different size fractions as well as by design. The presence of an extra volume of large particles results in larger porosity, which is advantageous for subsequent infiltration.

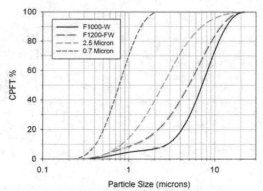

Figure 1 CPFT distributions of the four SiC powders used in this research.

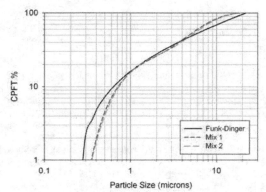

Figure 2. Plot of the CPFT distribution for Mix 1 and 2. They are compared with the Funk-Dinger distribution for D_L=22 microns, D_S=0.24 microns and n=0.37. Please note the CPFT scale is cut-off at 1%.

The SiC mixes were first dry mixed for 4 hrs. The mixing was done in a polymer bottle using a small charge of 6 mm alumina media. The mixing process incorporates the powders together and does not alter the particle size distribution. Four batches were prepared for with 0, 3, 5 and 7 wt. % of polycarbosilane (SMP-10, Star fire system, USA). The polycarbosiline was incorporated by stirring. Hexane was used as a solvent for dissolving the polycarbosilane and making the powder mixture more homogeneous. The slurry was dried under inert atmosphere and pressed uniaxially to produce rectangular bars with a green density kept constant at 1.8 g/cm^3 (~56% ρ_{th}). The pressure used was 12.1, 5.5, 3.5, and 1.7 MPa for 0, 3, 5, and 7 wt% polycarbosilane, respectively. The specimens were divided into 2 groups, i.e. with and without cold isostatic pressing at 250 MPa for 3 minutes. All the samples were then sintered at 1800°C for 2 h in argon in order to transform the polycarbosilane into silicon carbide.

The obtained preforms were infiltrated with 7075 aluminum alloy via gas-induced semi-solid (GISS) technique. The infiltration was conducted in static air. Details on the experimental setup are provided by Wannasin et. al[9]. During aluminum infiltration, the specimens have to withstand a temperature higher than 600°C under a threshold pressure of approximately 15 MPa. Thus the preforms were undergone thermal shock testing heating at a rate of 300°C/hr to a preset temperature (500°C or 700°C) and held for 20 min to equilibrate. The atmosphere was static air. The specimens were then submerged into a water bath at room temperature. The flexural strength of the shocked samples was measured and compared to those of control samples.

Measurement of three-point bending flexural strength was preformed using an Instron Universal system (model 4502). Specimens with dimensions of 3 x 4 x 30 mm were placed on a 20 mm span and loaded until fracture occurred. The crosshead speed was 1 mm/min and load at fracture was recorded. Young's modulus was measured by the fundamental resonant frequency technique by the ultrasonic pulse method (GrindoSonic, MK5 Industrial).

RESULTS

Microstructure

Figure 3 exhibit the fracture surface micrographs of the SiC preforms of two sample sets. Both sets contain samples with different amount of PCS at 0, 3, 5, and 7%. Samples in the left column; Fig. 3(a), 3(c), 3(e) and 3(g), were uniaxial pressed to controlled bulk density of approximately 60% theoretical density, and then pyrolyzed at 1800°C in argon. Small amount of in-situ SiC could be observed as tiny specs on surface of SiC particles. As the samples were pyrolyzed together in a graphite container within a graphite furnace, the volatile carbon and silicon species could condense and precipitate as in-situ SiC at high temperature.

On the other hand, samples in the right column, Fig. 3(b), 3(d), 3(f), and 3(h), were CIPped at 250 MPa before the pyrolysis and thus had approximately 10% higher bulk density than those in the left. It is obvious that in-situ SiC much more abundantly formed over the SiC particles. Especially the samples with 5% and 7% PCS, in-situ SiC fully covers the entire surface.

This is in agreement data in Fig. 4 which illustrates gradual increment of bulk density for CIPped samples, but insignificant increment for the uniaxial pressed samples. It should be noted that bulk density of samples without PCS are lower than those with PCS, either CIPped or not CIPped. Binding effect of the PCS does facilitate the forming process. In order to keep the bulk density of approximately 1.9 g/cm^3, samples with PCS 3%, 5%, and 7% were uniaxial pressed at 5.5, 3.5, and 1.7 MPa, respectively. It was not possible to keep the target bulk density for the sample without PCS because the sample would crack or delaminate. Maximum pressure used was 12.1 MPa and bulk density of only 1.7 g/cm^3 was obtained. By undergoing CIP at 250 MPa, bulk density would increase approximately 10%.

When exposed to oxidation during thermal shock treatment, edges of the SiC particles appeared rounded for both uniaxial pressed and CIPped preforms. Oxidation reaction would occur on SiC surface exposed to high temperature, forming oxide film viscous flow, especially at sharp edges. However, the most noticeable change observed was the diminishing amount of in-situ SiC formed on the CIPped preforms. Typically, the tiny in-situ SiC specs covering all over the surface of the CIPped sample (Fig.3(h) for instance) was almost vanished (Fig. 5(b)) when exposed to oxidation treatment, even at temperature lower than 600°C. This is evident that the tiny in-situ SiC specs from pre-ceramic polymer exhibit different oxidation mechanism from the SiC particles.

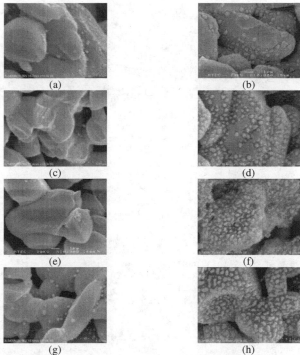

Figure 3. SEM image of porous SiC preforms with and without PCS-derived in-situ SiC for various amount of PCS (a) – (b) 0%, (c) – (d) 3%, (e) – (f) 5%, and (g) – (h) 7%. Samples on the left column were uniaxial pressed but those on the right were CIPped.

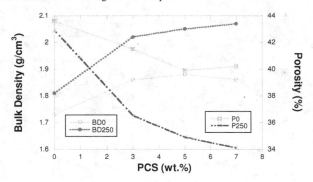

Figure 4. Bulk density (BD0 and BD250) and porosity (P0 and P250) change in the SiC preform as a function of added PCS.

(a)   (b)   (c)

Figure 5. Fractured surface of (a) SiC preform with 7% PCS and thermal shocked at 500°C, (b) SiC preform similar to (a) but was CIPped at 250 MPa, and (c) SiC-Al MMC.

Effect of the residual stress from CIP process can be observed by comparing Fig.5a and 5b. After thermal shocked at 500°C, the SiC preforms exhibited mainly two different type of damage topography on the SiC grain. Rounded edge was observed on the non-CIPped sample (Fig.5a). Cracks mostly propagate along this rounded boundary. In contrast, cleavage step are dominant on CIPped sample (Fig.5b). Intense stress concentration induced in SiC particles during cold isostatic press is enhanced by surface markings. These local disturbances become favoured sites for sub-surface deformation during fracture, especially at the tip of a propagating crack.

With the infiltrated preforms with GISS aluminum, forming SiCp/Al MMC, damage evolution of the composite is radically altered. The fracture surface of the composite showed typical ductile failure mode, features with big dimples corresponding to SiC particles and many other small dimples in the matrix. Crack initiation has been found from both reinforcement/matrix interfaces and broken particles. The majority of the SiC particles had fractured but intensive distortion of big dimples by much matrix deformation revealed the presence of some SiC particles decohesion within the matrix.

Strength & modulus

The mechanical strength of porous ceramics is related not only to the volume fraction and geometrical nature of the pores but also to the size and type of the interparticle connections [13] [14]. Fig. 6 shows the change of bending strength and elastic modulus for SiC preforms with various PCS content. Strength of uni-axial pressed samples with PCS addition are in the range of 25 – 40 MPa, but it deteriorated to lower than 10 MPa after thermal shock treatment. In contrast, strength of CIPped samples with PCS addition was considerably increased up to 70 MPa, or nearly double that of un-CIPped ones. Strength after the thermal shock treatment deteriorated by one third, but the values were still higher than 20 MPa, which meets the target of this study. It can be seen that the flexural strength is directly dependent on amount of in-situ SiC remained on the SiC particles. For example, comparing Fig. 3(h) and Fig. 5(b) which illustrates that the in-situ SiC was volatized by the heat treatment, the flexural strength dropped from 60 MPa down to 20 MPa.

Added in-situ SiC in the preform could significantly influence the elastic modulus. As shown by Fig. 6(b), elastic modulus of the CIPped samples increased with the increase of PCS content. However, the modulus of elasticity was not sensitive to its microstructure, so thermal shock treatment induced slight effect on elastic modulus of SiCp/Al composites.

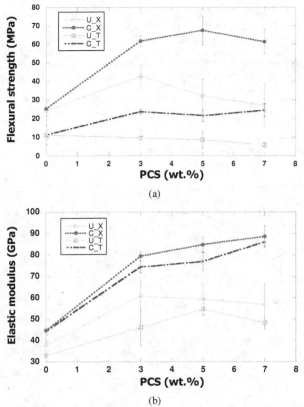

(a)

(b)

Figure 6. (a) Flexural strength and (b) Elastic modulus of the SiC preform comparing uniaxially (U) and cold isostatically (C) pressed, with (T) and without (X) thermal shock treatment, as a function of added polycarbosilane (PCS).

DISCUSSION

Preform: porosity VS strength & modulus

There are limitations to Funk-Dinger approach. Abrasive grades of SiC powder are only available in a limited number of size fractions near or below a 10 micron average. The Funk-Dinger distribution represents a continuous distribution of powder, which is hard to achieve by mixing three different size fractions together. In addition, aspects of the powders and processing, such as the morphology, the use of internal lubricants and the actual mixing details of the blends themselves, are not accounted for in the Funk-Dinger distribution. Thus it is only a guideline and may not represent the ideal distribution. Therefore, both CIP and PCS addition were applied to increase bulk density of the preform.

As explained by Lee[15], the PCS is first converted into an amorphous, glassy form of SiC, designated as α-SiC. This transformation process likely begins around 315°C-370°C and releases hydrogen gas. The reaction continues as the temperature is gradually increased, above the 1093°C-1204°C range, the α-SiC ceramic begins to crystallize and convert into its face-centered cubic beta form, β-SiC. The structural changes associated with these conversion effects are illustrated in Fig.7.

It is interesting to consider only physical characteristics, like porosity, without concern on the PCS addition or CIP condition. Fig. 8(a) exhibits that elastic modulus of the preform is linearly dependent on its porosity. The modulus – porosity relationship is identical for both samples with and without thermal shock treatment. This implies that the modulus is not sensitive to its thermal shock treatment.

Although porosity inversely affects the flexural strength as does the Young's modulus, the thermal shock severely deteriorates the strength. Linear fit equations of the plot show that extrapolation of the flexural strength of the SiC preform to zero porosity would be 223 MPa and 68 MPa for the samples with and without thermal shock treatment, respectively. Only one third of the strength, or 20 MPa of CIPped samples, will be available to withstand the threshold pressure of molten aluminum infiltration. This is considered sufficient comparing to compacts containing SiC with 0.72 micron mean particle diameter and 53% volume fraction which has the threshold pressures for infiltration of 12 MPa.[16]

While SiC is inert in almost any environment up to > 1650°-1700°C, under extreme situations, it may begin to break down, degrade, transform into a less stable form and sometimes even volatilize. In case of non-CIP SiC with low density, no in-situ SiC, passive oxidation dominate. 'passive' oxidation of the outer SiC layer begins to form as a thin, dense (nonporous), vitreous film of SiO_2

$$SiC(s) + O_2 \xrightarrow[\text{passive-oxidation-of-SiC}]{} SiO_2(s) + C(s)$$

amourphous silica   amorphous carbon

Prolonged exposure may lead to interface reactions between the oxide and the in-situ SiC. Under certain oxidation levels, some of the in-situ SiC may undergo degradation with volatilization.

$$2SiO_2 + SiC \xrightarrow[\text{Degradative-interface-oxidation}]{} 3SiO \uparrow + CO \uparrow$$

In case of CIPped samples, overall rate of oxidation exceeds that of oxide replenishment, oxidation converts from 'passive' to 'active' as exposed SiC surfaces begin to degrade and volatilize.

$$SiC(s) + O_2(g) \xrightarrow[\text{Active-oxidation-of-SiC}]{} SiO \uparrow + CO \uparrow$$

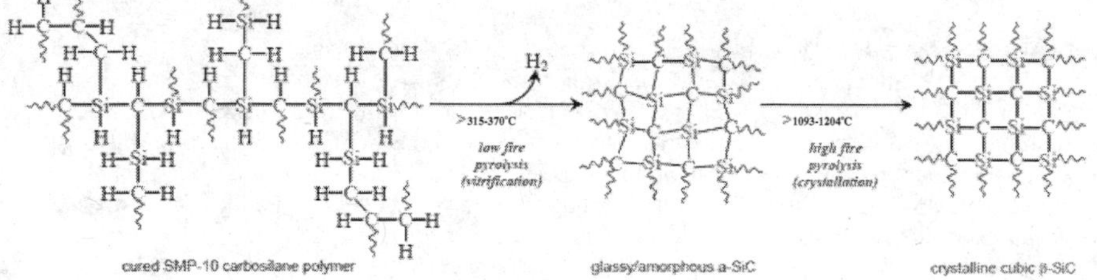

Figure 7. Schematic diagram for the pyrolysis of polycarbosilane (SMP-10) and transformation to SiC.[15]

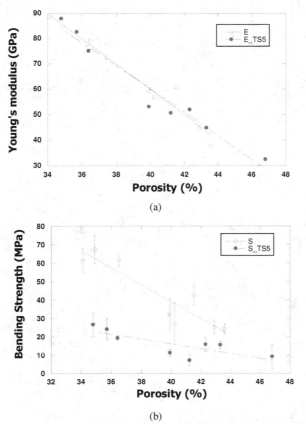

(a)

(b)

Figure 8. Thermal shock effect on (a) Young's modulus and (b) Flexural strength of the SiC preform as a function of porosity.

To achieve strong and reliable porous ceramics, a homogenous microstructure with fine grains and enhanced necks is expected. Flexural strength of the preform increases exponentially with bulk density (Fig.9). Comparing to porous SiC preform using other types of preceramic polymer as necking agent, i.e. polysilazane or polysiloxane, it is intriguing to see the strength limit of each in-situ SiC neck. It is obvious that bulk density of polysiloxane > polysilazane > polycarbosilane. Porous SiC with in-situ SiC from PCS is strongest and lightest.

In case of samples from this study, S0_0 and S250_0 which were not undergone thermal shock treatment, fit in well with strength of porous SiC with PCS as precursor for necking SiC. But the thermal shocked samples, S0_500 and S250_500, the strength dropped and the bulk density increased closer to characteristics plot of polysilazane.

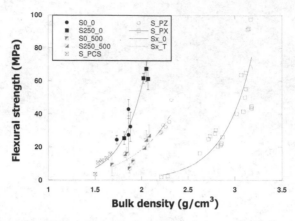

Figure 9. Flexural strength of the SiC preform with various bulk density comparing effect of in-situ SiC derived from polycarbosilane (PCS)[17], polysilazane (PZ)[18], and polysiloxane (PX)[19][20].

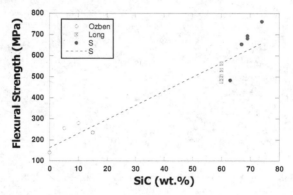

Figure 10. Flexural strength of the Al-SiCp MMC comparing effect of SiC content from this study (●) to those from Ozben[21] (□), and Long[22] (○)

MMC strength & modulus

Fig. 10 demonstrates that flexural strength of SiCp-Al MMC linearly increases with increasing SiC reinforcement. Composites with low SiC loading (10-20 wt%) are made by extrusion or casting. Typical strength value would be 200-300 MPa.[21] Adding SiC loading higher than 50% requires fabrication technique with high pressure, such as squeeze casting or hot pressing. Flexural strength of composite from 60% SiC and AlCu4Mg1Ag alloy of approximately 500 MPa was achieved by Long and co-workers.[22]

In this research, one of the new rheocasting processes, so-called, gas induced semi-solid (GISS) process is utilised. The GISS process utilizes the principle of rapid heat extraction and vigorous local extraction using the injection of fine gas bubbles through a graphite diffuser. Semi-solid slurries with different solid fractions can be obtained simply by varying the diffuser immersion times. Since supercooled temperature of the slurry (~600-640°C) is lower than temperature (>750°C) used for making low viscosity molten aluminum, defects form cooling shrinkage is prevented. Flexural strength approximately 480 MPa was attained form 60% SiC and 7075 Al alloy in this study. Samples with additional in-situ SiC, having total SiC approximately 70 wt%, exhibit impressive flexural strength close to 700 MPa (S in Fig.10).

CONCLUSIONS

MMC with high SiC loading and exceptional strength was successfully produced. The strength achieved, 655±104 MPa, was a result of several improved processing steps in porous SiC preform preparation and rheocasting aluminum infiltration technique.

Starting SiC mix was designed for high solid packing at 60% theoretical density. Cold isostatic press increased the bulk density by 10%. Added polycarbosilane assist both binding for handling in green state and bonding as in-situ SiC necks for aluminum infiltration. The optimum amount of PCS is 5-7 wt% in this study. The PCS transformation required high packing density to induced sufficient partial pressure of volatile species as well as low degree of freedom for in-situ SiC precipitation on SiC particle. When exposed the thermal shock at 500oC, the SiC particle and the in-situ SiC exhibited different oxidation mechanisms. SiC particles in green compact with lower bulk density (55 %TD) oxidized passively, forming rounded edged, while intensive cleavage cracks was observed on those with higher bulk density (65% TD). In contrast, the in-situ SiC undergone active oxidation and dwindling in amount under prolong exposure. Remaining strength of the preform should be high enough to withstand threshold pressure during aluminum infiltration. In this study, bending strength higher than 20 MPa was obtained after thermal shocked at 500oC.

Infiltration aluminum by GISS technique allows low viscosity and low temperature gradient. This make high solid loading SiCp/Al composite with minimum crack or damage during cooling cycle. Mixed ductile damage of the aluminum and Al/Si interface, brittle damage of the SiC particles are present all over fractured surface.

REFERENCES

[1] W.H. Hunt, Particulate reinforced MMCs. In: Kelly A, Zweben C, editors. Comprehensive composite materials, **3 (3.26)**. Amsterdam: Elsevier; 701–15 (2000)

[2] D. Chung and C. Zweben, Composites for electronic packaging and thermal management. In: Kelly A, Zweben C, editors. Comprehensive composite materials, **6 (6.38)**. Amsterdam: Elsevier;.701–25 (2000)

[3] W.H. Hunt, Metal matrix composites. In: Kelly A, Zweben C, editors. Comprehensive composite materials, **6 (6.05)**, Amsterdam: Elsevier; 57–66. (2000)

[4] M.L. Seleznev , J.A. Cornie, R.P. Mason, M.A. Ryals, High volume fraction ceramic reinforced aluminium matrix composites for automotive applications: mechanical properties and microstructural characterization. Technical paper, MMCC Inc. (1995) Available from: www.mmccinc.com..

[5] D.R. Dinger, One-Dimensional Packing of Spheres, Part I, *Ceram. Bull.*, **2**, 71-76 (2000)

[6] D.R. Dinger, "One-Dimensional Packing of Spheres, Part II", Ceram. Bull., **4**, 83-91 (2000)

7 L.V. Interrante, K. Moraes, Q. Liu, N. Lu, A. Puerta, L.G. Sneddon, Siliconbased ceramics from polymer precursors, *Pure Appl. Chem.* **74** [11] 2111–2117 (2002)

[8] S. Ren, X. Qu, J. Guo, X. He, M. Qin, X. Shen, Net-shape forming and properties of high volume fraction SiCp/Al composites, *Journal of Alloys and Compounds* **484**[1-2]256-62 (2009)

[9] J. Wannasin, S. Janudom, T. Rattanochaikul, R. Canyook, R. Burapa, T. Chucheep, and S. Thanabumrungkul, Research and development of gas induced semi-solid process for industrial applications, *Trans. Nonferrous Met. Soc. China* **20**, 1010-1015 (2010)

[10] T.S. Srivatsan, T.S. Sudarshan, and E.J. Lavernia, Discontinuously-reinforced metal matrix composites by rapid solidification, *Progress in Materials Science*, **39**, 317-409 (1995)

[11] J.A. Cheng, D. Apelian and R. D. Doherty. *Metallurgical & Materials Transactions* **17A**, 2049-2062 (1986).

[12] J. Wannasin and M.C. Flemings, Fabrication of Metal Matrix Composites by a High-Pressure Centrifugal Infiltration Process, *Journal of Materials Processing Technology*, **169**[2], 143-149 (2005).

[13] J.H. She and T. Ohji, Porous mullite ceramics with high strength, *J. Mater. Sci. Lett.*, **21**, 1833–1834 (2002)

[14] S.T. Oh, K. Tajima, M. Ando, and T. Ohji, Strengthening of porous alumina by pulse electric current sintering and nanocomposite processing, *J. Am. Ceram. Soc.*, **83**, 1314–1316 (2000)

[15] R.Lee, Carbosilanes: Reactions & Mechanisms of SMP-10 Pre-Ceramic Polymers, (2009) http://www.materialchemistry.com/DreamHC/Download/Reactions%20Pre-Ceramic.pdf

[16] J. Wannasin and M.C. Flemings, Threshold pressure for infiltration of ceramic compacts containing fine powders' Scripta Materialia 53 (2005) 657–661

[17] S. Zhu, S. Ding, H. Xi and R. Wang, Low-temperature fabrication of porous SiC ceramics by preceramic polymer reaction bonding, *Materials Letters* **59**, 595-97 (2005)

[18] J. Semen and J.J. Rogers, Preceramic composition and ceramic products *US.Patents* No.4,835,207 (1989)

[19] J. Eom, Y. Kim, I. Song, and H. Kim, Processing and properties of polysiloxane-derived porous silicon carbide ceramics using hollow microspheres as templates *Journal of the European Ceramic Society* **28**, 1029–1035 (2008)

[20] S. Chae, Y. Kim, I. Song, H. Kim, and M. Narisawa, Porosity control of porous silicon carbide ceramics, *Journal of the European Ceramic Society* **29**, 2867–2872 (2009)

[21] T. Ozben, E. Kilickap, and O. Cakir, Investigation of mechanical and machinability properties of SiC particle reinforced Al-MMC, *Journal of Materials Processing Technology*, **198**, 220-225 (2008).

[22] S.Long, O. Beffort, C. Cayron, and C. Bonjour, Microstructure and mechanical properties of a high volume fraction SiC particle reinforced AlCu4Mg1Ag squeeze casting,. *Mater Sci Eng A*, **269**, 175–85 (1999)

NOVEL HIGH TEMPERATURE WOUND OXIDE CERAMIC MATRIX COMPOSITES MANUFACTURED VIA FREEZE GELATION

Thays Machry, EADS Innovation Works, Munich, Germany
Christian Wilhelmi, EADS Innovation Works, Munich, Germany
Dietmar Koch, Advanced Ceramics, University of Bremen, Bremen

ABSTRACT

Ceramic Matrix Composites (CMC) based on oxide materials are of interest for high temperature applications not only due to their inherent oxidative stability, but also due to their damage tolerance and thermal shock resistance as well as their excellent thermo-mechanical performance.

The development of a new type of oxide CMC termed WITA-OX^{TM} with Nextel^{TM} 610 fibers and mullite matrix manufactured via filament winding and freeze gelation process is presented. The ceramic matrix is produced using the sol gel technique combined with the freeze gelation process. Solid nanoparticles are dispersed into a liquid and filler particles are added. The infiltration of the fiber bundles by the suspension is then conducted via filament winding. The nanoparticles from the sol form a three-dimensional network due to its gel transition which is forced by freezing of the solvent, resulting in ice crystals growth. After drying the green body was sintered to achieve a ceramic state. The effectiveness of fiber bundle infiltration were characterized by evaluation of interlaminar properties and fracture mechanical tests. Results show that slurry impregnation of the individual fiber bundles is successfully achieved and the mechanical tests on the CMCs indicate that the concept of this manufacturing route is promising.

The work was performed within the cooperative German BMBF HiPOC program (**Hi**gh **P**erformance **O**xide **C**eramics, consisting of 3 companies and 4 research institutes) dealing with the development of oxide/oxide CMC materials and components for the use in gas turbine applications.

INTRODUCTION

Ceramic Matrix Composites (CMC) manufactured with continuous oxide fibers and oxide matrix have the potential for combining high damage tolerance with high temperature resistance. As a consequence, the introduction of oxide CMC as structural components for e.g. gas turbine engines appears very promising due to their elevated thermo-mechanical performance and their low density compared to metal alloys or inter-metallic materials. Moreover oxide CMC are inherent resistant to high temperature oxidation which enables an acceptable lifespan of corresponding components. Thus the development of these sophisticated CMC materials and tailored cooling schemes is one of the main focuses for significantly increasing the operating temperatures in future gas turbines.

This scope has led to the German BMBF (German Federal Ministry of Education Research) collaborative HiPOC program (**Hi**gh **P**erformance **O**xide **C**eramics) started in February 2009 which main objectives is based on the development of different oxide/oxide CMC materials which may be used in gas turbines for power generation and aerospace propulsion, or as spin-off in space applications such as for Thermal Protection Systems and Hot Structures. In conjunction with an improved thermal management the main objectives are to minimise the fuel consumption and thereby reduce the CO_2 emission from the gas turbine. To achieve this goal, oxide CMCs and new design concepts for the attachment of the CMC component to the metal structure of the engine are developed focused on the combustion chamber and the turbine seal segment. As for precise design detailed experimental data are necessary, the different oxide/oxide CMC are tested in various loading modes (tension, compression, shear, off-axis loading) from room temperature to maximum application temperature. These studies indicate the high temperature potential of the CMC materials under investigation in this collaborative project [1, 12-16].

In CMC materials the interaction of the fibers and the ceramic matrix strongly influences the fracture behavior of the system, such as crack evolution and propagation. In order to determine the flaw tolerance of CMC it is important to understand the influence of the microstructure on the crack propagation. In this manner, the adjustment of the microstructure and the damage tolerance can be arranged in two different ways. The first is the use of fiber coatings for promoting crack deflection and frictional sliding along the fiber/matrix interface. These kinds of composites are called weak interface composites (WIC). The second approach is the use of a porous matrix, which mechanically decouples the fibers, producing inelastic strain during loading and allowing matrix crack evolution without damage of the fibers; this route does not need necessarily a fiber-matrix interface. These composites are named weak matrix composites (WMC) [2-5].

Nowadays oxide-oxide CMCs are manufactured e.g. at EADS Innovation Works (CMC standard material termed UMOX^{TM}), DLR or Pritzkow Ceramics (all Germany) and AFRL (Air Force Research Laboratory), COI Ceramics or University of Santa Barbara (all USA). These materials are mainly produced with oxide fibers known as Nextel^{TM} from 3M^{TM} Company (USA). These fibers are made of e.g. alumina (Nextel^{TM}610) or mullite (Nextel^{TM}720). The ceramic matrix developed for these materials are mainly manufactured using polymer based slurries, water based slurries or via sol based slurries [8]. In the work presented in this paper, the development of a new type of oxide CMC (termed WITA-OX^{TM}) by the combination of the sol-gel and freeze gelation technology with filament winding is investigated.

The sol–gel infiltration is a ceramic slurry infiltration process (CSI) and offers various advantages. The advantages over conventional ceramic processing routes include high matrix homogeneity (because the fine ceramic particles are intimately mixed in the colloidal state), the ability to prepare compositions which are difficult to be achieved by conventional methods, and relatively low sintering temperatures (as consequence of the high reactivity which arises from the very high surface area of colloidal particles) [6].

After infiltration of continuous fibers with the sol-gel matrix, the sample is frozen in order to force sol-gel transition. The sol is converted to a rigid matrix without significant shrinkage. In this step, the structure of the material is formed and the characteristics of the future porosity are determined once the continuous crystals of solvent are formed and grow in the slurry (Figure 1). Subsequent drying slightly above room temperature is conducted. This step results in bulk shrinkage often below 1%, owing to low capillary stresses associated with the relatively large and open porosity, typically 1-10µm in diameter, which results from the nucleation and growth of ice crystals during freezing. The green body is relatively weak and must, therefore, be sintered to achieve sufficient strength [7-8].

The main advantage of the freeze gelation process is the ability to form large or small complex-shaped components near to net shape with multidirectional fiber reinforcement either by simple casting for short fiber reinforcement or by filament winding and hand lay-up for continuous fiber reinforcement [9-11].

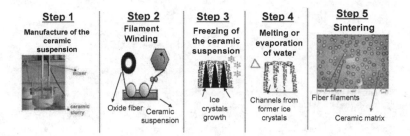

Figure 1 - Sketch of the manufacture of CMC via freeze gelation process of a sol-gel based suspension

EXPERIMENTAL SET-UP

Nextel™610 from 3M™ (coated) oxide fibers were impregnated with a sol-gel matrix via filament winding and subsequent lay-up on a quadratic tool. Afterwards three manufacturing variations were performed, one sample was taken directly to freezing after filament winding, another sample was taken into an autoclave and a third sample was pressure-less dried at ambient temperature (Figure 2). In a fourth variation uncoated Nextel™610 fibers were impregnated and frozen directly after lay-up.

For manufacturing of the ceramic matrix via sol-gel route water based suspension with dispersed SiO_2 nanoparticles was used as basis for the matrix. Mullite powder with average particle size of 3µm and additive agent were added to produce a homogeneous suspension with 75 wt%. The suspension was homogenized using a mechanical mixer from IKA.

The composites were then sintered at 1200°C using the same parameters for all of the samples. For characterization, microstructure, porosity, transversal tensile strength, and compressive shear strength were investigated; additionally single edge notched beam (SENB) tests were performed on sintered composites with (C) and without (U) fiber coating.

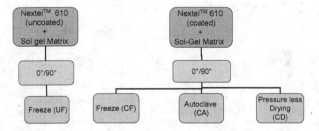

Figure 2 – Sketch of the manufacturing routes to produce four types of composites

RESULTS AND DISCUSSIONS

Microstructures of the four different materials were analyzed by optical microscope. The pictures are presented in Figure 3 and 4. In Figure 3 it is shown that all composites show a good microstructure without presence of delamination. In Figure 3a and 3b the former ice crystal structure is observed respected by diagonal oriented pore formation. The dark grey or black areas correspond to pores, light grey areas are the fiber filaments distributed in layers oriented in 0° and 90°. Figure 4 shows the same materials with a higher resolution. All variations are characterized by a very good and homogenous infiltration within the fiber bundles where nearly no regions with agglomerated filaments are observed. Fiber volume content of the samples was calculated between 25 and 30 vol%. The values for each composite are listed in the following Table 1.

Table 1 - Fiber Volume content of 4 different oxide CMC manufactured

Manufacture route	Fiber	Type	Fiber Volume Content (%)
Freeze Gelation	Nextel^{TM}610	CF	28,3
Freeze Gelation	Nextel^{TM}610	UF	23,4
Autoclave process	Nextel^{TM}610	CA	27,1
Dried in ambient temperature	Nextel^{TM}610	CD	24,7

The porosity of the composites was investigated using mercury intrusion technique. Total porosity is found between 30 and 40 %, Table 2 resumes the values of porosity and density for each composite. It can be observed that the highest porosity (39,4 %) is found after autoclave process. While, the composite dried in air presented lowest porosity (29,6 %).

Table 2- Porosity and density values of 4 different oxide CMC manufactured

Manufacture route	Fiber/ Coating	Type	Total Porosity (%)	Apparent Density (g/cm^3)
Freeze Gelation	Nextel^{TM}610	CF	33,5	2,75
Freeze Gelation	Nextel^{TM}610	UF	34,2	3,00
Autoclave process	Nextel^{TM}610	CA	39,4	3,08
Dried in ambient temperature	Nextel^{TM}610	CD	29,6	2,89

In Figure 5 the pore size distributions of the four different materials is shown. The CMC manufactured with the use of autoclave showed larger pore sizes and a mono-modal distribution, the application of pressure in the component may have caused the particles to move to the bottom of the material before the component was dried, leading to regions with higher porosity and to a larger total porosity. The pore distribution for the sample dried in air and frozen is similar, although the size generated from the ice crystal growth are uniformly distributed, in contrast to the samples dried in air.

Regarding the coated and uncoated composites, it becomes obvious (Figure 5b) that due to the higher hydrophility of fiber uncoated during processing and subsequent freezing, the composite was wetter compared to composite with coating. Therefore a higher amount and elongated ice crystals is created during freezing, leading to larger pore sizes and a more homogeneous distribution of the pores.

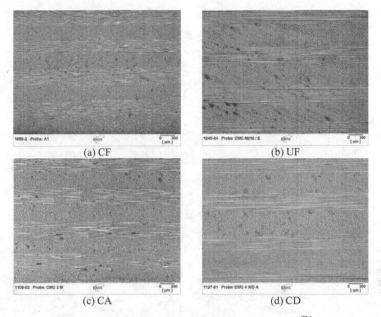

Figure 3 - Microstructure of CMCs manufactured by impregnation of Nextel^{TM}610 fibers with sol-gel matrix via filament winding and consolidated (a) via freeze gelation, (b) via freeze gelation (uncoated), (c) autoclave and (d) dried in ambient temperature

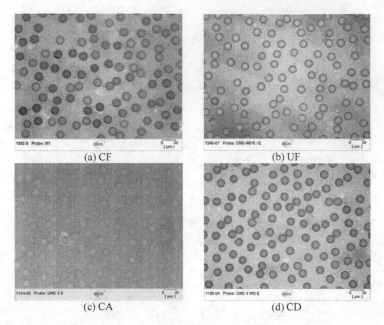

(a) CF   (b) UF

(c) CA   (d) CD

Figure 4 - Impregnation of fiber bundles with ceramic matrix of CMCs manufactured by infiltration of Nextel^{TM}610 fibers with sol-gel matrix via filament winding and consolidated (a) via freeze gelation, (b) via freeze gelation (uncoated), (c) autoclave and (d) dried in ambient temperature

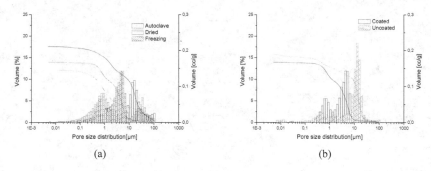

(a)   (b)

Figure 5 - Comparison of pore size distribution of CMCs manufactured via (a) freezing, autoclave and drying and (b) of CMCs made with and without fibre coating

In order to investigate the interlaminar properties of the composites manufactured with coated fibers, compressive-shear test and transversal tensile test were conducted. Compressive-shear test was followed by the norm DIN EN 658-4.

In Figure 6 the results from the compressive-shear test are presented. Composite manufactured using the autoclave process presented the lowest shear strength. After freezing or drying at ambient temperature similar strength values 8,1 and 7,2 MPa are measured. Although their strength values were similar the fracture mode was different. After freezing all samples failed by fracture across the center layers while, the dried samples showed pure interlaminar failure (Figure 7).

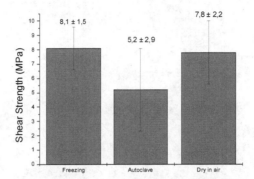

Figure 6 - Shear strength of composites made with Nextel™610 fibers via filament winding combined with freeze gelation, autoclave process or drying in air

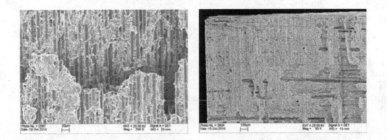

Figure 7 - Fracture surfaces of composites manufactured via freezing (left) and drying (right) after compressive-shear test

In Figure 8 the results from the transversal tensile tests are shown. The CMC manufactured via autoclave process showed lowest strength (0,23 MPa), all tested samples show interlaminar failure. The highest strength (2,32 MPa) is measured on samples manufactured via freezing. The dried composites present an intermediate strength (1,1 MPa) and interlaminar fracture behavior (Figure 9).

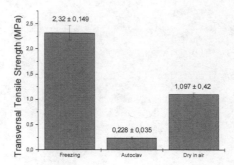

Figure 8 - Results from transversal tensile test of CMCs made with Nextel™610 fibers via filament winding combined with freeze gelation, autoclave process or drying in air

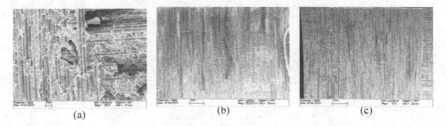

Figure 9 - Fracture surfaces of CMCs manufactured via freezing (a), autoclave (b) and drying in air (c) after transversal tensile test

In order to investigate the influence of the fiber coating, single edge notched beam (DIN 51 109) test was used to compare samples manufactured via freezing with coated and uncoated fibers. The results are presented in Figure 10. Fracture toughness of both composites are nearly the same (2,9 and 2,7 MPam^{0,5}). However, the fracture surfaces differ significantly. The coating of the fibers induces brittle fracture (Figure 11d) while significant fiber pull-out is observed with uncoated fibers which is a qualitative easure of enhanced fracture toughness (Figures 11 a-c).

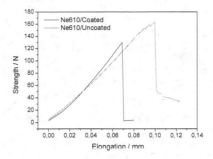

Figure 10 – Representative load displacement curves of CMCs manufactured with coated and uncoated fibers tested via SENB test

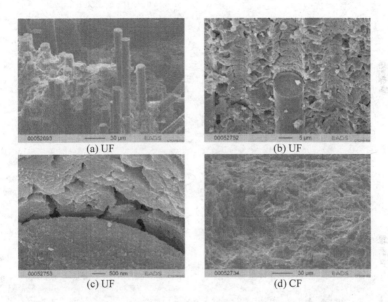

Figure 11 - Fracture surfaces of the CMCs manufactured with uncoated (a,b,c) and coated (d) fibers

CONCLUSIONS

The paper provides an overview of the development work on a new type of oxide based CMC material called WITA-OX^{TM} where sol-gel matrices and oxide fibers are combined using filament winding technique for fiber impregnation. The composites were manufactured with four different process variations using either freeze gelation route, autoclave or pressure-less drying at ambient temperature. Additionally, the freeze gelation technique was performed with coated and uncoated fibers.

All samples were infiltrated via filament winding with the result of an excellent and homogeneous impregnation of the sol-gel matrix within the fiber. Nevertheless it was observed that the different routes influenced the properties of the composites. The average pore size was found to be larger in the samples fabricated in the autoclave. A reason for this might be the autoclave process itself, where the stability of the slurry was disturbed leading to a precipitation of particles or water to the bottom of the material before the component was dried. This may explain the lower interlaminar properties of this material. The composites manufactured via freeze gelation show a porosity influenced by the growth of ice crystals during freezing and higher density of the matrix between the channel like pores. The drying route does not induce such oriented pores. Here the pores are distributed equally throughout the matrix and this result in lower interlaminar strength properties. Therefore, the use of the freeze gelation route proved to be the most promising route to manufacture CMCs with suitable properties when sol-gel technique and fiber bundle impregnation via filament winding is performed.

The comparison of CMCs with coated and uncoated fibers, respectively, shows that the coating induces brittle fracture while uncoated fibers are debonded easily which results in enhanced toughness. One reason for this might be the higher hydrophility of the uncoated fiber. During processing and freezing, the composite was probably still wetter compared to composite with coating, therefore a higher amount of ice crystals with elongated crystals was created during freezing, leading to easier fiber debonding.

In conclusion, the production of oxide based CMCs combining freeze gelation of sol-gel matrices and filament winding technique results in a composite with very homogenous microstructure and the use of uncoated fibers allows crack branching within the porous matrix and prevention of premature failure of fibers. Thus, fiber properties can be transferred to composite properties effectively.

After successful confirmation of the described manufacture process for oxide based CMC material the development work was continued. Further development activities e.g. concerning higher fiber volume contents are currently under investigation and show already promising results for this novel type of high temperature CMC material.

ACKNOWLEDGMENTS

This work was performed within the HiPOC project (**H**igh **P**erformance **O**xide **C**eramics) started in February 2009 (contract number 03X3528) within the WING framework. It is funded in part by the German Federal Ministry of Education and Research (BMBF) and administered by Projektträger Jülich (PTJ). Their support is gratefully acknowledged.

REFERENCES

[1] J. Lane, J. Morrison, S. Mazzola, and H. Grote, Oxide-Based CMCs for Gas Turbines, 5th International Conference of High-Temperature Ceramic Matrix Composites (2004).

[2] F.W. Zok and C.G. Levi, Mechanical properties of porous-matrix ceramic composites. Advanced Engineering Materials, 3(1-2), 15-23 (2001).

[3] C.G. Levi, J. Y. Yang, B. J. Dalgleish, F.W. Zok and A. G. Evans, Processing and Performance of an All-Oxide Ceramic Composite. J. Am. Ceram. Soc., 81 (8), 2077-2086 (1998).

[4] D. Koch, R. Knoche and G. Grathwohl, Multiphase Fiber Composites. Ceramics Science and Technology, 12 (1), 511-582 (2008).

[5] R.A. Simon and R. Danzer, Oxide fiber composites with promising properties for high-temperature structural applications. Advanced Engineering Materials, 8 (11), 1129-1134, (2006).

[6] M.K. Naskar, M. Chatterjee, A. Dey and K. Basu, Effects of processing parameters on the fabrication of near-net-shape fibre reinforced oxide ceramic matrix composites via sol-gel route. Ceramics International, 30 (2), 257-265 (2004).

[7] R.S. Russellfloyd, B. Harris, R.G. Cooke, J. Laurie and F.W. Harnmett, Application of Sol-Gel Processing Techniques for the Manufacture of Fiber-Reinforced Ceramics. J. Am. Ceram. Soc, 76, 2635-2643 (1993).

[8] T. Machry, D. Koch and C. Wilhelmi, Development of a new Oxide Ceramic Matrix Composite. 7th International Conference of High-Temperature Ceramic Matrix Composites, 435-445 (2010).

[9] R.S. Russellfloyd, B. Harris, R.W. Jones, R.G. Cooke, T.H. Wang, J. Laurie and F.W. Hammett, Sol-Gel Processing of Fibre-Reinforced Ceramic Shapes. British Ceramic Transactions, 92, 8-12 (1993).

[10] S. Deville, Freeze-Casting of Porous Ceramics: A Review of Current Achievements and Issues. Advanced Engineering Materials, 10, 155-169 (2008).

[11] R. Gillissen, J.P. Erauw, and A. Smolders, Gelcasting, a near net shape technique. Materials and Design, 21, 251-257 (2000).

[12] B. Newman and W. Schäfer, Processing and Properties of Oxide/Oxide Composites for Industrial Applications, In: High Temperature Ceramic Matrix Composites (Ed. W. Krenkel, R. Naslain, H. Schneider), Wiley-VCH, 600-609 (2001).

[13] W. D. Vogel and U. Spelz, Cost Effective Production Techniques for Continuous Fibre Reinforced Ceramic Matrix Composites, In: Ceramic Processing Science and Technology, 51, 225-259 (1995).

[14] D. Koch, K. Tushtev, K. Rezwan, C. Wilhelmi, S. Denis, and J. Göring, Effect of Microstructure on Room and High Temperature Properties of Oxide/Oxide Composites Developed for Gas Turbine Applications, *Proceedings of the 34ᵗʰ International Conference on Advanced Ceramics and Composites (ICACC)*, Florida, USA (2011).

[15] M. Gerendás, Y. Cadoret, C. Wilhelmi, T. Machry, R. Knoche, T. Behrendt, T. Aumeier, S. Denis, J. Göring, D. Koch, and K. Tushtev, Improvements of Oxide/Oxide CMC and Development of Combustor and Turbine Components in the HiPOC Program, in press: *Proceedings of the ASME Turbo Expo 2011: Power for Land, Sea Air*, GT2011-45460, Canada (2011)

[16] R. Knoche, E. Werth, M. Weth, J. Gómez García, C. Wilhelmi and M. Gerendás, Design and Development Approach for Gas Turbine Combustion Chambers made of Oxide Ceramic Matrix Composites, *Proceedings of the 34ᵗʰ International Conference on Advanced Ceramics and Composites (ICACC)*, Florida, USA (2011).

EFFECT OF REACTION TIME ON COMPOSITION AND PROPERTIES OF SIC-DIAMOND CERAMIC COMPOSITES

S. Salamone, O. Spriggs
M Cubed Technologies, Inc.
1 Tralee Industrial Park
Newark, DE 19711

ABSTRACT

Composites of reaction bonded silicon carbide (RBSC) and diamond have huge potential because of the high hardness, high thermal conductivity and high stiffness they exhibit. There is a strong drive to increase the quantity of diamond powder incorporated, thereby improving the properties of these composites. The addition of up to 20% diamond to the RBSC composites poses technical challenges that must balance higher final properties with increasing processing difficulties. The thermal stability of diamonds is such that high processing and/or longer reaction times can degrade their properties. X-ray diffraction of diamond powders has shown an increasing transformation to graphite during elevated processing temperatures. Analytical techniques (ICP-AES) have also shown decreasing silicon content with increasing reaction temperatures in SiC-Diamond composites. The current study relates isothermal reaction time to microstructure and properties of the SiC-Diamond composites. These physical properties are directly related to the final composition (e.g. remaining diamond and silicon content) of the composite.

INTRODUCTION

Diamond containing composites have potential use in applications ranging from wear products to electronic devices because of its extremely high hardness and high thermal conductivity.[1-2] The processing techniques commonly employed usually involve sintering at high temperatures and high pressures with various reactive infiltration alloys and diamond particle sizes.[3-6] In contrast, the reaction bonding process developed and discussed here results in a Si/SiC and diamond composite fabricated at lower processing temperatures with a low weight percent of residual metal. Low temperature processing of diamond composites is advantageous because diamond is a meta-stable compound (i.e., it converts to graphite at high temperatures and low pressures).[7] However, this transformation can be minimized and or eliminated by carefully controlling the reaction bonding processing conditions. In addition to the aforementioned advantages of diamond, it is also a very high density form of carbon. During the reaction bonding process, some of the diamond (e.g., the surfaces of the particles) will react with the infiltrating Si alloy to form -SiC.[7] Due to the high density of carbon in the diamond structure, the volume of SiC formed during this reaction is very high relative to the volume of diamond that is reacted. This allows the production of a reaction bonded ceramic with very low free Si content in the microstructure.[8-9] The amount of residual silicon in RBSC materials is important because it has been shown to decrease the desired high stiffness, hardness, strength and toughness of this class of materials.[10-11] With that in mind, silicon content has been measured for the infiltrated materials currently under investigation.

The reaction bonding process has two key steps. First a preform is made that consists of ceramic particles (e.g., SiC and Diamond) and carbon. Second, the preform is infiltrated with molten silicon. During the infiltration process, a $Si + C \rightarrow SiC$ reaction occurs. Thus, in the case where the starting preform consists of SiC and diamond particles, and carbon, the finished ceramic should contain four basic constituents, namely the original SiC and diamond particles, reaction formed SiC and residual silicon.

These following experiments are intended to give a better understanding of how processing can affect the physical properties of reaction bonded based materials. Controlling residual silicon content and suppressing graphite formation in diamond containing RBSC composites will be discussed.

EXPERIMENTAL PROCEDURES

Powder compacts consisting of 12 m SiC particles and synthetic, high pressure high temperature manufactured diamond particles, with an average particle size of 22 m, were combined with specific levels of additional carbon. Preforms were consolidated using these SiC/diamond/C mixtures, and then infiltrated with molten Si to yield a Si/SiC and diamond ceramic.

The physical properties of the infiltrated composites were measured using several common techniques summarized in Table I. All the microstructures were characterized by examining fracture surfaces using a JEOL JSM-6400 Scanning Electron Microscope. The scanning electron microscope (SEM) images were taken in Back-Scattered Mode to differentiate the phases present (compositional differences) - e.g., Si metal (brightest phase), SiC (intermediate/gray phase), and Diamond (dark phase). All images are static fracture surfaces due to the difficulty of polishing diamond containing specimens.

Samples were sent to independent laboratories to identify compositional changes occurring over the various thermal conditions. The residual silicon metal content was determined using inductively coupled plasma atomic emission spectroscopy (ICP-AES) and the amount of graphite formation was detected using a Bruker D4 diffractometer with Cu radiation at 45KV/40mA.

Table I: Summary of Properties and Techniques Used to Quantify the Various Composites.

Property	Technique	Standard
Density	Immersion	ASTM B 311
Elastic Modulus	Ultrasonic Pulse Echo	ASTM D 2845
Silicon Content	ICP-AES	
Phase Analysis	Powder XRD	

RESULTS AND DISCUSSION

Initial experiments were conducted with RBSC samples containing 14 volume percent of 22 m diamond particles. The preforms were infiltrated at three different temperatures and held for a short period of time to allow for furnace equilibrium. The density and Young's modulus is plotted as a function of the temperature in Figure 1. The density appears fairly independent, however, the modulus which is highly dependent on the amount of diamond remaining, drops roughly two percent over a range of less than 100 °C. Two processes could be responsible for the degradation. The first is the reaction of silicon and diamond (carbon) to form SiC. The second is the formation of a graphite phase (density 2.25 g/cc) from the diamond (density 3.51 g/cc) due to thermal instability.

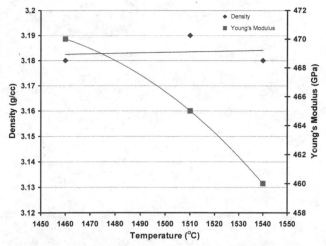

Figure 1: Density and Young's modulus as a function of infiltration temperature.

When processed in the preform and before infiltration, the diamond particles are protected with an in situ formed carbon coating to ensure that they survive the infiltration with silicon metal. However, an experiment was performed using uncoated diamond powder exposed to the same infiltration temperature/vacuum cycle as the isothermal experiments conducted in this study. Figure 2 is a SEM micrograph showing the size (22 m average) and morphology of the starting diamond powder. After processing the diamond particles (uncoated) in the standard infiltration cycle, X-Ray diffraction was performed to determine what, if any, reactions might have occurred. Figure 3 is a plot of the diffraction pattern, with the matching stick patterns, for the processed diamond powder. Graphite phase formation was detected (approximately 8.5 wt%) along with some SiC (5.5 wt%) from the ubiquitous Si vapor present in the furnace. Several avenues can be taken to increase the final diamond content and reduce the graphite transformation and SiC reaction. From the data in Figure 1, decreasing the temperature is clearly one way to accomplish this. Since the goal is to increase the amount of diamond and hence improve the final properties of the composite, samples containing 21 vol% of the 22 m diamonds were developed for the isothermal experiments.

Figure 2: Representative particle morphology of the 22 m diamond powder.

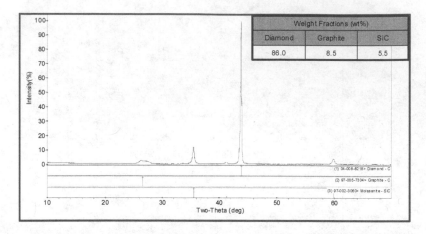

Figure 3: XRD analysis of diamond powder after exposure in vacuum at temperature.

Physical Properties as a Function of Reaction Time

Table II contains the physical properties and compositional data for the 12 m RBSC with 21 vol% diamond composites, infiltrated at $1470°$ C, as a function of isothermal reaction time. Also shown, for comparative purposes, is data for a standard reaction bonded material sans diamond particles. There is a large increase in density and Young's modulus due to the addition of diamond particles. The residual silicon content, as measured by wet chemistry analysis (ICP-AES), changes dramatically when diamond is added to the matrix. This is most likely due to the reaction of silicon and carbon to form silicon carbide, as discussed previously. Figure 4 is a plot of the density and Young's modulus as a function of reaction time for the diamond composites. There is a drop in both density and modulus over time. From the data, the optimal condition for maintaining high density and modulus composites, under these conditions, appears to be short infiltration times.

Table II: Properties and Compositional Analysis of 12 m RBSC with 21 vol% Diamond

Isothermal Temperature – 1470° C				
Isothermal Time (hr)	**Density (g/cc)**	**Young's Modulus (GPa)**	**Silicon (wt %) (ICP)**	**Graphite (wt %) (XRD)**
1.5	3.21	490	1.3	0
3	3.20	481	.88	0
6	3.19	475	.95	.6
12	3.17	463	.86	1.2
Baseline Sample				
12 m SiC –No diamond	3.07	396	11.4	----

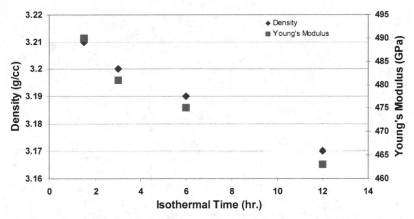

Figure 4: Density and Young's modulus as a function of time for RBSC matrix with 21 vol% diamond particles. (average diamond particle size – 22 m)

XRD Analysis

To determine the thermal stability of the diamond as a function of processing time, powder diffraction was performed after every experiment. Each sample was ground in a WC ball mill to produce a fine powder. This introduced some WC contamination and these peaks have been identified in the following diffraction patterns. Figure 5 shows the formation of graphite after 6 hours of infiltration. The accompanying pattern for the sample held for only 1.5 hours does not have any graphite peaks. The diffraction data verifies that some of the diamond transforms to graphite between three and six hours of processing. Rietveld analysis of the phase compositions determined that the weight percent graphite present increases (doubles) as the isothermal infiltration time doubles from six to twelve hours, as shown in Table II.

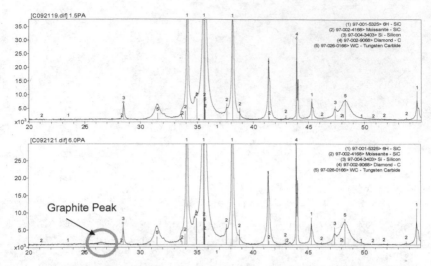

Figure 5: Powder diffraction plots comparing samples 1.5 hr and 6.0 hr. The formation of a graphite phase starts to develop within 6 hours.

Microstructure

The physical properties (e.g. Young's modulus, density and measured silicon content) change as the infiltration time increases. Microstructural comparisons may help to corroborate the measured values. The first step is to determine the various visual differences (i.e. phases) in the micrographs. Elemental analysis (EDS) was done to aid in the identification of the SiC matrix, diamond particles and silicon alloy. The following figures were taken at two levels of magnification to elucidate any similarities or differences that were developed during the changing heat treatments and processing variables. As noted earlier each micrograph was taken in back scattered mode to accentuate the different phases, the darkest phase is the diamond (when applicable), the gray phase is SiC and the white (or lightest) phase is residual silicon alloy.

Figure 6 consists of a SEM image of the 1.5 hr sample and the spectra from the three colored circular regions highlighted on the micrograph. The blue circular region (B) contains a diamond particle and the spectra shows only a carbon peak. This combined with the presence of diamond peaks in XRD data and the enhanced physical properties (modulus and density) leads us to believe these are diamond particles that survived the thermal cycle. The red circular region (A) contains the silicon alloy used to infiltrate the preform. The SiC matrix particles are identified in the green region (C). The spectra contains only silicon and carbon peaks. When compensated for absorption and intensity behaviors, the ratio is nearly stoichiometric.

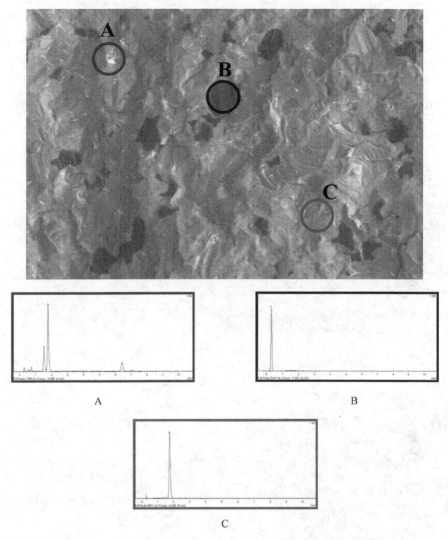

Figure 6: SEM image and EDS spectra of the 12 m SiC with 21 vol% diamond particles processed at temperature for 1.5 hours. Spectra A is infiltration alloy, Spectra B is diamond particle and Spectra C is SiC matrix

Figure 7 compares the RBSC composites with and without the diamond particles. At both magnification levels (100X & 500X), the difference in silicon content is very noticeable. The light regions (silicon alloy) appear throughout the baseline sample (a), but the sample containing diamond particles (b) appears to have a diminished amount of the silicon phase. It appears to contain almost no alloy and looks very much like an all-ceramic composite. This finding is consistent with the density, Young's modulus and chemical analysis results in Table II.

The distribution of diamond particles in image (b) is fairly homogeneous. The fracture surface is predominately transgranular, with many grains displaying rough cleavage planes across the surface of the particle.

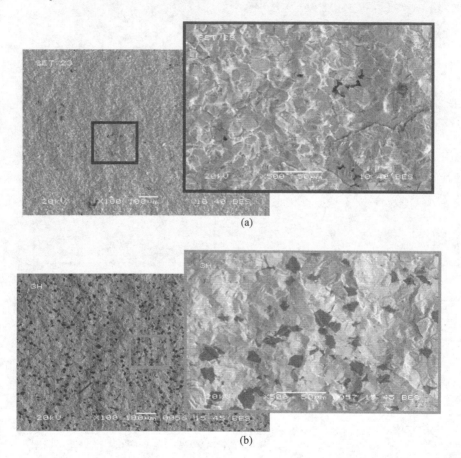

(a)

(b)

Figure7: Micrograph of (a) the 12 m SiC matrix with no diamonds and (b) the sample containing 21 vol% of 22 m diamond particles.

SUMMARY

The ability to produce RBSC composites with a high (21 vol%) diamond content has been demonstrated. It was found that the physical properties of these highly loaded Si/SiC + diamond composites can be greatly influenced by the time held at infiltration temperature. The density and Young's modulus of the samples decrease over time as the reaction time continues. XRD and property measurements have shown that diamond particles become unstable at high processing temperatures. In addition, the formation of a graphite phase (from the diamond phase) occurs during extended isothermal times. Thus the processing conditions for optimal properties were found to be at low temperature and short thermal exposure times. Knowing and understanding these effects can lead to a wide range of physical properties, making this class of materials extremely tailor-able.

REFERENCES
1.   I. E. Clark and P. A. Bex, "The Use of PCD for Petroleum and Mining Drilling," Industrial Diamond Review, 43-49 (1999).
2.   W. Zhu, G. P. Kochansky, S. Jin, "Low-Field Electron Emission from Undoped Nanostructured Diamond", Science 282 1471 (1998).
3.   Y. S. Ko, T. Tsurumi, O. Fukunaga, and T. Yano, "High Pressure Sintering of Diamond-SiC Composite," Journal of Material Science, vol. 36, 469-475, (2001).
4.   J. Qian, G. Voronin, T. W. Zerda, D. He, and Y. Zhao, "High-Pressure, High-Temperature Sintering of Diamond-SiC Composites by Ball-Milled Diamond—Si Mixtures," J. Mater. Res., vol. 17, No. 8, 2153-2160, (2002).
5.   S. K. Gordeev, S. G. Zhukov, L. V. Danchukova, and T. C. Ekstrom, "Low pressure Fabrication of Diamond-SiC-Si Composites," Inorganic Materials, vol. 37, No. 6, 579-583, (2001).
6.   G. Bobrovnitchii, A. Skury, A. Osipov, R. Tardim, "SiC Infiltrated Diamond Composites" Materials Science Forum Vols., 591-593, 654-660, (2008).
7.   C. Pantea, G. A. Voronin, T. Zerda, J. Zhang, L.Wang, Y. Wang, T. Uchida, Y. Zhao, "Kinetics of SiC formation during high P–T reaction between diamond and silicon", Diamond & Related Materials 14, 1611 – 1615, (2005).
8.   P. Karandikar, S. Wong, G.Evans, M.Aghajanian " Microstructural Design for Si-B4C-Diamond System", submitted 35th International Conference & Exposition on Advanced Ceramics & Composites
9.   S. Salamone, R. Neill, M. Aghajanian, "Si/SiC and Diamond Composites: Microstructure-Mechanical Properties Correlation", CESP Vol. 31 [2] 97-106 (2010).
10.   S. Salamone, P. Karandikar, A. Marshall, D. D. Marchant, M. Sennett, "Effects of Si:SiC Ratio and SiC Grain Size on Properties of RBSC", in Mechanical Properties and Performance of Engineering Ceramics and Composites III, Ceram. Eng. Sci. Proc., 28, E. Lara-Curzio et al. editors, 101-109, (2008).
11.   O. P. Chakrabarti, S. Ghosh, and J. Mukerji, "Influence of Grain Size, Free Silicon Content and Temperature on the Strength and Toughness of Reaction-Bonded Silicon Carbide," Ceram. Int., 20 [5] 283 – 86 (1994).

PRESSURELESS SINTERING OF BORON CARBIDE USING AMORPHOUS BORON AND SiC AS ADDITIVES

(1) Célio A. Costa, Ph.D – POLI/COPPE/UFRJ – Centro de Tecnologia, Bl. F/210 – Cidade Universitária – Rio de Janeiro, RJ - CEP: 21949-900 – BRAZIL – celio@metalmat.ufrj.br
(2) Victor Manuel Domingues – POLI/COPPE/UFRJ – Centro de Tecnologia, Bl. F/210 – Cidade Universitária – Caixa Postal 68505 - Rio de Janeiro, RJ - CEP: 21949-900 – BRAZIL
(3) José Brant de Campos, D.Sc. – Universidade do Estado do Rio de Janeiro, Departamento de Engenharia Mecânica – R. Fonseca Teles 121 – CEP: 20940-200 - Rio de Janeiro – RJ - Rio de Janeiro/RJ – Brazil - brant@uerj.br;
(4) Pedro Augusto de S.L. Cosentino, D.Sc., Lt Col Military Engineer – Brazilian Army Evaluation Center – 9140, Barra de Guaratiba Rd - Rio de Janeiro/RJ – BRAZIL – CEP: 23020-240 - pacosen@gmail.com.

ABSTRACT

The B_4C and B_{am} was high energy milled with SiC spheres during four different times. The SiC was added to the composition, possibly as nanoparticles. The materials were pressed and sintered at 2150 °C for 30 minutes. The results showed that a direct correlation between milling times, density and the elastic modulus, which were measured to be 276, 333, 365 and 420 GPa as the milling time varied from zero, 15, 30 and 60 minutes, respectively.

INTRODUCTION

Boron carbide (B_4C) is a structural ceramic material with applications in both structural and functional fields. The combinations of very low density with mechanical and nuclear properties enable it for application from ballistic armour to x-ray transmission windows, for instance [1,2].

To process B_4C and to obtain high density is a difficult task, since the nature of the covalent bonds leads to very low diffusion rate. Typically, high density materials are obtained through pressure application, atmosphere control and proper sintering additives, which the commonly used are: AlF_3, Be_2C, TiC, TiB_2, W_2B_5, SiC, Al, Mg, Ni, Fe, Cu, Si and C; with C being the most effective among all of them [1,3,4].

The present study evaluated the processing of B_4C with addition amorphous boron (B_{am}) and SiC, which might have been added as nanoparticles. The results showed that milling time had a strong effect on the density and on the elastic modulus, the density was higher than 84% for all milling time and the elastic modulus measured was 276, 333, 365 and 420 GPa as the milling time varied from zero, 15, 30 and 60 minutes, respectively.

EXPERIMENTAL

The raw materials used in this study, boron carbide (B_4C) and amorphous boron (B_{am}), are described in Table I. Silicon carbide (SiC) became part of the mixture as intentional contaminations from the high energy milling process.

Table I – Raw material used, supplier and specification

Stock	Supplier	d_{50} µm (Laser Diffraction)	Superficial Area m^2/g (BET)
Boron Carbide Grade HP	H. C. Starck, Germany	2,5	6,0 – 9,0
Amorphous Boron Grade I	H. C. Starck, Germany	1,0 – 2,0	>10,0

The starting composition was 98.5 wt% of B_4C and 1.5 wt% B_{am}. This composition was homogenized via ball milling, with IPA and zirconia spheres for 12 hours. This same composition was high energy milled with isopropyl alcohol (IPA) in the presence of SiC spheres, which diameter were in the range from 3 to 5 mm, in a planetary mill (250 rpm) during 15, 30 and 60 minutes. The high energy milling processed added SiC particles to the starting composition, which were quantified by Rietveld method. Four compositions were generated, codified as follows: i) only homogenized (B_4C+B_{am}-0; no high energy milling), ii) high energy milled for 15 minutes (B_4C+B_{am}-15), iii) high energy milled for 30 minutes (B_4C+B_{am}-30) and iv) high energy milled for 60 minutes (B_4C+B_{am}-60).

The compositions were dried, desaglomerated and uniaxially pressed with 10 MPa in rectangular plates with 60 x 65 x 7 mm. The green plates were then isostatically pressed with 150 MPa, resulting in a green density of about 60% for all the mixtures, based on B_4C theoretical density.

All compositions were sintered at 2150 °C, during 30 minutes, in argonne with nitrogen gas mixture, and -12 psi pressure (no gas flux).

The sintered plates were surface grinded in both sides to guarantee parallelism, and the final thickness was 6.8 mm. The elastic modulus was then measured by ultrasound; generator (Krauftkramer USIP40), oscilloscope (Agilent Tech), longitudinal wave transducer of 5 MHz (Krauftkramer G5KB) and transversal wave transducer of 2.25 MHz (Panametrics V154).

The materials were then cut to be analyzed by X-ray diffraction and to have the density measured by Archimedes density.

RESULTS AND DISCUSSION

The particle size distribution of the B_4C+B_{am}-0 is shown in Figure 1. It is observed a bimodal distribution with the average centered in 2.5 µm and a large particles size in between 10 and 12 µm. After the high energy milling for three different times (15, 30 and 60 minutes) the particle size distribution became as shown in Figure 2, where it is observed a change from bimodal to monomodal distribution, with all particles below 10 µm. The change in the particle distribution was the result of particle fracture, since the large particles (> 10µm) in the B_4C+B_{am}-0 composition were eliminated. However, there was a twofold addition that must be accounted in these distributions, namely, the amorphous boron and the very small SiC particles from the milling spheres, since the comminution processed used is quite aggressive.

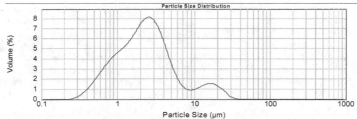

Figure 1 – Particle size distribution of the as received B_4C+B_{am}-0.

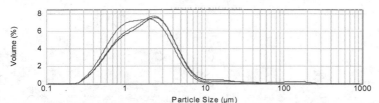

Figure 2 – Particle size distribution after planetary milling. The blue, the green and the red lines are for 15, 30 and 60 milling time, respectively.

The particle size distribution parameters (d_{10}, d_{50} and d_{90}) and resulting properties of the four materials studied are summarized in Table 2. The comminution process was most effective in reducing the large particle sizes, where values of 20%, 31% and 48% were obtained for d_{10}, d_{50} and d_{90} for 60 minutes of milling, compared to the homogenized composition. The green density was also improved by the particle size reduction process, but as the distribution became narrowed, as the milling time increased, the density slightly decreased (30 and 60 minutes), as expected.

The sintered density increased from 84% to almost 100%, in therms of theoric density, as the B_4C particles were reduced and SiC particles were introduced. Even though the microstructure characterization has not been done yet, it quite reasonable to assume that SiC particles did not impair the sintering process, since high density and elastic modulus has been obtained. It is worth mention that the SiC particles contributed to density measured above 100%, since its density is 3.22 g/cm^3.

Table 2 - Parameters of the milled powder, amount of SiC present in the composition, and the respective properties of the green and sintered compositions.

	B_4C+B_{am}-0	B_4C+B_{am}-15	B_4C+B_{am}-30	B_4C+B_{am}-60
d_{10} (µm)	0,743	0,648	0,641	0,592
d_{50} (µm)	2,255	1,902	1,816	1,553
d_{90} (µm)	7,48	5,003	4,493	3,897
Amount of SiC [%]	0,13	1,55	3,07	7,01
Green body density [%]	58,7	63,4	61,1	61,0
Sintered density [%]	84,2	88,9	93,7	101,3
Poisson Ratio [ ]	0,20	0,19	0,18	0,19
E [GPa]	276	333	365	420

X ray diffraction was performed for all compositions and the phases were quantified by the Rietveld method, as shown in Table 2 and exemplified in Figure 3 for the B_4C+B_{am}-60 condition. In Figure 3, the blue line represents the experimental data and the red one the calculated results, showing a very good agreement between the experimental and calculated data. For all compositions, it was identified the presence of $B_{13}C_2$, pure graphite and SiC, this last one in amount proportional to the milling time. In fact, the amount of SiC added per minute of milling was about 0.11%. It was also possible to identify the polytype of SiC being added, 3C and 6H. Graphite 2H was also identified for all compositions, but its amount was 1.15% and did not change for all milling times.

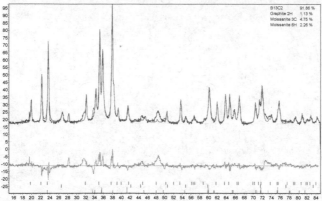

Figure 3 – X-ray diffraction and Rietveld analysis for the B_4C+B_{am}-60 sample. About 7% of SiC was added to the original composition.

The properties of the sintered compositions are shown in Table 2, where it clearly observed that density increased in the milling time and the elastic modulus too. The density varied from 84% for the homogeneized B_4C+B_{am}-0 composition, with and elastic modulus of 276 GPa, to 101% with a corresponding elastic modulus of 420 GPa. The Poisson ration for composition was about 0.19. The elastic modulus of the pressureless composition B_4C+B_{am}-60 is comparable to the hot pressed boron carbide reported in the literature, which is in the range of 450 to 476 GPa [5], showing a promising material.

CONCLUSIONS

High energy milling of B_4C and amorphous boron with silicon carbide spheres during 60 minutes resulted in high density (about 100%) and high elastic modulus (420 GPa) material, comparable to the hot pressed ones. Further studies are needed to characterize the microstructure, specially the location and size of the SiC particles, and to measure the MOR and fracture toughness.

ACKNOWLEDGEMENT

The authors wish to thanks the Brazilian Petroleum Agency (ANP-PRH 35), the Brazilian Army and CBPF for supporting and collaborating with this research.

REFERENCES

1- H. Lee; R. F. Speyer; *J. Am. Ceram. Soc.,* 86 [9] 1468–73 (2003)
2- V. M. D. Menezes; *Monography*, Department of Metallurgical and Materials Engineering, Universidade Federal do Rio de Janeiro, 2009
3- T. K. Roy, C. Subramanian, A. K. Suri; *Ceramics International*, 32 (2006) 227–233
4- L. S. Sigl; *Journal of the European Ceramic Society*, 18 (1998) 1521-1529
5- Champagne, B; Angers, R.; J. Am. Ceram. Soc., 62 [3-4] 149-153 (1979)

EFFECT OF REACTIVE HEAT TREATMENT ON PROPERTIES OF Al-Mg-B$_4$C COMPOSITES

M. K. Aghajanian, A. L. McCormick, W. M. Waggoner
M Cubed Technologies, Inc.
1 Tralee Industrial Park
Newark, DE 19711

ABSTRACT

Metal matrix composites consisting of boron carbide (B$_4$C) particles in an Al alloy matrix have been well studied due to their utility in various structural and nuclear (neutron absorbing) applications. In addition, prior work exists where such composites have been reactively heat treated to yield lightweight and hard ceramic-like materials consisting of Al-B-C ternary phases. The present work expands on this prior work and examines the quantnary Al-Mg-B-C system. The advantages of this system are reduced density due to the presence of Mg, and the ability to form the ultra-hard AlMgB$_{14}$ phase under the correct processing conditions. Composites were made by infiltrating preforms of B$_4$C particulates with molten Al-Mg alloy. Key variables were Al:Mg ratio of the alloy and residence time at temperature (i.e., reaction time) after infiltration. The results demonstrate that composites with properties ranging from metal-like to ceramic-like can be produced in the experimental space studied, and that significant opportunities for additional optimization exist. In particular, longer reaction times led to increased formation of hard reaction products, and the use of an alloy of nominally 75Al – 25Mg resulted in the greatest reaction rate (i.e., greatest formation of hard phases).

INTRODUCTION

The Al-B-C system has been studied extensively [1-3]. It was found that the properties of Al/B$_4$C metal matrix composites (MMCs) could be significantly modified via a reactive heat treatment during which lightweight and hard Al-B-C ternary ceramic compounds were formed. By controlling the heat treatment conditions (e.g., time and temperature), free metal content could be varied from nominally 20 vol. % to zero. The work evaluated a wide range of heat treatment temperatures (e.g., 800 to 1400°C) and times. At the lower temperatures, where processing is most favorable, reaction rates were slow, requiring hundreds of hours to completely react away the free metal to yield a ceramic-like material [1].

More recently, researchers at Iowa State have formed ultra-hard and lightweight ceramics in the Mg-Al-B system [4, 5]. These compounds (e.g., AlMgB$_{14}$) are harder than SiC, yet have a very low density of 2.66 g/cc. The Mg-Al-B compounds were produced by hot pressing stoichiometric mixtures of the elements at 1350 to 1400°C. Moreover, increases in hardness were observed by the addition of doping elements (e.g., Si) to the system.

In the present work, B$_4$C-containing preforms were infiltrated with Al-Mg alloys. Then, a reactive heat treatment was conducted with the goal of forming lightweight Al-B-C and Mg-Al-B compounds. The use of an Al-Mg alloy provides the benefits of lower processing temperatures (i.e., lower melting temperatures), higher reaction rates, and enhanced tailorability. Herein, the effects of Al:Mg ratio and reaction time on microstructure, chemistry, and properties are presented.

EXPERIMENTAL PROCEDURES

Five Al-Mg alloy compositions were selected for the study. The two end points were commercially available Al and Mg alloys, with the other alloys specifically produced for the present study. The specific alloys were (by wt. %):

- Commercial Alloy AA518 (nominally 92Al – 8Mg)
- 75Al – 25Mg
- 50Al – 50Mg
- 25Al – 75Mg
- Commercial Alloy AZ91A (nominally 8Al – 92Mg)

For all infiltration experiments a preform consisting of 50 vol. % 50 micron B_4C particles was utilized. An SEM photomicrograph of the reinforcement is provided in Figure 1, and shows a nominally 20 to 120 micron distribution of blocky-shaped particles.

Figure 1: SEM Photomicrographs of B_4C Powder

Infiltration of the matrix metal occurred by coupling the ceramic powder compacts with the Al-Mg alloy in an atmosphere of flowing nitrogen at 700°C (Figure 2). After the infiltration was complete, the furnace atmosphere was switched to Ar and the composites were held for heat treatment times of 10 and 40 hours.

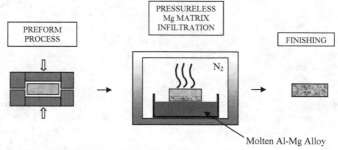

Figure 2: Schematic of Pressureless Infiltration Process for Al-Mg/B_4C Composites [6, 7]

Phase composition was characterized by semi-quantitative x-ray diffraction (XRD) using a Panalytical X'pert Pro diffractometer. The phases were identified by the use of the Powder Diffraction File published by the International Centre for Diffraction Data.

Physical and mechanical properties were measured with the test methods shown in Table II. Density was measured only once per material on the bulk billet of composite. Young's modulus was measured at three locations on each billet. Flexural strength and fracture toughness were measured with five to ten samples using a Sintech universal test frame in conjunction with Test Works materials testing software. Hardness was measured on a ground surface using a minimum of eight indents. Microstructures were evaluated using a Leica D 2500 M optical microscope and the Clemex Vision PE imaging software.

Table II: Test Methods

Property	Test Description	Test Procedure
Density	Water Immersion	ASTM B 311
Young's Modulus	Ultrasonic Pulse-Echo	ASTM E 494
Flexural Strength	Four-Point Bend	ASTM C 1161
Hardness	Rockwell C	ASTM E 18
Fracture Toughness	Four-Point Bend Chevron Notch	ASTM C 1421

RESULTS AND DISCUSSION

A summary of property data is provided in Table III. In general, density decreases as the content of the low density Mg phase increases (densities of Al and Mg are 2.70 and 1.74 g/cc, respectively). Moreover, density increases as reaction time increases, suggesting conversion of Mg to higher density binary and ternary compounds. Despite the fact that Al has a higher Young's modulus than Mg (70 vs. 44 GPa), Young's modulus initially increases as the Mg content increases, with a maximum value obtained with the 75Al-25Mg alloy. In addition, Young's modulus increases with reaction time. This behavior is shown graphically in Figure 3. These findings indicate that the addition of Mg to Al alloys increases reactivity with B$_4$C, thus allowing more rapid formation of high modulus binary and ternary reaction products. Although less consistent, the hardness data follow the Young's modulus results. Increased hardness is expected as a result of reduced metal content and increased content of hard reaction products. Thus, as expected, the higher reaction time led to higher hardness values (with the exception of the 75Al-25Mg alloy). Furthermore, the highest hardnesses were obtained in the 75Al-25Mg alloy system, again suggesting rapid reaction kinetics between this alloy and B$_4$C.

The strength and toughness data follow the indication of higher reaction rates with the Al-Mg mixtures, as opposed to the Al and Mg alloy end points of the study. Specifically, the composites made with the Al-rich alloy (AA518) and the Mg-rich alloy (AZ91A) have relatively high strengths and high fracture toughness values. This suggests a high residual metal content. On the other hand, the composites made with the Al-Mg mixtures have lower strengths and ceramic-like fracture toughness values, indicating a low residual metal content. Finally, consistent with previous findings, fracture toughness dropped as reaction time was increased, indicating reduced metal content through reaction to harder phases.

Table III: Summary of Property Data

	10 hr. Reaction Time				
	Density (g/cc)	Young's Modulus (GPa)	Hardness (R$_C$)	Flexural Strength (MPa)	Fracture Toughness (MPa-m$^{1/2}$)
92Al - 8Mg	2.58	194	30	305	11.1
75Al - 25Mg	2.59	250	47	157	4.9
50Al - 50Mg	2.48	208	---	---	---
25Al - 75Mg	2.31	169	29	178	6.0
8Al - 92Mg	2.21	158	21	283	10.0

	40 hr. Reaction Time				
	Density (g/cc)	Young's Modulus (GPa)	Hardness (R$_C$)	Flexural Strength (MPa)	Fracture Toughness (MPa-m$^{1/2}$)
92Al - 8Mg	2.64	228	40	274	9.4
75Al - 25Mg	2.66	273	44	179	5.0
50Al - 50Mg	2.56	240	---	---	---
25Al - 75Mg	2.44	195	39	193	5.4
8Al - 92Mg	2.21	164	23	239	8.7

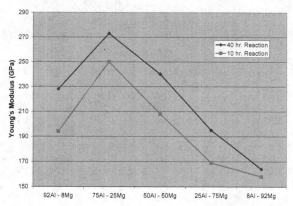

Figure 3: Effects of Al:Mg Ratio and Reaction Time on Young's Modulus of Al-Mg-B$_4$C Composites

The XRD results are provided in Table IV. These results are semi-quantitative and are valuable for comparison. However, the results cannot be taken as fully quantitative (e.g., the residual B$_4$C content is over-estimated as the starting content was only 50 vol. %). Multiple different ternary compounds were formed in the various experiments. Thus, to ease comparison, the data were simply segregated as Al-B-C, Al-Mg-B and Mg-B-C. With that said, the desired Al-B-C and Al-Mg-B phases seen in prior art [1-5] were successfully formed.

The XRD results follow the property data very well. In virtually all cases, the increased reaction time led to an increase in reaction products (Al-B-C, Mg-B-C, Al-B and Al-Mg-B) and a decrease in starting constituents (B$_4$C, Al, Mg). Moreover, the reaction rate with the commercial Al (92Al-8Mg) and Mg (8Al-92Mg) alloys was much slower than with the Al-Mg mixtures (75:25 and 25:75). In the case of the mixtures, a far greater content of reaction products formed during the allotted 10 and 40 hr. reaction times. In particular, the 75Al-25Mg alloy that resulted in the highest Young's modulus and hardness values also shows the highest content of reaction products.

Table IV: Semi-Quantitative XRD Results (data in wt. %)

	92Al - 8Mg		75Al - 25Mg		25Al - 75Mg		8Al - 92Mg	
	10 hrs.	40 hrs.	10 hrs.	40 hrs.	10 hrs.	40 hrs.	10 hrs.	40 hrs.
Al	20	11	5	6			2	
Mg				2	13	9	12	14
Al-Mg		1	9	2	7	6		1
B$_4$C	69	65	51	44	68	50	79	70
Al-B-C	5	11	20	23	1	1		3
Mg-B-C		1			12	32	4	10
Al-B	2	2					3	2
Al-Mg-B	4	11	16	23		2		

Representative optical photomicrographs are provided in Figures 4 and 5. The commercial Al alloy (92Al-8Mg) system is shown in Figure 4. Per Table IV, this alloy resulted in little formation of reaction products, particularly at the short reaction time. The photomicrograph clearly shows the B$_4$C particles and the metallic matrix, with a small content of reaction products. The level of reaction products increases with reaction time. The images in Figure 5 are very different. This figure shows the 75Al-25Mg alloy system that generated the most reaction products per the XRD data (Table IV). The photomicrographs show very high levels of reaction products, both in the matrix and at the surfaces of the B$_4$C particles. Again, a significant increase in the level of reaction products is seen at the higher reaction time. If the goal is the development of a fully ceramic-like composite, it should be possible to remove all the metal phase via reaction by increasing reaction time, increasing reaction temperature, decreasing B$_4$C particle size (i.e., more surface area for reaction) and/or increasing green density of the starting B$_4$C preform.

Figure 4: Optical Photomicrographs of 92Al-8Mg System (dark phase is B₄C particles, bright
phase is metallic matrix, and remaining phases are reaction products)
Left: 10 hr. Reaction
Right: 40 hr. Reaction

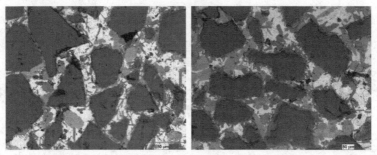

Figure 5: Optical Photomicrographs of 75Al-25Mg System (dark phase is B₄C particles, bright
phase is metallic matrix, and remaining phases are reaction products)
Left: 10 hr. Reaction
Right: 40 hr. Reaction

SUMMARY
      Composites in the Al-Mg-B₄C system were fabricated and exposed to various reactive
heat treatments. Key variables studied were Al:Mg ratio and reaction time. Microstructure,
properties, and chemistry were evaluated. Key observations can be summarized as follows:
   1. Al-Mg/B₄C composites with a wide range of Al:Mg ratios could be successfully
       fabricated with the pressureless molten metal infiltration process.
   2. Low temperature (700°C) reactive heat treatments were successful in driving Al-Mg-B-C
       reactions, leading to a conversion from metal-like to ceramic-like properties. The desired
       Al-B-C and Al-Mg-B phases from the prior art were formed.
   3. The use of Al-Mg mixtures as opposed to commercial Al-rich and Mg-rich alloys led to
       much greater reaction rates. This provides the opportunity to fully remove all metallic
       phases through reaction.

Future variables for study include B$_4$C particle size, B$_4$C loading in original preforms, the addition of additives to Al-Mg alloy (e.g., Si), the addition of additives to B$_4$C preform (e.g., B), and reactive heat treatment temperature.

REFERENCES
1. D.C. Halverson, A.J. Pyzik, I.A. Aksay and W.E. Snowden, "Processing of Boron Carbide-Aluminum Composites," *J. Am. Ceram. Soc.*, **72** [5] 775-80 (1989).
2. A.J. Pyzik and D.R. Beaman, "Al-B-C Phase Development and Effects on Mechanical Properties of B$_4$C/Al-Derived Composites," *J. Am. Ceram. Soc.*, **78** [2] 305-12 (1995).
3. A.J. Pyzik, R.A. Newman, A. Wetzel and E. Dubensky, "Composition Control in Aluminum Boron Carbide Composites," *Mechanical Properties and Performance of Engineering Ceramics II: Ceramic Engineering and Science Proceedings*, **27** [2] 457-72 (2008).
4. B.A. Cook, J.L. Harringa and A.M. Russell, "Superabrasive Boride and a Method of Preparing the Same by Mechanical Alloying and Hot Pressing," United States Patent No. 6,099,605, Aug. 8, 2000.
5. B.A. Cook, J.L. Harringa and A.M. Russell, "Superabrasive Boride and a Method of Preparing the Same by Mechanical Alloying and Hot Pressing," United States Patent No. 6,432,855, Aug. 13, 2002.
6. M.K. Aghajanian, J.T. Burke, D.R.White and A.S. Nagelberg, "A New Infiltration Process for the Fabrication of Metal-Matrix Composites," *SAMPE Q,* **20**(4) 43-47 (1989).
7. M.K. Aghajanian, M.A. Rocazella, J.T. Burke, and S.D. Keck, "The Fabrication of Metal Matrix Composites by a Pressureless Infiltration Technique," *J. Mater. Sci., 26 447-54 (1991).*

COHESIVE STRENGTH OF DRY POWDERS USING RHEOLOGY

Nicholas Ku, Sara Reynaud, and Rich Haber
Rutgers University
Piscataway, NJ, USA

Chuck Rohn
Malvern Instruments Inc.
Westborough, MA, USA

ABSTRACT
    In many industries, powders are a vital step in the processing toward manufacturing the final product. A challenge arises in the ability to predict the transport properties of these powders. Poor powder flowabilty results in difficulties in processing steps such as mixing. The solution is the development of a test to simulate the flow behavior of dry powders. Powder flowabilty is controlled by interparticle cohesion. Particle size, shape, and bulk density will all affect the flowabilty of the powder. The use of a rheometer allows for the collection of viscoelastic data of the cohesive powders. From this data, cohesion is predicted by determining the shear strength of the powder. In this study, talc, kaolin clay, and alumina powders were measured, due to their importance to cordierite processing. A method was developed to allow for the measurement of the shear stress of a powder while having precise control over the axial load present. The cohesive energy densities (CED), a measure of flowabilty, of the three powders were measured at two axial loads: 20kPa and 40kPa. The results show talc having the highest CED, followed by kaolin clay, with alumina showing the lowest value.

INTRODUCTION
    Fine powders are extensively used in industrial processing and manufacturing. However, the cohesive powder behavior is not yet fully understood. Because of the tendency of a powder to aggregate, the behavior of powder is unpredictable[8]. There is a great need to understand the fundamental rheological properties of particles in order to predict the flow behavior and define basic constitutive relationship of stress and strain of bulk powders. Many theoretical models[1-4] have been proposed in the literature to explain certain bulk powder properties, but these models are not suitable to describe flow behavior. The development of a constitutive powder rheology model able to describe the powder behavior at the microscale would be very useful for the scale-up of industrial processes.
    Fine powders are not able to flow freely due to interparticle cohesion. This cohesive force between the powder particles is mainly van der Waals forces. When the size of the particles is very small, the impact of van der Waals forces on their behavior becomes significant. Adhesion between particles is of infinitely short range and acts only over the contact area. Particle cohesiveness is difficult to measure as it is influenced by many factors such as particle size distribution, particle shape, surface roughness, surface free energy, hardness and elasticity[5].
    When viewed in bulk, a powder can support a finite stress without continuously deforming. This static state occurs provided a critical stress state is not exceeded. Once the yield stress is exceeded, the powder will continuously deform without any further increase in stress. At this high shear rate regime, individual powder particles are energized enough for the powder assembly to act as a freely flowing liquid. The kinetic energy of the powder is then sufficient to overcome the cohesive binding energy[6].
    The interparticle cohesive forces can be divided into frictional forces, which affect moving particles sat contact points between particles, and adhesion forces, which bind particles together and provide a pulling resistance as particles move in relation to each other. Both these forces will be dependent on the number of contact points with each particle's nearest neighbors as well as the normal

force acting on each particle at those contact points[7]. Therefore, the cohesive binding energy of a bulk powder is dependent upon the compaction pressure exerted on the powder. As the yield strength of a bulk powder is measured as a function of compaction pressure, strength is gained with consolidation pressure but levels off at high pressures. This is due to the fact that as compaction pressure increases, the packing fraction of the powder changes, increasing the number of contact points between neighboring particles. With more contact points, the cohesive forces increase. Also, with increased pressure, the normal force between the particles would increase, increasing the frictional force between particles. The yield strength of the powder levels off at high pressures as the powder reaches its maximum packing volume and change in number of nearest neighbors remained constant[8].

Angle of repose is another technique used to characterize the flowability of powders. The method measures the angle of inclination of the free surface to the horizontal of a bulk solid pile. Many of the flowability measurement methods offer a well-defined scale of measuring flowability but the powder under processing parameters may act differently than what is described by the different methods. One of the main problems occuring during the diagnosis of the flowability of powders is the compressibility of the material. For example, an angle of repose test does give a measurement for flowability, but under different conditions, such as large quantities of powder in a hopper, where powders are under compaction, the flowability of these powders can be different. The reason for this change is that powder flowability is affected by the cohesive strength of the powder, which changes with compaction, or the axial load on the powder. To fully understand the flow behavior of a powder, the relationship between the cohesive strength of the powder as a function of the consolidating pressure must be developed[9].

To understand the flowability of powders, an understanding of the interaction between the particles is needed. In order for a bulk powder to exhibit fluid motion, the individual particles must break the cohesive forces between itself and its nearest neighbors. The purpose of this measurement is to provide insight into the cohesive force between particles in a bulk powder. By using parallel plate rheometer, powder is sheared and the critical shear stress and cohesive energy density of the powder can be found. The cohesive forces within a powder increase as the pressure on the particles increase, so the gap between the plants are reduced to increase the axial load on the powder. By finding the critical shear stress and cohesive energy density of the powder for varying axial loads, a relationship can be developed between the pressure placed on a powder and its flowability. The purpose of this paper is to develop the methodology of such a measurement to study the flowability of powders.

EXPERIMENTAL

Three different ceramic powders were measured in this study. Talc, kaolin clay, and alumina were chosen for their importance to the ceramic industry as well as there difference in morphology. To prevent ambient moisture from affecting the results of the measurement, the powder was first dried. The powders were placed in a 100°C oven for at least 24 hours to insure they are bone dry before being tested.

The measurement was conducted on a Malvern Instruments - Bohlin Gemini HRnano Rheometer, equipped with a Peltier plate system which allows for accurate temperature control during the measurement. The geometry used was a 15-mm diameter smooth parallel plate. The volume under the parallel plate and the displacement of 600 microns was set as the standard volume, calculated to be a volume of 1.06E-7 m^3. The bulk density of the powder was found using the tapped bulk density procedure (ASTM D7481-09). By multiplying the bulk density of the sample by the standard sample volume, the mass of sample to be used was found.

A ring with 15.06mm diameter and 3mm height was made to fit around the geometry and contain the powder under the geometry, but at the same time allow enough clearance to not add additional frictional forces to the rheometer geometry. The containment ring was centered on the plate so that the geometry freely spins without being obstructed by the ring. The correctly measured amount of the powder sample was then placed inside the ring and the upper plate was slowly rotated and lowered onto the powder. The powder sample after being placed between the plates is not flat, and contains many hills and valleys. Rotating the upper plate at the point of contact shifts the powder surface and produces a test specimen that is uniform in height and distributes the load evenly across the sample. The geometry is lowered until the powder is pre-compacted to a pressure of about 25MPa. The plate is then raised until the axial load on the sample is removed, then relowered until contact is made with the powder. At this point, the sample is ready for oscillation sweeps to be run. Oscillatory tests were performed at very low strains ranging form 1E-06 to 1E-04. The measurements were made at 1 Hz and a temperature of 25°C.

The goal of this technique is to find a relationship for the strength of the powder. A simplified approach directly correlates the mechanical properties of the powder to the theoretical cohesive strength[11]. The theoretical cohesive energy density, CED, is calculated as the product of shear modulus and the square of the critical strain. The CED represents the energy density per unit volume developed by mechanical loading for deformation failure.

$$CED = G' * \gamma_c^2 \tag{1}$$

where G' is the elastic modulus, $_c$ is the critical strain, and CED is a function of pressure. The CED is therefore a value representing the energy needed to separate particles from one another.

RESULTS

A study was conducted to find the CED of three different powders: talc, kaolin clay, and alumina. To understand the difference in morphology of the powders, SEM images were taken of the powder. As particle shape has an impact on the cohesive strength of a powder, SEM images of the powder were taken to visually observe the powder in detail. The SEM used was a Zeiss Sigma Field Emission SEM. The powder was adhered to the sample stud using a colloidal Ag paste to prevent charging. Images were taken at 2,000x using a beam current of 3.00kV with the Everhart-Thornley Detector.

Figure 1. SEM image of talc powder at 2,000x magnification.

From viewing the SEM image in Figure 1, insight into the morphology of the talc is present. The talc is microcrystalline talc, with a large particle size distribution. The largest powders are easily 20 microns in length. Talc, being a sheet-silicate, has a tablet-like structure. Each tablet is not only different in length, but also thickness, with some particles seemingly comprised of only a few layers.

Figure 2. SEM image of kaolin clay powder at 2,000x magnification.

The kaolin clay, seen in Figure 2, also a sheet-silicate, has a similar particle shape to the talc. The particles are tablet s comprised of multiple layers. The kaolin overall has a much smaller particle size than the talc, but the same occurrence of particles with varying layer thickness does occur.

Figure 3. SEM image of alumina powder at 2,000x magnification.

Figure 3 shows the alumina particle, and the image is of a single large aggregate. The particle is relatively spherical compared to the tablet-shaped talc and kaolin. By having a such a different particle shape, it is expected the packing fraction to be very different. As cohesive strength of a powder is very dependent on the packing fraction, it can be expected that the CED of the alumina should show a difference from the other two powders.

The amplitude sweeps for the three powders were conducted. The log of the elastic modulus was then graphed as a function of the shear stress. Using Equation 1, the cohesive energy density of the powder was then calculated. The sweep was made at two different axial pressures, 20 kPa and 40 kPa, to help gain an insight into the effect of compaction on the CED.

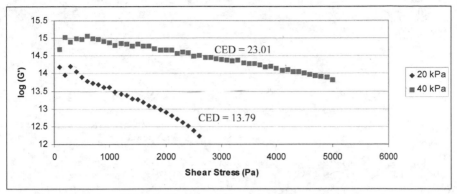

Figure 4. Elastic modulus curves of talc powder.

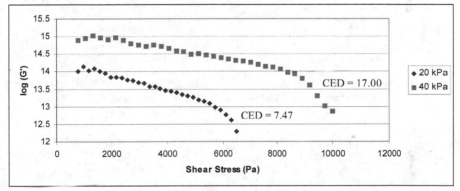

Figure 5. Elastic modulus curves of kaolin clay powder.

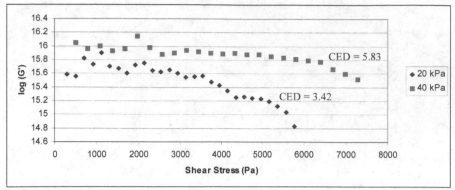

Figure 6. Elastic modulus curves of alumina powder.

Figures 5 to 6 show the rheological data taken from cohesive strength measurement, the CED, measured at the critical point of each sweep, is shown next to each curve. There is a trend that higher axial pressures yield a higher CED. As stated in Johanson[8], the increased axial pressure would decrease flowability of a powder, which correlates to in increase in CED. The CED of a powder can be used as a means to measure the flowability of a powder, with a high flowability corresponding to a low CED.

CONCLUSIONS
The cohesive energy density of a powder is the energy needed to separate to particles in contact with one another. By using the presented rheological measurement method, the CED can be used as a means to measure the flowability of a powder. Factors that affect flowability of a powder, such as particle composition, shape, size distribution, and compaction, all show an effect on the measured CED. Future work is needed to better understand the relationship of these factors, but the measurement provides a method to quantitatively measure the flowabilty of a powder.

REFERENCES
[1]Aydin I, Briscoe BJ, Sanliturk KY. The internal form of compacted ceramic components: A comparison of a finite element modelling with experiment. Powder Technology 1996;89:239.
[2]Cazacu O, Jin J, Cristescu ND. A new constitutive model for alumina powder compaction. KONA Powder and Particle 1997;15:103.
[3]Feise H, Schwedes J. Investigation on the behavior of cohesive powder in the biaxial tester. Partec 95, 3rd European Symp. on Storage and Flow of Particulate Solids. Nurnberg, Germany, 1995. p.119.
[4]Gethin DT, Lewis RW, Ariffin AK. Modeling compaction and ejection processing in the generation of green powder compact. ASME IMECE, vol. 216: AMD, 1995. p.27.
[5]Puri VM, Tripodi MA, Manbeck HB, Messing GL. Constitutive model for dry cohesive powders with applications to powder compaction. KONA Powder and Particle 1995;13:135.
[6]Chen YH, Jallo L, Quintanilla MAS, Dave R. Characterization of particle and bulk level cohesion reduction of surface modified fine aluminum powders. Colloids and Surfaces A: Physiochem. Eng. Aspects 2010;361:66.
[7]Klausner JF, Chen DM, Mei RW. Experimental investigation of cohesive powder rheology. Powder Technology 2000;112:94.

[8]Johanson K. Effect of particle shape on unconfined yield strength. Powder Technology 2009;194:246.

[9]Ileleji KE, Zhou B. The angle of repose of bulk corn stover particles. Powder Technology 2008;187:110.

[10]Bumiller M, Carson J, Prescott J. A preliminary investigation concerning the effect of particle shape on a powder's flow properties. Malvern Instruments, Inc.

[11]Rohn CL. Analytical Polymer Rheology: Structure-Processing-Property Relationships. Cincinnati, OH: Hanser Gardner Publications, 1995.

EFFECT OF HEAT TREATMENT ON THERMAL PROPERTIES OF PITCH-BASED CARBON FIBER AND PAN-BASED CARBON FIBER CARBON-CARBON COMPOSITES

[a]Sardar S. Iqbal[1], [b]Ralph Dinwiddie, [b] Wallace Porter, [b] Michael Lance, [a]Peter Filip

[a] Center for Advanced Friction Studies, Mech Engg. & Energy Processes, Southern Illinois Univ. Carbondale, IL 62901

[b] Oak Ridge National Labs, Oak Ridge, TN

ABSTRACT

Thermal properties of two directional (2D) pitch-based carbon fiber with charred resin and three directional (3D) PAN-based carbon fiber with CVI carbon matrix C/C composite were investigated for non-heat treated (NHT) and heat treated (HT) materials through the thickness (z-direction). Heat treatment was performed at 1800, 2100 and 2400 °C for 1-hr in inert argon atmosphere. Thermal diffusivity, heat capacity and bulk density were measured to calculate thermal conductivity. Thermal diffusivity and conductivity was the highest for 3D C/C heat treated at maximum temperature with non-heat treated one exhibiting the lowest thermal conductivity. Similarly, 2D C/C heat treated at maximum temperature exhibited the highest thermal diffusivity and thermal conductivity. Polarized light microscopy (PLM) images of HT C/C show a progressive improvement in microstructure when compared to NHT C/C. However, HT 2D and 3D C/C composites exhibited extensive shrinkage of charred resin and CVI carbon matrix, respectively, from fibers resulting in intra and inter-bundles cracking when compared to NHT one. Raman spectroscopy and XRD results of NHT and HT C/C indicated increased ordering of structure. A progressive improvement in thermal properties was observed with increased heat treatment temperatures.

INTRODUCTION

Carbon-carbon (C/C) composites are the most valuable ceramic composites. They possess excellent specific thermal, mechanical properties. They exhibit high thermal properties, strength and modulus, high friction coefficient and low wear from room temperature to high temperatures [1]. Carbon-carbon composites retain their properties at temperatures where most of the high end alloys give in. The properties of C/C are dependent on various factors such as fiber orientation distribution, fiber types, matrix type, microstructure, defects, fiber/matrix interface, and pores/voids [2].

Due to the excellent specific mechanical, thermal, wear, and frictional properties, C/C composites have been the best candidates in today's brake industries in aviation as well as in selected automotive industries. The use of C/C as brake materials also includes high speed train, racing cars, motorcycles and tanks [3,4]. For multidirectional composites such as 2D and 3D, the major interest lies in the mechanical and thermal properties of C/C composites in order to survive the high pressure and temperature seen in use as brake materials. The mechanical and thermal properties are governed by factors such as microstructure, fiber orientation, architecture, matrix type, fiber/matrix bonding, density, defects and porosity, and strongly depend on heat treatment temperature and oxidation effects [2]. All these factors strongly influence the tribological behavior of the materials.

The present study includes the effect of heat treatment on thermal properties of 2D pitch-based carbon fiber, and 3D PAN-based carbon fiber C/C composites.

---

[1] Corresponding author: sarwatiq@yahoo.com, +16183193963

EXPERIMENTAL PROCEDURES

Two different types of C/Cs, a two-directional "2D" C/C (material A) consisting of randomly oriented chopped pitch-based carbon fibers and charred resin matrix (sample A) , and a three-directional "3D" C/C (material D) made of continuous-non-woven ex-PAN (polyacrylonitrile based) carbon-fibers and CVI-matrix were studied. The samples from the discs were obtained through the thickness direction to the disc.

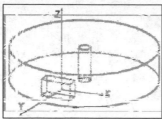

Fig.1 Schematic showing orthogonal reference axes X, Y and Z for tangential, radial and through the thickness directions, respectively.

The samples were heat treated at 1800, 2100 and 2400 °C in graphitization furnace (Model: TP-4X10-G-G-D64A-A-27, Centorr Associates Inc). Heating was performed at the rate of 10°C/min for 1 hour in an Argon atmosphere. The samples were designated corresponding to their heat treatment temperature.

Table I. C/C sample designation according to heat treatment temperature.

Sample name		Heat treatment temperature (C°)
2D C/C	**3D C/C**	
ANHT	DNHT	Non-heat-treated
A18	D18	1800
A21	D21	2100
A24	D24	2400

The microstructure and extinction angles (A_e) of different pyrocarbons were inspected using Nikon Microphot-FX and Nikon Eclipse LV 150 polarized light microscopes. Samples were ground using different grit sizes of sand papers (120-1200 μm) followed by polishing in diamond slurries with grain size down to 1 μm. Final polishing was performed by using 50 nm particle size alumina on micro cloth.

Raman spectra were acquired with a Dilor XY800 Raman Microprobe (Horiba, Edison, NJ) with an Innova 308c Ar ion laser (Coherent, Inc., Santa Clara, CA) operating at 5145Å and an attached microscope. Four peaks (and a baseline) were fitted to each spectrum; the D and D' bands (at ~1350 and ~1620 cm^{-1}, respectively) the G band (at ~1580 cm^{-1}) and one for the background oxygen vibrational mode at 1555 cm^{-1}.

Thermal properties were measured according to ASTM C1470 to measure thermal diffusivity, specific heat capacity, and thermal conductivity.

RESULTS AND DISCUSSION

Fibers act as a reinforcing agent. Fiber orientation causes a considerable anisotropy in a number of physical properties. Histograms indicating fiber orientation distribution of materials A and D are shown in Fig. 2a and b, respectively.

Histogram of sample A along z-direction is not shown due to its 2D architecture and the absence of fibers oriented in the z- direction. Histogram for fiber oriented through thickness direction (z-direction) of material D shows that there are close to 35% of fibers oriented at lower angles, with a relatively large proportion (~ 55%) oriented at higher angles ($60° < \theta < 90°$) (Fig. 4-1b).

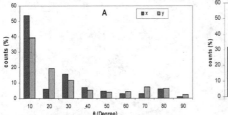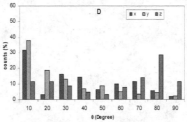

Fig.2 Histograms of material A along x and y directions (a), and material D along x, y and z directions (b).

Figure 3 shows the PLM images of material A before and after heat treatment at 2400 °C. Material A before heat treatment (HT) has fewer intra and inter-bundle cracks. The microstructure of carbon matrix after HT is still isotropic (ISO) and the areas adjacent to fibers exhibit graphite-like microstructure enveloping the fibers. The 'graphitization' of 'non-graphitizing carbon' is assumed to occur due to high stresses leading to 'stress graphitization' of charred resin matrix [5-8]. After HT at 2400 °C, the intra/inter bundle crack density increases. Also, the enlargement and expansion of pre-existing cracks and voids/porosity was detected. Apparently, the matrix and fiber shrink and retract with respect to each other after heat treatment. Another pertinent effect of HT is the inter-connection of the pre-existed voids which have expanded into larger voids.

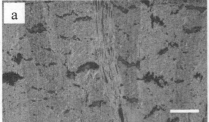

Fig.3   PLM images of material A before (a), and after (b) heat treatment at 2400 °C showing the cracks appeared after HTT. The micron bar is 800 μm.

Figure 4 shows the PLM images of material D before and after heat treatment (HT) at 2400 °C. The rough laminar (RL) microstructure of CVI becomes highly anisotropic with extinction angle, A_e, increased to 22° when compared to the non-heat treated (NHT) C/C material D exhibiting 18°. An extensive intra and inter-bundle crack initiation and propagation is evident after HT in Fig. 4b.

Additionally, the voids and porosity become larger after HT. The intra bundle cracks as well as the inter bundle cracks are obvious after HT in PLM images. Also the pores and voids become larger after HT. Mirasol et al. [9] reported an increase in the porosity with heat treatment and attributed it to the widening of micro pores. Apparently, the shrinkage of matrix and related stress generation at fiber/matrix interface rendered the material with higher crack density. The shrinkage of the CVI matrix is more pronounced when compared with PAN fibers.

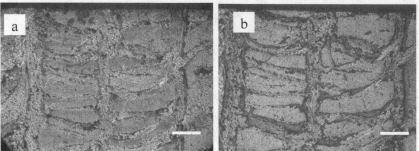

Fig.4 PLM images of material D before (a), and after (b) heat treatment at 2400 °C showing the cracks appeared after HTT. The micron bar is 800 µm.

Figure 5 shows the Raman spectra for non-heat treated (NHT) and heat treated (HT) C/C materials A and D. The Raman spectra show three distinct peaks: one at 1354 cm^{-1} (I_D), a second at 1554 cm^{-1}, and third one at 1580 cm^{-1}. The peak at 1354 cm^{-1} (I_D) is related to the less ordered carbon corresponding to the finite size of the crystallites and the to the defects in the carbon fibers and matrix. The peak at 1580 cm^{-1} (I_G) corresponds to highly organized graphite-like carbon. The peak at 1554 cm^{-1} is not related to the structural changes, but is due to oxygen.

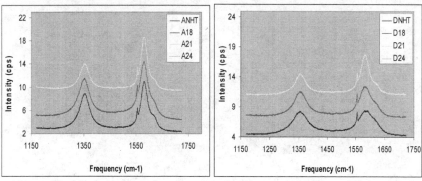

Fig.5 Raman spectra of non-heat treated and heat treated C/C materials A and D.

Figure 6 gives I(D)/I(G), the ratio of the intensity of D and G bands, for C/C plotted as a function of heat treatment temperature. The symbols indicate the mean values of 10 measurements on each sample and error bars denote the standard deviations. The ratio of the intensity of D to G band shows a progressive decrease with increasing heat treatment temperature, which indicates that the heat treatment is increasingly organizing the carbon microstructure into a turbostratic, graphite-like structure. However, D24 appeared more organized and ordered (heat treatment temperature of 2400 °C) than material A24, which could be attributed to the presence of "graphitizeable" rough laminar CVI carbon in material D when compared with the isotropic microstructure of charred resin matrix in material A. The microstructure of CVI carbon changes more readily from the glassy to trubostratic carbon or to graphite-like microstructure [10], with micro crystallites growing in size [11], however, charred resin matrix is less likely to organize after heat treatment.

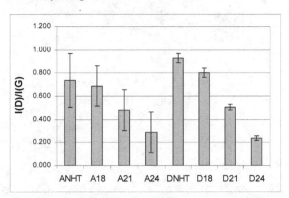

Fig.6 Ratio of the I(D)/I(G) of the D and G bands for C/C materials A and D as a function of heat treatment temperature. Error bars represent the standard deviation of ten measurements at one specimen.

Figure 7 shows thermal diffusivity of material A through the thickness direction (z). Sample A21 exhibited the highest value of thermal diffusivity with respect to the temperature, followed by A24. ANHT exhibited the lowest thermal diffusivity. It appears that high heat treatment temperature (HTT) has a beneficial effect on materials A. A24 has increased crack density, and dimensions of voids and pores when compared to ANHT as obvious from the PLM images. The heat treatment may have resulted in the interconnection of micron and submicron pores and voids and introduced extensive network of cracks. Additionally, A21 may have less pores, voids, and crack density than A24 which provides for the higher thermal diffusivity and conductivity.

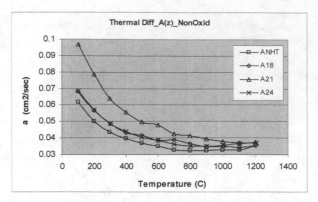

Fig.7 Thermal diffusivity of non-heat treated and heat treated C/C material A.

Figure 8 shows thermal diffusivity of materials D. Thermal diffusivity of D24 through the thickness direction (z) is the highest followed by D21 with DNHT and D18 exhibiting the lowest value of thermal diffusivity. Heat treatment at 2400 °C has resulted in the introduction of increased crack density and enlargement and expansion of pores/voids.

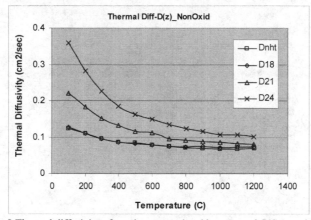

Fig.8 Thermal diffusivity of non-heat treated and heat treated C/C material D.

Figure 9 shows the specific heat capacity of material A and D for samples ANHT, A24, DNHT and D24. It is obvious that the heat capacity of materials A and D does not change with the rise in temperature. The heat capacity of A24 is approximately 2% higher than ANHT. Similarly heat capacity of D24 is only 1% higher than DNHT.

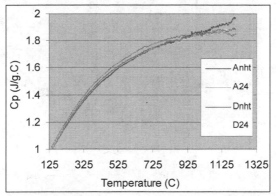

Fig.9 Specific heat capacity of non-heat treated and 2400 °C heat treated C/C materials A and D.

Figure 10 shows thermal conductivity of non-heat treated and heat treated C/C materials A. The increase in thermal conductivity is due to the increase in the crystallinity of the HT materials when compared to the NHT material A. It is pertinent to note that the A24 after heat treatment experiences more intra and inter bundle cracks, expansion and enlargement of the pre-existing cracks and pores. Thermal conductivity of ANHT drops exponentially as temperature increases. At the maximum temperature, thermal conductivity of A24 is higher than that of ANHT with A21 and A18 in intermediate range. The charred resin matrix is less likely to organize after high temperature heat treatments [1]. However, isotropic nature of charred resin matrix in comparison to anisotropic nature of pitch-fiber in expansion and shrinkage, the bonding at temperatures 2100 and 2400 °C may have decreased at the interface. However, intra-bundle cracks, more evident after heat treatment than the cracking of the matrix as shown in PLM images of Fig. 3b.

Although, the space between the pitch-fiber is filled with isotropic microstructure, the difference in coefficient of thermal expansion (CTE) between matrix and pitch-fiber thermal behavior at higher temperature become pronounced resulting in a sharp decrease in thermal conductivity. Since heat transfer mechanism in C/C is highly dependent on microstructure of matrix, fibers and its orientation [11-13]. Mirasol et al. [9] reported about the effect of heat treatment (1000 and 2900 °C.), and attributed an increase in the development of narrow constriction (closed porosity) to increasing temperatures.

It is also obvious that the charred resin carbon matrix is neither a major factor nor does play a significant part in transmitting heat energy when compared to the fiber orientation distribution [14, 15], despite the fact that pitch based carbon fibers are known to possess highly ordered graphitic structure, and lower out-of-plane (z-direction) thermal diffusivity than in-plane thermal diffusivity demonstrates the indispensible importance of fiber orientation distribution on thermal properties.

Too much cracking due to shrinkage of fiber and matrix may have also occurred incase of A24 in comparison to A21, however, the isotropic microstructure which does not graphitize at temperatures up to 3000 °C [1, 16].

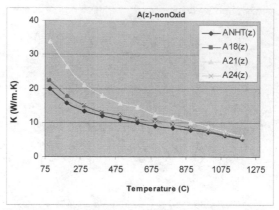

Fig.10 Thermal conductivity of NHT and HT C/C materials A.

Figure 11 shows thermal conductivity of non-heat treated and heat treated C/C materials D.

Thermal conductivity of DNHT and D18 are the lowest. Thermal conductivity increases substantially with heat treatment temperature. Thermal conductivity of D24 is the highest followed by D21. The increase in thermal conductivity is strongly dependent on fiber orientation distribution, fiber and matrix types. It is pertinent to note that D24 exhibits the highest value of thermal conductivity despite having a larger crack density, increase and enlargement of pores and voids after heat treatment as evident from PLM images (Figs. 4b). It is also obvious that the heat treatment has developed the microstructure of C/Cs with interplanar distance decreasing, crystallite thickness and size increasing, which have caused the materials to exhibit improvement in thermal conductivity with highest heat treated C/C.

It has also been reported that graphitization can be triggered in the matrix when the carbon–carbon composite is heated to significantly high temperatures [9, 17]. This transformation was suggested to be related to the accumulation of stress in the matrix during carbonization, when the matrix and fibers undergo different extents of shrinkage, with the former experiencing a larger shrinkage than the latter [9]. In effect, a highly stressed region around the reinforcing fillers develops. These stresses become relieved when the composite is treated at higher temperatures and is accompanied by the rearrangements of the graphene sheets into a more ordered structure, i.e. formation of graphite or near-graphite (or anisotropic) structure around the reinforcing fibers [10, 18, 19]. Setton et al. [20] reported that properties of CVD based carbons are highly anisotropic with thermal properties highest along the graphene plane and weakest perpendicular to the planes. They attributed it to the preferred orientation of the layers parallel to the fiber surface. It is evident for the 3D C/C with CVI carbon matrix. The RL CVI matrix possesses the longest and the narrowest crystallites, whereas the SL matrix contains less ordered smaller crystallites.

Jie et al. [21] reported the bonding between CVI pyrocarbon and carbon as an important factor on thermal behavior. They attributed a strong interface boding between fiber and matrix to higher thermal properties, in addition to higher crystallinity of CVI carbon matrix C/C.

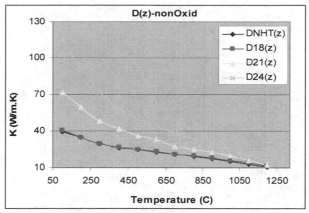

Fig.11 Thermal conductivity of NHT and HT C/C materials D.

Overall, thermal conductivity of 3D material D along z-direction is the highest than 2D material A, and it is attributed to the higher percentage of fibers oriented along z-direction when compared to materials A which has null fibers oriented along z-direction. Also, it is interesting to note that the 2D material A having pitch-based carbon fibers exhibited much lower thermal conductivity than 3D PAN-based C/C material D, since PAN-based fibers are turbostratic with no organized structure when compared to pitch-based C/Cs which are reported to exhibit increasing order of microstructure with heat treatment.

The heat transfer mechanism in C/C is controlled by interaction of phonons with structural and lattice defects. Therefore, a good heat transfer requires not only highly oriented fibers but also highly oriented graphite-like or graphitic phases (consisting of highly aligned graphene sheets) within fibers and matrix [10, 12, 13, 23, 24].

CONCLUSIONS

The impact of heat treatment on thermal conductivity of two different C/Cs, a 2D pitch-based C/C and 3D PAN-based C/C materials were studied in comparison to non-heat treated ones.

Microstructure of C/Cs developed with increasing heat treatment temperature. Heat treatment increased crack density and pre-existing pore size. Specific heat capacity did not show any change due to heat treatment. In general, thermal conductivity of pitch-based carbon fiber C/C increased with heat treatment temperature. Material A heat treated at 2100 °C exhibited highest value of thermal conductivity followed by A24, which could be attributed to the introduction of increased crack density and probable shrinkage of matrix and fiber with respect to each other since 2D material A is composed of charred resin carbon matrix. Thermal conductivity of 3D C/C material D increased with heat treatment temperature in comparison to DNHT. However, substantial increase in thermal conductivity is observed for C/C D24 heat treated to the maximum temperature, and could be attributed to the highly ordered structure development with increasing heat treatment temperature.

ACKNOWLEDGEMENTS
Authors would like to acknowledge the Honeywell Aircraft Landing Systems (ALS) for material. This research was sponsored by the National Science Foundation (Grant EEC 3369523372), State of Illinois

and a consortium of 12 industrial partners of the Center for Advanced Friction Studies, Southern Illinois University at Carbondale (http://frictioncenter.siu.edu).

REFERENCES

[1]   Sardar Iqbal, Peter Filip, Through the thickness properties in relation to the fiber orientation distribution and microstructure. Structure-Property relationships in multi-functional materials, MS&T 2008, p.1206-21

[2]   Sardar S. Iqbal, Peter Filip, Tensile and compressive properties of 2D pitch-based and 3D PAN-based C/C composites in relation to fiber orientation distribution and microstrucutre, Processing and Properties of Advanced Ceramics and Composites II: MS&T 10, 205-217

[3]   Krenkel W, Berndt F. C/C–SiC composites for space applications and advanced friction systems. Mat Sci Eng A 2005; 412: 177–81.

[4]   G. Savag, Carbon-Carbon Composites, p197-223, 1st ed, 1997, Chapman and Hall, UK.

[5]   Xintao Li, Kezhi Li, Hejun Li, Jian Wei, Chuang Wang, Microstructures and mechanical properties of carbon/carbon composites reinforced with carbon nanaofibers/nanotubes produced in situ, Carbon., 45, 1662-1688, (2007).,

[6]   E Yasuda, S Kimura, Y Shibusa. Tans. Jap Soc. Com. Mat. 6. p.14

[7]   S Kimura, Y Tanabe, N Takase, E Yasuda. *J. Chem. Soc. Jap.* (1981), 9: pp1474.

[8]   R Sexena, RH Bragg. *Carbon* (1978); 16: 373.

[9]   JR. Mirasol, PA Thrower, LR Radovis, On the oxidation resistance of carbon-carbon composites: Importance of Fiber Structure for Composite Reactivity. Carbon (1995), 33(4), 545-554.

[10]   E Fitzer. Carbon fibers – Present state and future applications in carbon fibers, filaments and composites (eds) J L Figueiredo et al (Kluwer Academic).1989, pp 3–41

[11]   LIAO XiaoLing1, LI HeJun1†, XU WenFeng2 & LI KeZhi, The effect of applied stress on damage mode of 3D C/C composites under bend-bend fatigue loading. Science in China Series E-Technological Sciences | February 2007 | vol. 50 | no. 1 | 97-102.

[12]   Li H J. Carbon/carbon composites. New Carbon Mater, 2001, 16 (2): 79—80.

[13]   Pierson H O. Handbook of carbon, graphite, diamond and fullerenes, 1993, pp. 67-98 (Noyes).

[14]   G. Zak., Estimation of three-dimensional fiber-orientation distribution in short-fiber composites by a two-section method, J. of Comp. Mat., 35 (4), 316-339, (2001)

[15]   A Vaxman, N Narkis. A. Siegmann, S. Kenig. Polymer Composites. 10(84), 1989.

[16]   G. Savage, Carbon-carbon Composites. Chapman & Hall, USA, 178-237, (1993).

[17]   P.L. Walker, Hah. McKinstry, JV Pustinger. X-Ray diffraction studies of carbon gasification,. Industrial & Engineering Chemistry 46(8), 1651-1658, (1954)

[18]   Jun-Ming Su , Jun Yang, Zhi-Chao Xiao, Shao-Jian Zhou, Zhi-Gang Peng, Jian-Guo Xin, Rui Li, Mei Han, Sheng-Li Zhao, Li-Min Gu Xi "Structure and properties of carbon/carbon composite materials for aircraft brake discs", New Carbon Materials, 2006, 21(1): 81–89.

[19]   CR Thomas. Essentials of carbon-carbon composites, The Royal Society of Chemistry, Cambridge, 1993. pp 4-21].

[20]   Ralph Setton. Carbon, A fundamental element for research and its applications: In Carbon Molecules and Materials, Taylor & Francis, 2002;1-51

[21]   Chen Jie, Xiong Xiang, Xiao Peng, Zhang Hong-Bo, The catalytic effect of boric acid on polyacrylonitrile-based carbon fibers and the thermal conductivity of carbon/carbon composites produced from them,. Carbon 48; 2010: 2341-46

[22]   Burchell T D 1999 Carbon materials for advanced technologies (Pergamon)

[23]   Peter Morgan, Carbon Fibers and Their Composites, Taylor and Francis, FL, 2005, 586-591

Tribological Properties of Ceramics and Composites

CERAMIC FOAM / ALUMINIUM ALLOY INTERPENETRATING COMPOSITES FOR WEAR RESISTANCE APPLICATIONS

J. Liu, J. Binner, R. Higginson, C. Munnings
Department of Materials, Loughborough University
Loughborough, UK

ABSTRACT

Metal-ceramic interpenetrating composites (IPCs) consist of 3D-continuous matrices of discrete metal and ceramic phases. The goal is to develop superior multifunctional properties compared with traditional MMCs with a view to using these composites in applications including wear resistance, electrical components and light weight structural parts in industries including aerospace, automotive and defence, amongst others. In this paper, a pressureless infiltration technique was used to produce IPCs by infiltrating molten aluminium alloys into a range of gel-cast ceramic foams produced from alumina, mullite, and spinel. The wear rates of the composites were measured under dry sliding conditions. The wear mechanism was investigated and the effects of the foam density and cell size on the wear properties were determined. It was found that the alumina and spinel reinforcement-based IPCs performed better than those based on mullite. It was also found that all the foam-based IPCs were up to twice as wear resistant as MMCs made by infiltrating a bed of ceramic powder.

INTRODUCTION

Ceramic reinforced aluminium matrix composites (AMCs) have received substantial attention from the automotive and aerospace industries due to their light weight, high elastic modulus, improved strength and good wear resistance. The latter is one the most attractive properties of the AMCs in applications including automotive brake systems, callipers, pistons, cylinder liners, connecting rods and turbine compressors[1-4].

In contrast to other mechanical properties, the study of wear behaviour is quite empirically based as there are a wide variety of wear mechanisms resulting from the huge number of possible material combinations and processing techniques, and the test conditions are also an important parameter that influences the wear behaviour[5-9]. However, it is generally agreed that with an increase in the ceramic volume content, the wear resistance of the composites improves[5-11]. The wear mechanisms are broadly categorised into those that involve cracking, or essentially brittle fracture, or are dominated by plastic deformation[5]. The bonding between the metal and ceramic phases, which is affected by the processing method, plays an important role in determining the wear mechanism. Strong bonding improves the wear resistance as the wear is dominated by plastic deformation[8] and the material's hardness becomes the main factor that influences wear behaviour[12]. With weak bonding, the ceramic reinforcements are easily removed from the matrix leading to severe material loss[13].

Conventional AMCs are fabricated by adding ceramic particles or fibres into Al alloys, however, there is a volume restriction on how much can be added and dispersed uniformly. In contrast, 3-3 interpenetrating composites (IPC) consisting of 3 dimensionally continuous matrices of discrete metal and ceramic phases may provide advantages for wear resistance applications as they can offer a higher load bearing capacity than the conventional AMCs. Some researchers[2,5,14,15] have studied the wear behaviour of IPCs and compared it with conventional AMCs, finding that the former is significantly better. Imbeni et al.[2] reported that IPCs yielded a much lower wear rate than conventional AMCs when abraded by soft silica particles. Furthermore, a crossover was found with the conventional AMCs in that there was a load above which the wear rate of the AMC was higher than that of the unreinforced metal[6]. A transition from mild to severe wear in an AMC was also observed by

Zhang et al.[16], whilst Ceschini et al.[14] and Wang et al.[15] claimed that such a transition does not occur in IPCs.

As potentially attractive candidates for high performance wear resistance applications, metal-ceramic IPCs have been produced by a range of fabrication methods. One approach used a displacement reaction between a precursor and molten aluminium to form a continuous interpenetrating ceramic phase within a metal matrix, however the volume content of the ceramic phase was limited to ~70% with average pore sizes of ~5 μm[2,14]. Other researchers have infiltrated molten Al alloys into porous ceramic preforms with or without the assistance of pressure[15,17,18,19]. The fabrication of porous ceramic can be relatively straightforward, e.g. by making a green body by die-pressing or vacuum forming and then sintering to the desired density[19], or by foaming a ceramic slip[17] to achieve a very porous green body that is then sintered.

Chang et al.[20] used gel-cast alumina foams as the basis for making Al_2O_3/Al(Mg) interpenetrating composites and assessed their wear performance under dry sliding conditions. The results indicated that foam density and cell size were important parameters in determining the wear behaviour with composites with larger cell sizes offering better wear resistance. They also found that with 15% ceramic content, the IPC exhibited ploughing wear throughout the process with no obvious ceramic struts protruding out; whilst with 27% ceramic foam density, a transition from "ploughing" to "protective" wear of the IPC occurred after the initial ploughing stage, the ceramic struts protruding from the worn surface providing very effective protection to the soft metal.

In the present work, the objectives were to study the wear mechanisms of IPCs, and to investigate further the relationships between the wear properties and the characteristics of the precursor ceramic foams. In addition, IPCs have been made from a range of different ceramic foams including alumina, mullite and spinel, and the results compared with IPCs made from ceramic powders and fibres.

EXPERIMENTAL

Al-Mg alloys containing approximately 10 wt% Mg were produced from commercially pure Al and a Mg-Al master alloy, AZ81, by casting. Alumina (Al_2O_3), spinel ($MgAl_2O_4$) and mullite ($Al_6Si_2O_{13}$) foams produced by gel casting were produced by Dysons Thermal Technology (Dysons TT) Limited, UK[21]. These foams had spherical, open pores inter-connected by circular windows. The density of the foams varied from 15% to 40%, and the cell size was in the range of 100 to 500 μm for each of the different density foams. 30% dense alumina powder-based preforms were also provided by Dysons TT. They were made from a mixture of alumina fibres (SAFFIL) and alumina powder by first vacuum forming and then sintering.

Following the pressureless infiltration process developed by Chang[18], ceramic/metal assemblies were placed in a tube furnace as illustrated in Figure 1. Note that the assemblies were contained in an alumina boat; this prevented the molten alloy from making contact with the tube furnace walls in the event of problems during infiltration. When the infiltration was complete, as observed through a quartz glass window at one end of the tube furnace, the composite was cooled.

The wear resistance of the composites was assessed using the same method as Chang[20]; this involved using a linear reciprocating ball-on-flat method at ambient temperature in a dry environment, Figure 2. Composite samples measuring 35±2 mm × 15±2 mm × 8±2 mm (length × width × thickness) in size were ground on SiC paper from 120 to 1200 grit and then fixed in a sample holder. Following calibration to ensure the counter balance weight and the moveable weight (Figure 2) were level, the load was applied vertically downward onto the horizontal, flat sample through a 12 mm diameter tungsten carbide ball, which slid on the sample surface with a speed of 50 mms[-1] and a 25 mm stroke length. After the desired sliding distance, an electronic balance with a sensitivity of 0.1 mg was used to measure the weight loss of the sample and the wear rate was calculated by the equation:

$$Wear\ rate = \frac{m_{loss}}{\rho L} \qquad (1)$$

where m_{loss} is the weight loss, ρ is the density of the sample, and L is the sliding distance. At least three tests were carried out on each kind of sample under each set of conditions. Samples that have been tested are shown in Appendix.

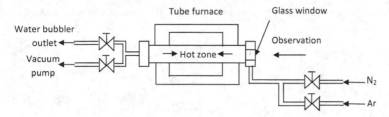

Figure 1: Schematic of the tube furnace setup.

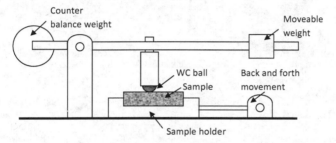

Figure 2: Schematic of the linear reciprocating wear test equipment[20].

To determine the bulk hardness of the interpenetrating composites, a standard Rockwell Hardness tester was used with a 1.6 mm (1/16 in.) steel ball. A pre-load of 10 kg was applied followed by a full load of 150 kg (corresponding to the G scale), and then the depth of penetration was noted. At least ten measurements were made on each sample.

A LEO VP 1530 FEGSEM was used to examine the microstructures of the ceramic preforms, the IPCs and the worn surfaces. The interpenetrating composite samples were metallographically polished using 6 μm and then 1 μm diamond paste before being examined in the FEGSEM.

RESULTS AND DISCUSSION

Microstructure of the ceramic preforms and the interpenetrating composites

SEM micrographs of 27% dense alumina foam with 200 μm average cell diameter and the 30% dense alumina powder/fibre preform are shown in Figure 3(a) and (b) respectively. The alumina foam had the typical structure produced using the gel casting technique, with open spherical cells that were connected by circular windows. The alumina powder/fibre preform also had open porosity, but its pores were much smaller than those in the foam and the bonding between the alumina particles were weaker than those in the foams. Examination of both types of preform revealed macroscopically isotropic microstructures, thus isotropic properties were anticipated in the final composites. The spinel and mullite foams had fundamentally the same structure as the alumina foams, although different features were found in the ceramic foam struts at higher magnifications, Figure 4. It can be seen that whilst the alumina foam strut was dense, which yielded good mechanical properties, the spinel foam strut was more open and made of larger particles. The mullite strut was also dense with few pores and a glassy phase can also be observed, which is believed to be residual liquid phase present during sintering.

Figure 3. SEM micrographs of (a) 30% dense alumina foam with 160 μm average cell size; (b) 30% dense alumina powder/fibre preform.

Typical SEM micrographs of the ceramic foam / Al-Mg alloy IPCs and those made from infiltrating the alumina powder / fibre-based preform with the alloy are shown in Figure 5, with the metal phase appearing brighter and the darker phase being the ceramic. It can be seen that effectively full infiltration has occurred with both types of perform, despite the different microstructural characteristics. No bulk secondary phases were found in these composites, although some defects were observed in the metal phase due to the shrinkage of the metal during cooling. The powder / fibre reinforced composite showed more porosity compared with the foam-based IPCs, although the level was still negligible. The reason is likely to be the narrower channels that will have existed in the higher packing density perform, which the molten alloy could not infiltrate. From the micrographs, it can also be seen that good bonding appears to exist between the aluminium alloy and the ceramic phases; this was due to the good wetting between the two phases. Extensive characterisation of the interface between the metal and ceramic in alumina foam / Al-Mg alloy IPCs has been undertaken previously[18,22] and it is believed to be due to the formation of Mg_3N_2 and AlN at the interface during the infiltration process[22].

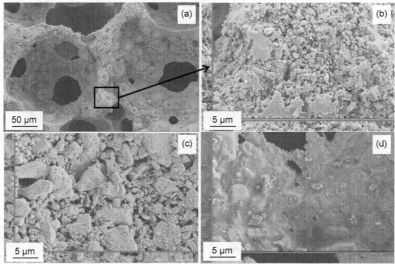

Figure 4. SEM micrographs of (a) and (b) alumina foam strut, (c) spinel foam strut and (d) mullite foam strut .

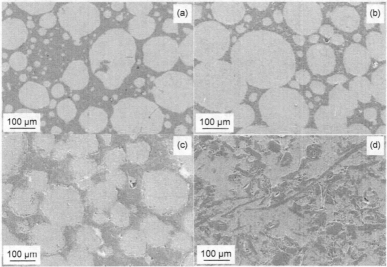

Figure 5. SEM micrographs of interpenetrating composites made from Al-10Mg alloy and (a) 27% dense alumina foam, 200 μm average cell size, (b) 20% dense spinel foam, 200 μm average cell size, (c) 30% dense mullite foam, 100 μm average cell size, and (d) 30% dense alumina powder/fibre preform.

Parameters of wear resistance of ceramic foam reinforced IPCs

1. Foam density

The wear rates of Al(Mg)/mullite and Al(Mg)/spinel interpenetrating composites with a range of foam densities after 250 m sliding are shown in Figure 6. It may be observed that the wear rate of the IPCs increased with load. However, the effect of foam density on the wear rates was not significant until the load increased to 20 N, despite the use of different cell sizes or ceramic materials. The worn surfaces of the Al(Mg)/mullite composites made from 20%, 30% and 40% dense mullite foam with 300 μm average cell size after 5000 m sliding are compared in Figure 7. It can seen that with an increase in the foam density, more and stronger ceramic struts protruded out of the surface, providing more effective resistance to the ploughing of the counter ball, hence less and finer grooves were formed.

The Rockwell hardness of interpenetrating composites with a range of ceramic content, Table I, indicates that composites with higher ceramic content yielded higher hardness, in agreement with the wear data and confirming their higher load bearing capacity.

Table I: Rockwell G hardness of the Al(Mg)/mullite composites.

Ceramic content	20% mullite			30% mullite			40% mullite		
Cell size / μm	100	300	500	100	300	500	100	300	500
Rockwell G hardness	66±4	68±6	62±3	78±2	81±3	72±5	94±2	96±3	90±3

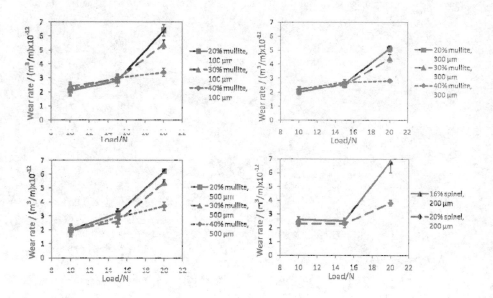

Figure 6. Variation of wear rates after 250 m sliding distance as a function of load and ceramic foam density.

2. Foam cell size

The variation in the wear rates of Al(Mg)/mullite interpenetrating composites as functions of load and foam cell size after 250 m sliding is shown in Figure 8. Consistent with Chang's results[20], there was a marginal effect of foam cell size on the wear performance. However, a common point of the three graphs in Figure 8 is that, for composites made from the same dense ceramic foam, those with 300 μm average cell size performed better than those with smaller or larger cell sizes. In addition, results in Table I consistently show the 300 μm cell size to be the hardest. Note that the size of indents, examined by FEGSEM, is bigger than 1 mm diameter.

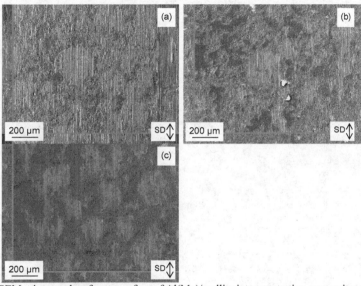

Figure 7. SEM micrographs of worn surface of Al(Mg)/mullite interpenetrating composites made from (a) 20%, (b) 30%, and (c) 40% dense mullite foam with 300 μm average cell size after 5000 m sliding.

Comparing samples made from 20% dense foams with 100 μm and 300 μm average cell sizes, the latter has thicker, hence harder and stronger, ceramic struts, so that it can protect the alloy from wear more effectively than the former. It is consistent with Chang's[20] results that the Al(Mg) / 27% Al$_2$O$_3$ foam IPC with 150-200 μm average cell size exhibited a lower wear rate than a composite based on a foam with a 100-150 μm average cell size. This is also similar to the results of Kök[10] who found, in ceramic particle reinforced Al matrix composites, that the larger the ceramic particles, the more wear resistant the final composites. However, with the same foam density, the composites with average cell sizes of 500 μm have even thicker and stronger ceramic struts than the former two composites, hence it should have yielded the best wear resistance. One possible explanation is that with an average cell size of 500 μm, the counter ball had a chance to penetrate deeper into the metal, resulting in more material loss from the metal phase. Therefore, the relationship between cell size and wear resistance of the foam-based interpenetrating composites is not linear, but shows a maxima; the best wear property is yielded by composites with moderate cell sizes. Furthermore, the consistent relationship between the wear resistance and hardness of IPCs has been shown again. There may be other possible confounding factors that make the 300 µm IPCs more wear resistant, but no evidence has been found to show them yet.

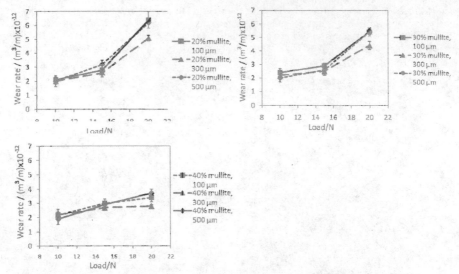

Figure 8. Variation of wear rates after 250 m sliding as a function of load and ceramic foam cell size.

3. Ceramic materials

The wear rates of the ceramic foam-based IPCs made from Al(Mg) alloy and a range of ceramic materials are given in Table II. The wear rate value of the spinel-based composite has overlap with that of the alumina composite when the scatter is considered, revealing similar wear resistance of the two composites. It can be seen that among the composites made from alumina, spinel and mullite foams, the mullite-based composites yielded the highest wear rates. Note that the alumina and spinel foam only has 200 µm average cell size which, as just discussed before, is not the best, whilst data of mullite IPCs is for those which have the 'best' 300 µm average cell size. Figure 9 shows the worn surfaces of the Al(Mg) / 27% alumina and Al(Mg) / 30% mullite composites with similar cell sizes after 5000 m sliding under a 20 N load. It can be seen that the alumina struts protruding on the surface appear to have suffered only localised damage at the edges of the struts and there are no grooves formed in them, whilst in mullite composites, not only the metal phase but also the mullite struts experience ploughing with the evidence of grooves everywhere. This is probably a consequence of the lower hardness of the mullite IPC made from liquid phase sintered mullite foam compared with the alumina IPC (Table I and III).

Table II. Wear rates of the foam-based interpenetrating composites made from Al(Mg) alloy and different ceramic materials after 5000 m sliding, in m³/m ($\times 10^{-12}$).

Foam density	15%	20%	27%	30%	40%
Alumina, 200 µm	5.0±0.3	-	1.8±0.1	-	-
Spinel, 200 µm	6.2±0.9	3.8±1.0	-	-	-
Mullite, 300 µm	-	5.3±0.3	-	5.7±1	3.2±0.2

Table III. Rockwell G hardness of the Al(Mg)/Al₂O₃ composites.

Materials	Al(Mg)/27% Al₂O₃ foam, 200 μm	Al(Mg)/30% Al₂O₃ powder/fibre
Rockwell G hardness	87±3	66±2

The degree of sintering of the ceramic foam is also an important factor that affects the wear resistance of the final IPC. Table IV gives the wear rates of IPCs made from mullite foams sintered at different temperature. SEM examination showed little difference of foam cell size but influence of the sintering temperature on the ceramic struts that the lower the sintering temperature, the more the pores observed on the struts. Therefore, with a higher degree of sintering resulting from higher sintering temperature, the ceramic foam played the better role of the reinforcement in the IPC during the wearing process.

Table IV. Wear rates of mullite IPCs made from 20% mullite foams sintered at different temperatures after 250 m and 5000 m sliding under 20 N load, in m³/m ($\times 10^{-12}$).

Sintering temperatures	1000°C	1200°C
250 m	16.6±0.8	5.1±0.2
5000 m	9.7±0.5	5.3±0.3

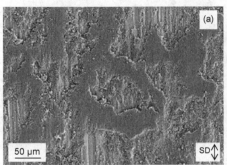

Figure 9. SEM micrographs of worn surfaces of (a). Al(Mg)/27% Al₂O₃ interpenetrating composite with 200 μm average cell size, and (b). Al(Mg)/30% mullite interpenetrating composite with 100 μm average cell size after 5000 m sliding.

4. Comparison of ceramic foam reinforced IPC and ceramic powder/fibre reinforced IPC

The results of wear testing 27% alumina foam reinforced IPCs and the 30% alumina powder/fibre reinforced composite under the 20 N load for 250 m (5000 cycles) and 5000 m (100,000 cycles) are shown in Table V. From the table, the foam-based IPC is about twice as wear resistant after both sliding distances, despite having slightly less ceramic content. This could be explained by the results of Rockwell hardness test, as Table III shows, that the former is significantly harder than the latter suggesting that the former provides higher load bearing capacity.

Table V. Wear rates of the Al(Mg)/Al₂O₃ composites after 250 m and 5000 m sliding under 20 N load, in m³/m ($\times 10^{-12}$).

Materials	Al(Mg)/27% Al₂O₃ foam, 200 μm	Al(Mg)/30% Al₂O₃ powder/fibre
250 m	3.7±0.3	7.0±1.5
5000 m	1.8±0.1	4.7±0.2

The worn surfaces and wear debris of the two kinds of composites were examined by SEM and the micrographs are shown in Figure 10. Severe grooves along the sliding direction can be observed in Figure 10(a) only in the metal phase which appears bright in the image; the dark Al_2O_3 struts protruded and contributed to the wear resistance of the composite. The wear debris of the foam-based composite, Figure 10(b), mainly consisted of Al alloy thin flakes that were cut from the composite by the counter ball, and some dispersed fine Al_2O_3 chips found on the flake surfaces. In Figure 10(c), the ceramic phase protruding out the worn surface was observed as the brighter region. The wear debris of the powder/fibre reinforced composite, shown in Figure 10(d), contained metal flakes as well, but thicker debris was observed which might contain ceramic particles inside as stronger peak of oxygen was identified by EDX than the former wear debris. It also can be seen in Figure 10(d) that some Al_2O_3 powder/fibres were ground off along with the metal.

Figure 10. SEM micrographs of (a) worn surface of Al(Mg)/Al$_2$O$_3$ foam interpenetrating composite after 5000 m sliding under 20 N load with its wear debris in (b)[20]; (c) worn surface of Al(Mg)/Al$_2$O$_3$ powder/fibre composite after 5000 m sliding under 20 N load with its wear debris in (d).

In the ceramic foam reinforced Al alloy composite, the 3-dimensionally connected ceramic struts were dense and strong due to sintering, hence didn't show fracture or breakage unlike in AMCs made by adding ceramic powders or fibres into Al alloy, or infiltrating a bed of ceramic powder or by fibre. For the ceramic foam-based IPCs, the counter ball was grinding the ceramic phase and only chipping off some fine ceramic debris. The main material loss was from the soft metal phase, which suffered ploughing damage. The ceramic struts protruding out of the surface bore most of the wear load and effectively resisted the penetration and cutting into the surface, protecting the soft metal phase and reducing the material loss. Compared with the foam, the ceramic phase in the powder/fibre reinforced composite was based on weaker bonding between the ceramic particles which could be more easily broken during wear. Therefore, as Kök[10] and Roy[23] found in their work, the wear resistance of the ceramic reinforced composites was not only dependent on the bulk hardness of the composites, it is also dependent on characteristics and properties of the ceramic phase itself.

CONCLUSIONS

Under dry sliding wear test conditions, the ceramic foam/Al(Mg) interpenetrating composites exhibited a mechanism whereby initially the metal was smeared across the surface of the wear region of the composite. Subsequently, this metal layer was removed, causing significant material loss. After this the ceramic struts became exposed and protruded from the surface, resisting further wear and leading to a steady state situation. The characteristics of the ceramic foams appear to be very important in determining the wear properties of the composites. The denser the ceramic foam, the stronger the foam struts, and hence the more effective the composites were in resisting wear. However, a non-linear relationship between the foam cell size and the wear rate was observed; the composites with moderate mean foam cell sizes of about 300 µm exhibited better properties than composites with smaller (100 µm) or larger (500 µm) cell sizes. In addition, the interpenetrating composites made from either alumina foam or spinel foam yielded lower wear rates than composites made from mullite foams, mainly due to the former's strong foam struts and higher hardness. Finally, compared with ceramic particle reinforced Al matrix composites, the foam-based interpenetrating composites showed significantly better performance in the wear tests.

FUTURE WORK

To get a better comparison and understanding of the resistance of IPCs made of alumina, spinel and mullite, the wear resistance of dense ceramics and fracture strength of ceramic foams will be measured and compared, as well as the ligament hardness of the ceramic foams. In addition, other possible confounding factors that make the IPCs with 300 µm average cell size the most wear resistant will be investigated in the future.

REFERENCES

[1]EuroTrends Research Ltd, *Materials in Cars: Options for Change*, New Barnet, 56-78 (2000).
[2]V. Imbeni, I. M. Hutchings, M. C. Breslin, Abrasive Wear Behaviour of an Al_2O_3-Al Co-continuous Composite, *Wear*, **233-235**, 462-467 (1999).
[3]P. Wycliffe, Friction and Wear of Duralcan Reinforced Aluminium Composites in Automotive Braking Systems, *SAE Technical Paper Series*, **930187** (1993).
[4]R. Dwivedi, Performance of MMC Rotors in Dynamometer Testing, *SAE Technical Paper Series*, **940848** (1994).
[5]M. Sternitzke, M. Knechtel, M. Hoffman, E. Broszeit, J. Rödel, Wear Properties of Alumina/Aluminium Composites with Interpenetrating Networks, *J. Am. Ceram. Soc.*, **79**, 121-128 (1996).
[6]A. K. Mondal, B. S. S. Rao, S. Kumar, Wear Behaviour of AE42+20% Saffil Mg-MMC, *Tribol. Int.*, **40**, 290-296 (2007).
[7]H. Ahalatci, T. Kocer, E. Candan, H. Cimenoğlu, Wear Behaviour of Al / (Al_2O_3P+SiCp) hybrid composites, *Tribol. Int.*, **39**, 213-220 (2006).
[8]F. M. Hosking, F. Folgar Portilllo, R. Wanderlin, R. Mehrabian, Composites of Aluminium Alloys: Fabrication and Wear Behaviour, *J. Mater. Sci.*, **17**, 477-498 (1982).
[9]T. Miyajima, Y. Iwai, Effects of Reinforcements on Sliding Wear Behaviour of Aluminium Matrix Composites, *Wear*, **255**, 606-616 (2003).
[10]M. Kök, Abrasive Wear of Al2O3 Particle Reinforced 2024 Aluminium Alloy Composites Fabricated by Vortex Method, *Composites A*, **37**, 457-464 (2006).

[11]M. K. Surappa, S. V. Prasad, P. K. Rohatgi, Wear and Abrasion of Cast Al-Alumina Particle Composites, *Wear*, **77**, 295-302 (1982).
[12]R. W. Rice, Micromechanics of Microstructural Aspects of Ceramic Wear, *Ceram. Eng. Sci. Proc.*, **6** [7-8] 940-58 (1985).
[13]C. L. Lust, L. F. Yallard, Wear Characteristics of An Alumina Silicon Carbide Whisker Composite at Temperatures to 800°C in Air, *Tribol. Trans.*, **35**, 331-339 (1989).
[14]L. Ceschini, G. S. Daehn, G. L. Garaghani, C. Martini, Friction and Wear Behaviour of C4 Al_2O_3/Al Composites under Dry Sliding Conditions, *Wear*, **216**, 229-238 (1998).
[15]S. Wang, H. Geng, J. Zhang, Y. Wang, Interpenetrating Microstructure and Properties of Si_3N_4/Al-Mg Composites Fabricated by Pressureless Infiltration, *Appl. Compos. Mater.*, **13**, 115-126 (2006).
[16]J. Zhang, A. T. Alpas, Wear Regimes and Transitions in Al_2O_3 Particulate-reinforced Aluminium Alloys, *Mater. Sci. Eng. A*, **161**, 273-284 (1993).
[17]D. R. Clarke, Interpenetrating Phase Composites, *J. Am. Ceram. Soc.*, **75** [4], 739-59 (1992).
[18]J. G. P. Binner, H. Chang, R. L Higginson, Processing of Ceramic-metal Interpenetrating Composites, *J. Eur. Ceram. Soc.*, **29**, 837-42 (2009).
[19]J. Lee, J. Ahn, J. Shim, Z. Shi, H. Lee, Control of the Interface in SiC / Al Composites, *Scrip. Mater.*, **41** [8], 895-900 (1999).
[20]H. Chang, J. Binner, R. Higginson, Dry Sliding Wear Behaviour of Al(Mg) / Al_2O_3 Interpenetrating Composites Produced by Pressureless Infiltration Technique, *Wear*, **268** [1-2], 166-171 (2009).
[21]P. Sepulveda, J. G. P. Binner, Processing of Cellular Ceramics by Foaming and In situ Polymerisation of Organic Monomers, *J. Eur. Ceram. Soc.*, **19**, 2059-66 (1999).
[22]H. Chang, R. L. Higginson, J. G. P. Binner, Interface Study by Dual Beam FIB-TEM in A Pressureless Infiltrated Al(Mg) / Al_2O_3 Interpenetrating Composites, *J. Microsc.*, **233**, 132-129 (2009).
[23]M. Roy, B. Venkataraman, V. V. Bhanuprasad, Y. R. Mahajan, G. Sundararajan, The Effect of Particulate Reinforcement on the Sliding Wear Behaviour of Aluminium Matrix Composites, *Metall. Trans. A*, **23**, 2833-47 (1992).

APPENDIX

Ceramic /Al-12Mg interpenetrating composites that were tested by the linear reciprocating dry sliding wear test:

Materials	Preform density	Average cell size / μm	250 m sliding			5000 m sliding
			10 N	15 N	20 N	20 N
Mullite foam	20%	100	√	√	√	√
		300	√	√	√	√
		500	√	√	√	√
	30%	100	√	√	√	√
		300	√	√	√	√
		500	√	√	√	√
	40%	100	√	√	√	√
		300	√	√	√	√
		500	√	√	√	√
Spinel foam	15%	200	√	√	√	√
	20%	200	√	√	√	√
Alumina foam	27%	200	x	x	√	√
Alumina powder/fibre	30%	N/A	x	x	√	√

AN EXPERIMENTAL STUDY ON THE EFFECTS OF SIC ON THE SINTERING AND MECHANICAL PROPERTIES OF CR₃C₂-NICR CERMETS

Ali OZER[a,b,#], Waltraud M. KRIVEN[b], Yahya Kemal TUR[a]

a: Gebze Institute of Technology, Material Science and Eng. Dept, 41400, Gebze-KOCAELI, TURKIYE
b: University of Illinois at Urbana-Champaign, Material Science Engineering Dept, Urbana, 61801, IL, USA

ABSTRACT

Cr_3C_2-NiCr type cermets have been used in cutting industry since World War II for their relatively low sintering temperatures and relatively low densities. Their high corrosion resistance at high speeds and temperatures increases interest to these materials. In this study, the powder mixture of 75wt% Cr_3C_2-25wt% NiCr with a Ni-Cr weight ratio of 4/1 was produced as master composite at 1350°C for investigating its microstructure, micro/nano hardness and elastic modulus. SiC was added as reinforcement by decreasing the ceramic amount to study the effects on nanomechanical properties of Cr_3C_2-NiCr cermets. SEM and EDX analysis performed for identifying microstructures indicated that full densification occurs at 1350°C for master cermets. By increasing SiC ratio in cermets, the densification and relative density of doped cermets were not decreased even increased except for 8wt% SiC addition. Nanoindentation and elastic modulus tests were performed to characterize the mechanical properties of cermets.

INTRODUCTION

Transition metal carbides such as VC, WC, ZrC and Cr_3C_2 have been paid attention by scientists because of their high melting points, high chemical stability, high hardness and high temperature resistances. Cr-C phases (Cr_3C_2, Cr_7C_3 and $Cr_{23}C_6$) were used on steels to interfere with chemical attacks [1-3]. Cr_3C_2 phase known as tongbaite has superior properties such as high temperature oxidation resistance (up to 900°C) and high temperature degradation properties (up to 1800°C), to be used for thermal barrier coatings for steels or any other metal. Besides, chromium carbide is introduced into other carbide sintering cycles to inhibit the grain growth [2-5].

A number of production methods for Cr-C phases were examined in literature that include reduction from oxides [1, 6, 7], sputtering as coating [2, 6], carbothermal reduction [3, 7], mechanical-thermal synthesis [4].

The most important issue in metal cutting and drilling is the removal of binder that causes sudden damage of the tool. Metal phase acting as binder should be chosen appropriately for the purpose (e.g. not using nickel in stainless steel machining tools). This can cause damage of both tool and working part hence they contain same elements to show adhesive and severe wear in high speeds and elevated temperatures. Oxidation of metal phase decreases the fracture toughness drastically and damages the tool as well [8-10].

SiC has been a very attractive ceramic material due to its superior mechanical, chemical and even electrical properties for structural applications [11]. These properties made SiC a very good reinforcement material beyond its monolithic form for both metals and ceramics to increase properties especially hardness, wear resistance, strength, chemical stability by producing a composite [12].

In this study, an experimental study was attempted to examine the effects of SiC particulate addition on sintering behavior, and nanomechanical properties such as nanohardness, and elastic modulus of Cr_3C_2 based cermets.

EXPERIMENTAL PROCEDURE

Cr_3C_2, Ni, Cr, and SiC powders were purchased from Atlantic Equipment Eng. Inc., NJ-USA with median particle sizes of 4.7 μm, 5 μm, 4.3 μm and 5.6 μm, respectively. Fig.1 shows the scanning electron microscopy (SEM) micrographs of starting powders. The Cr_3C_2 and SiC particles were pre-milled in a planetary mill for 2 h at 500 rpm prior to powder mixing in order to decrease the initial particle sizes. The average sizes of Cr_3C_2 and SiC powders were obtained as 1.7 μm and 0.7 μm, respectively as seen at the bottom row of Fig.1. Studied powder mixtures and compositions were given in Table 1. As seen in Table 1, the metal weight percentage was held constant while increasing the SiC particulate reinforcement in total ceramic amount to obtain certain wt. % ratio of SiC in composite.

Table1. Chemical compositions of studied composites by wt. %.

Materials \ Compounds	Cr_3C_2	Ni	Cr	SiC
75wt%Cr_3C_2-25wt%NiCr(4:1)	75	20	5	-
73wt%Cr_3C_2-2wt%SiC-25wt%NiCr	73	20	5	2
71wt%Cr_3C_2-4wt%SiC-25wt%NiCr	71	20	5	4
67wt%Cr_3C_2-8wt%SiC-25wt%NiCr	67	20	5	8

Powder mixtures were mixed and milled in a Fritsch "pulverisette 7" planetary ball mill at 300 rpm for 2 h in ethanol medium. 3Y-TZP balls were used as grinding media. After drying, the powder mixtures were uniaxially pressed under 60 MPa and green compacts were obtained by cold isostatical pressing under 350 MPa. The dimensions prior to CIP were 5mm x 5mm x 26mm, prior to sintering, the dimensions were obtained as 4.9mm x 4.9mm x 25.5 mm which corresponds to a volume shrinkage of 6% after CIP. Green compacts were between 67-72% of relative density of composites' theoretical densities which is calculated from rule of mixtures with respect to volume ratio of powders. The sintering cycles were carried out in a Lindberg™ tube furnace with Ar feeding atmosphere at 20 ml/min flow rate. The heating rate and dwell time were held constant in all experiments as 5°C/min and 30 min., respectively. 1350°C was chosen as to study the effect of consolidation of composites on the mechanical properties.

Particle size analysis was conducted by Horiba Laser Particle Sizer LA950 using water as wet medium. Relative densities were measured geometrically after rough (400 grit SiC paper) grinding. Samples that have relative density over 96% were taken into consideration for all tests. XRD analyses were performed by Siemens Bruker D5000 diffractometer with Cu Kα radiation that has a wavelength of 1.5406 Å to identify the phases in powder mixtures and in sintered samples, and to evaluate primary crystallite sizes. Scan speed was held constant as 0.3°/min. Jeol JSM 6060LV scanning electron microscope (SEM) was employed for surface and structural analysis of powders in sintered samples.

Fig.1. SEM micrographs of starting powders

Nanoindentation hardness and elastic modulus tests were carried out via Hysitron TI950 Tribo-Indenter to investigate the nanomechanical properties of master and doped cermets. Nanoindenter was equipped with low load transducer that is capable of 10mN and a Berkovich type diamond triangular pyramid indenter which has an elastic modulus of 1141 GPa and a Poisson's ratio of 0.007.

RESULTS AND DISCUSSION

SiC was proposed to be ground as finer as possible to achieve a good distribution in matrix and interact with Cr_3C_2 ceramic phase and Ni(Cr) metal binder phase as much as possible by increasing relative surface area. For this purpose, planetary ball mill was used for 2 hours at 500 rpm since the submicron range (0.7-1.1 microns) for SiC powders was reached as seen from Fig 1. Peak broadenings for SiC's moissanite phase after 2 hours of planetary ball milling showed that, the average crystallite size was decreased from 500 nm down to 75 nm which proves the decrease in particle size from 5.6 μm down to 0.7 μm. Cr_3C_2 phase was also pre-milled at same conditions as SiC prior to mixing to increase the interaction area of particles and 1.7 μm particle size with a narrow size distribution was achieved as can be seen from Fig.1. The FWHM(full width at half maximum) peak fitting method was used to calculate the primary crystallite sizes of as received and milled powders.

XRD result graph in Fig.2 shows that by increasing SiC wt. % in cermet, the diffractions at 34°, 36°, and 60° 2θ angles which belong to SiC (moissanite) phase were increased due to increasing vol% of SiC in cermets while Cr_3C_2 phase's volume ratio was decreasing in cermet and also because of SiC's much lower density than other materials forming cermet. By increasing SiC ratio in composites, the relative volume ratio of Ni that decreased gradually in cermets resulted a gradual decrease at peak intensities especially on Ni(111) peak at 44° and Ni(200) peak at 52°. Since no densification was observed for 8 wt% SiC doped samples, XRD result doesn't show the actual phases for sintered bodies. For 2wt% and 4 wt.% SiC doped cermets, there occurs a peak splitting at 44° (2Theta) (100% main peak for both Ni and Cr_7C_3) which corresponds to a slight increase in Cr_7C_3 phase due to the accelerated C diffusion, increasing core-rim structure and relative density up to 99.5% and 98.5% for 2wt% and 4wt% SiC doping, respectively.

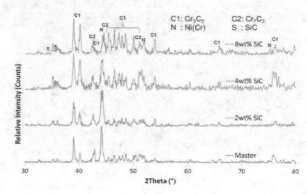

Fig.2. XRD analysis of SiC doped samples sintered at 1350°C for 30 min.

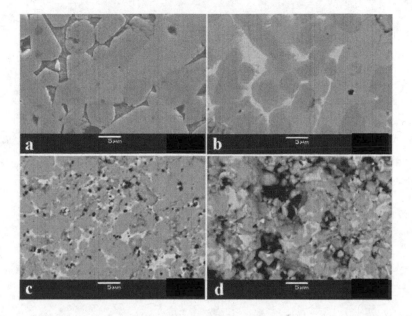

Fig.3. SEM micrographs of a) master, b) 2wt%, c) 4wt%, d) 8wt% SiC added cermets sintered at 1350°C.

Fig. 3 represents the SEM backscattered electron contrast (BEC) images of cermets which are liquid phase sintered at 1350°C. Relative densities were measured geometrically as 97%, 99.5%, 98.5% and 76% for master, 2wt%, 4wt% and 8wt% SiC reinforced cermets, respectively. It's obvious to see the entrapped porosities in grains and porosities at triple-quadruple grain junctions in Fig 3a and 3b. Fig.3b shows a core-rim structure and 3D connected grains with binder phase among them. Since the grain growth is very fast, entrapped porosities within grains are visible. In Fig.3c, almost all black irregular shaped spots were determined by EDS as SiC reinforcements which show a uniform distribution in matrix. A dramatic decrease in particle size by increasing SiC wt.% is well observable with the micrograph (a: 12μm; b: 8 μm; c: 3.5 μm; d:1.5 μm) and it's expected to increase the strength as well as hardness. Fig.3d shows the 8wt% SiC added Cr$_3$C$_2$-NiCr cermet with a fine particle size distribution which seems to be as same as starting particle size ranges except metal binder phase (Ni20Cr alloy) that begins melting at 1350°C. Increase in relative density from 67% (green density) to 76% (1350°C sintered) can be attributed to the partial viscous melting of metal phase and surrounding some grains to initiate neck formation at that temperature.

Since 8wt% SiC added samples didn't show any densification at 1350°C, only XRD analysis was carried out for them, no mechanical evaluation could have been done, as it's mentioned, samples under 96% relative densities were not taken into consideration.

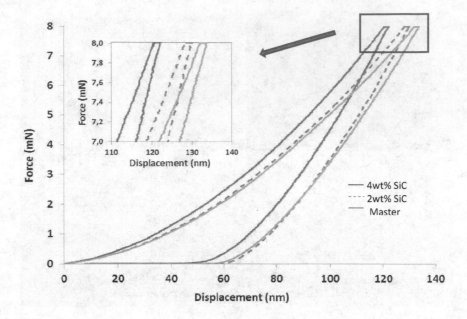

Fig.4. Average nanoindentation-elastic modulus plots for master and doped cermets; arrow indicates the detailed view of the data points in rectangle.

Fig.4 represents the average plots obtained from nanoindentation tests for master, 2wt% SiC and 4wt% SiC doped cermets. A detailed view can be seen for upper portion of data for ease of sight by an arrow. At least 10 measurements were averaged to obtain the resultant graph. Elastic modulus slope in Fig.4 (detailed view shown by arrow) can be seen to be steeper by SiC addition that leads elastic modulus to increase. Another possible phase to contribute to elastic modulus could be Cr_7C_3 phase which can be seen in XRD analysis (Fig.2) by increasing SiC wt. %. Cr_7C_3 was reported to have an elastic modulus of 374 MPa which is a bit higher than that of Cr_3C_2 (367 MPa) theoretically [13], but in this case a small increase in Cr_7C_3 was not considered to have a significant role on increasing elastic modulus when compared with increasing SiC amount.

As seen from Fig.5, elastic modulus which were measured by curve fitting method developed by Oliver and Pharr [14] using data from 95% to 20% of unloading curve by TriboScan software increased by increasing SiC addition into cermets which were densified over 98%. Elastic modulus was measured 310±9 GPa for master cermet, and increased up to 323±11 GPa and 363±14 GPa for 2wt%SiC and 4wt%SiC doped cermets, respectively. Nanohardness was also evaluated for all densified cermets and found to be increasing from 23.9±0.2GPa; lowest data point for master cermet to 24.6±0.3GPa; highest data point for 4wt% SiC doped cermets. Increase in elastic modulus and hardness can be attributed to the contribution of SiC's high hardness and high elastic modulus by increasing SiC ratio without decreasing the density. It's seen from Fig.5, estimation could be done that increasing hardness decreased the displacement (penetration depth).

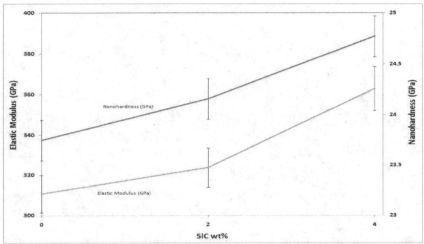

Fig.5. Elastic modulus and nanohardness graph versus SiC wt.%.

CONCLUSIONS

2wt% and 4wt% SiC doped Cr₃C₂-NiCr cermets was sintered up to near theoretical densities even at 1350°C for 30 min. 8wt% SiC doped samples wasn't able to densify due to increased volume ratio of SiC in composite and therefore constrained sintering at that temperature.

SiC showed a very good distribution in composites and decreased the average grain size by interfering with the Cr₃C₂- Cr₃C₂ and Ni- Cr₃C₂ grain boundary interactions. Hardness increased so that total displacement was decreased by increasing SiC ratio. Elastic modulus and hardness were increased from 310 GPa to 360 GPa and from 23 GPa to 25 GPa by increasing SiC wt.% ratio in composites, respectively.

ACKNOWLEDGEMENTS

The authors would gratefully like to acknowledge that various steps of this work were carried out in part in the Frederick Seitz Materials Research Laboratory Center of Microanalysis of Materials facility, University of Illinois at Urbana-Champaign and in Materials Research Laboratories in Gebze Institute of Technology. The support of Turkish Higher Education Council for Ali Ozer's visit is appreciatory acknowledged. We also acknowledge US National Science Foundation (Grant no: 490918-919000-191100) for supporting this research.

REFERENCES

[1] Reza Ebrahimi-Kahrizsangi, Hossein Monajati Zadeh, Vahid Nemati, "Synthesis of chromium carbide by reduction of chromium oxide with methane", Int. Journal of Refractory Metals & Hard Materials, 28(2010) 412–415.

[2] Pau AL, Lim J., "Effect of deposition parameters on the properties of chromium carbide coating deposited on the steel by sputtering", Mater. Sci Eng A, 332(2002) 123–128.

[3] Sheng-Chang Wang, How-Tung Lin, Pramoda K. Nayak, Shin-Yun Chang, Jow-Lay Huang, "Carbothermal reduction process for synthesis of nanosized chromium carbide via metal-organic vapor deposition", Thin Solid Films, 518(2010) 7360–7365.

[4] Osvaldo M. Cintho, Eliane A.P.Favilla, Jos´e D.T. Capocchi, "Mechanical–thermal synthesis of chromium carbides", Journal of Alloys and Compounds, 439(2007) 189–195.

[5] Håkan Engqvist, Gianluigi A. Botton, Niklas Axe´n, and Sture Hogmark, "Microstructure and Abrasive Wear of Binderless Carbides", J. American Ceram. Soc., 83[10] (2000) 2491–2496.

[6] A. Paul, Jongmin Lim, Kyunsuk Choi, Chongmu Lee, "Effects of deposition parameters on the properties of chromium carbide coatings deposited onto steel by sputtering", Materials Science and Engineering A, 332 (2002) 123–128.

[7] L.-M. Berger, S. Stolle, W. Gruner, K. Wetzig, "Investigation of the carbothermal reduction process of chromium oxide by micro- and lab-scale methods", Int, J. of Refr. Metals and Hard Mat, 19(2001) 109-121.

[8] C.-A. Jeng, J.-L. Huang, "The influence of oxidation on crack resistance in injection moulded Cr_3C_2/Al_2O_3 composites", Journal of the European Ceramic Society 23 (2003) 1477–1484.

[9] V. A. Maslyuk and S. G. Napara-Volgina, "Wear-Resistant And Corrosion-Resistant Carbide-Steel-Type Materials Having Different Matrices", Powder Metallurgy and Metal Ceramics, 38(1999) 9-10.

[10] C. Duran, S. Eroglu, "Liquid-phase sintering and properties of Cr$_3$C$_2$/NiCr cermets", J. of Mat. Pro. Tech., 74(1998) 69-73.

[11] Kay Andr´e Weidenmann, Georg Rixecker, Fritz Aldinger, "Liquid phase sintered silicon carbide (LPS-SiC) ceramics having remarkably high oxidation resistance in wet air", Journal of the European Ceramic Society 26 (2006) 2453–2457.

[12] Sang-Kee Lee, Wataru Ishida, Seung-Yun Lee, Ki-Woo Nam, Kotoji Ando, "Crack-healing behavior and resultant strength properties of silicon carbide ceramic", Journal of the European Ceramic Society 25 (2005) 569–576.

[13] B. Xiao, J.D. Xing, J. Feng, Y.F. Li, C.T. Zhou, W. Su, X.J. Xie and Y.H. Chen, "Theoretical study on the stability and mechanical property of Cr$_7$C$_3$", Physica B: Condensed Matter, 403(13-16), 2008, 2273-2281.

[14] W.C. Oliver, G.M. Pharr, "Measurement of hardness and elastic modulus by instrumented indentation: Advances in understanding and refinements to methodology", J. Mater. Res., 19(1), (2004) 3-20.

INCREASING THE OPERATING PRESSURE OF GASOLINE INJECTION PUMPS VIA CERAMIC SLIDING SYSTEMS

C. Pfister, H. Kubach, U. Spicher
Institut fuer Kolbenmaschinen, Karlsruhe Institute of Technology (KIT),
Kaiserstrasse 12, D-76131 Karlsruhe, GERMANY

ABSTRACT

Spray-guided gasoline direct injection (GDI) already shows strong potential to reduce both fuel consumption and pollutant emissions of modern combustion engines. This concept can be significantly improved by increasing the fuel injection pressure; higher pressures lead to smaller droplets being injected into the combustion chamber, thus accelerating mixture formation.

The materials used in the sliding systems of modern gasoline pumps wear severely at injection pressures above 20 MPa, whereas laboratory tests show substantial improvements in the performance of GDI engines at higher pressures. The use of ceramic parts in the sliding systems of the fuel pump should help to overcome these difficulties.

As part of the Collaborative Research Centre "High performance sliding and friction systems based on advanced ceramics", investigations have been performed at the Karlsruhe Institute of Technology on a high-pressure pump built with ceramic sliding systems. The pump is fitted with several sensors to assess the performance of the sliding parts.

The investigations performed within this project demonstrate great potential to increase injection pressures in GDI engines by using advanced ceramics in the sliding systems of the high-pressure gasoline pump.

INTRODUCTION

The development of reciprocating engines focuses on reducing both fuel consumption and pollutant emissions. For gasoline engines, such reductions can be realized with a spray-guided gasoline direct injection concept (GDI). In comparison to conventional port fuel injection engines, reductions in fuel consumption of up to 50% can be achieved, as shown in figure 1.

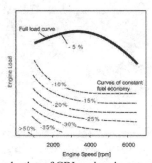

Figure 1. Fuel consumption reduction of GDI engines in comparison to port fuel injection[1].

The time available for mixture preparation in spray-guided GDI engines is very short, as the fuel is injected directly into the combustion chamber. In order to ensure stable ignition and combustion conditions, high injection pressures are required. Indeed, increasing injection pressure leads to smaller droplets sprayed into the combustion chamber and to a higher droplet surface area-to-volume ratio.

This enables a faster evaporation of gasoline[1, 2, 3]. Experiments performed at the Institut fuer Kolbenmaschinen at the Karlsruhe Institute of Technology have shown that increasing the injection pressure from 20 MPa to 50 MPa reduces the mean injected droplet size from 10 μm to approximately 6 μm[4]. The benefit of these smaller droplets is a strong reduction of the soot concentration in the exhaust[5], as shown in figure 2.

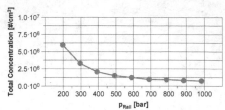

Figure 2. Particulate Emissions of a GDI engine with injection pressures up to 1000 bar[5].

However, modern high-pressure pumps for gasoline injection systems typically reach pressures of 10-20 MPa. The tribological stresses in their sliding contacts are very high because of the low viscosity and lubricity as well as the reduced sulphur content of gasoline. As a result, the use of conventional materials, such as steel, limits the pump delivery pressure because of high friction and wear[2, 3, 6, 7]. A pump with high friction forces requires more driving power and could offset the benefits of high-pressure injection. Moreover, high wear shortens the lifetime of the pump.

Advanced materials are required to increase the delivery pressure of high-pressure gasoline injection pumps. Due to their unique properties such as hardness and chemical stability, advanced ceramics promise to overcome this engineering challenge[8, 9]. In order to investigate the performance of various material pairs under realistic operating conditions, a model pump operating at up to 50 MPa and fitted with several sensors has been designed. Its main characteristics are described in the next section.

MODEL PUMP

The model pump designed at the Institut fuer Kolbenmaschinen in Karlsruhe is a fuel-lubricated single-piston pump based on the design of a conventional 3-piston radial pump, which is considered to be the best compromise between size, efficiency and costs[3]. The kinematics of this type of pump are shown in figure 3.

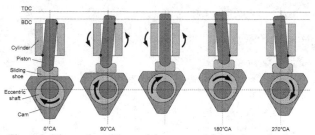

Figure 3. Kinematics of the model pump used for the investigations.

The eccentric shaft of the pump is driven by a variable speed asynchronous motor. Its rotation induces a circular translation of the cam which drives the piston via the sliding shoe. The piston/sliding shoe assembly maintains contact with the cam via retaining springs (not shown in the figure). The relative displacement and speed in both piston/cylinder and cam/sliding shoe contacts are shown in figure 4.

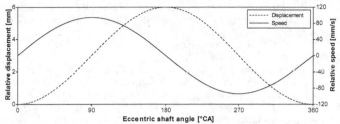

Figure 4. Relative displacement and speed in both sliding contacts at 300 rpm.

A very small clearance (5 to 7 µm) between the piston and the cylinder serves to lubricate this sliding contact while maintaining low levels of fuel leakage. This design ensures the purity of the gasoline delivered because there is no need for oil lubrication and guaranties a satisfactory lifetime of the pump. The clearance causes rocking of the piston in the cylinder at approximately 90 °Crank Angle (°CA) and 270 °CA after bottom dead centre (BDC), as the sliding sense of the cam/sliding shoe system is reversed at these shaft angles. Consequently, the piston is in contact with the cylinder in only two points (B and C or B* and C* in figure 5).

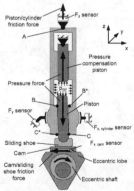

Figure 5. Schematic view of the force measurement and axis convention.

As the objective of this model pump is to assess the behaviour of various material pairs in the two main fuel-lubricated sliding systems (piston/cylinder and cam/sliding shoe), several parameters are measured on a shaft-angle-resolved basis in the test rig:
- Cylinder pressure
- Rail pressure

• Forces applied to the cylinder in the x, y, and z direction
• Forces applied to the cam in the x direction (sliding direction)

Figure 5 shows how the forces applied to the cylinder and to the cam are measured. The $F_{x,cam}$ sensor indicates the friction force in the cam/sliding shoe system. The contact force in this sliding system can be calculated by summing the force applied to the piston by the fuel pressure in the cylinder and the value supplied by the F_z sensor. This sensor measures the friction force in the piston/cylinder system, as a second piston compensates the pressure force in the z direction (pressure compensation piston in figure 5). As the cylinder assembly is articulated at A, it is possible to calculate the contact force applied in B and C (respectively B* and C*) with the value supplied by $F_{x,cylinder}$ as a function of the distance between these points.

The results presented in this paper focus on the cam/sliding shoe system, which is subjected to highest tribological stresses[10]. Figure 6 gives an example of highly time-resolved measurements against eccentric shaft angle in this system. The first diagram shows the contact force and the second one shows the friction force in the cam/sliding-shoe contact at 50 MPa delivery pressure and 300 rpm. The black curves are physical measurements performed with the model pump whereas the gray curves are calculated values with a friction coefficient set to 0.1. As a convention, the crank angle of the bottom dead centre of the piston is set to 0.

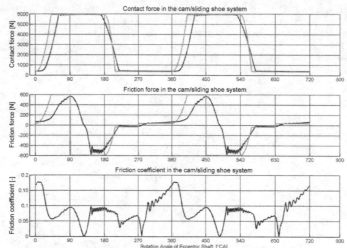

Figure 6. Plots of force measurements (black) and theoretical values (grey) in the cam/sliding shoe contact against eccentric shaft angle during 2 strokes. Material pair: self-mated silicon carbide; fuel: isooctane; rotation speed: 300 rpm; delivery pressure: 50 MPa; theoretical friction coefficient: 0.1.

Some differences between the theoretical and real forces can be observed because of the leakage between the piston and the cylinder which delays the achievement of maximal pressure during the compression (between 0 °CA and 90 °CA) and precedes the theoretical loss of pressure at approximately 180 °CA. Elastic deformation due to the pressure variations also influences the measured forces. Moreover, the real instant friction coefficient in the cam/sliding shoe system (third diagram in figure 6) is time variant and depends on various parameters such as contact force and sliding speed, which are not constant during pump operation.

In addition to the highly time-resolved measurements, the test rig supplies some other measurements, which are listed below:
Pressure in the low pressure circuit
Low pressure circuit temperature
High pressure circuit temperature
Fuel temperature in the tank
Output mass flow (Coriolis mass flow meter)

DESIGN OF THE SLIDING SYSTEMS

Geometry
Shown in figure 7 are the dimensions of the piston and cylinder system as well as those of the cam and sliding shoe system. The piston has a nominal diameter of 12 mm and a displacement of 6 mm for a theoretical displaced volume of 678 mm³. The ceramic cylinder is shrunken into a steel jacket to keep it under compression stress for an optimal reliability[9]. A metal jacket is glued to the bottom of the piston for its integration in the pump.
The cam and the sliding shoe are both glued into a steel holder, which facilitates the replacement of the material pairs in the pump. The amplitude of their relative displacement is the same as in the piston/cylinder system. The sliding shoe is ring-shaped in order to keep more fuel in the contact area and to facilitate lubrication and cooling. Both surfaces are flat ground and there is a light chamfer (0.5 mm x 15°) on the ring. The maximal contact pressure at 50 MPa pump delivery pressure is approximately 13 MPa.

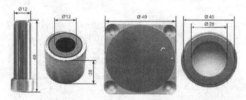

Figure 7. Dimensions in millimetres of the sliding parts in the model pump. From left to right: piston, cylinder, cam, sliding shoe.

Materials
For all the investigations, the material used for the cylinder is sintered silicon carbide (SSiC). For the piston, two materials are used: silicon carbide and hardened bearing steel (equivalent to AISI 52100). Former tests on this sliding system have shown very good results for these material combinations[10]. No change in the behaviour of the cam/sliding shoe system has been measured when changing the material of the piston.
Two ceramic materials have been investigated for the cam/sliding shoe system: commercial silicon carbide (self mated or in combination with bearing steel) and SiAlON. Both ceramics have very good characteristics regarding hardness, strength and chemical stability for use in a fuel-lubricated high-pressure pump[11, 12, 13]. The SiAlON parts used for the tests have been produced at the Institut fuer Keramik im Maschinenbau at the Karlsruhe Institute of Technology. They offer the advantage of

optimizing their mechanical characteristics to the system by varying the proportion of α and β phases of the material.

The main characteristics of these materials are listed below:

Silicon carbide (EKasic F from ESK Ceramics):
> Sinter type: sintered SiC
> Hardness (Vickers): 2500 HV 10
> Bending strength (S_0): 405 MPa
> Weibull modulus: 7.2
> Fracture toughness: 3.65 MPa.m$^{1/2}$
> Elastic modulus: 430 GPa

SiAlON (produced at the Institut fuer Keramik im Maschinenbau, Karlsruhe Institute of Technology):
> Proportion α/β: 60/40
> Hardness (Vickers): 1780 HV 10
> Bending strength (S_0): 748 MPa
> Weibull modulus: 6.1
> Fracture toughness: 10.4 MPa.m$^{1/2}$
> Elastic modulus: 316 GPa

Bearing steel
> Type: 100Cr6 (equiv. AISI 52100)
> Hardness (Vickers): 800 HV 10
> Elastic modulus: 210 GPa

Roughness and texture

The piston and the cylinder have a surface roughness of approximately $R_a = 0.05$ µm. For the cam and the sliding shoe, different roughnesses have been investigated, as shown in Table I.

Table I. Roughness of surfaces in the cam/sliding shoe sliding system.

Material	Roughness Ra [µm]		Roughness Rq [µm]	
	Cam	Sliding shoe	Cam	Sliding shoe
100Cr6 (fine ground)	-	0.13	-	0.16
SSiC (polished)	0.006	0.006	0.008	0.008
SSiC (fine ground)	0.09	0.1	0.11	0.16
SSiC (rough ground)	0.2	0.5	0.3	0.7
SiAlON (fine ground)	0.06	0.16	0.08	0.2

In order to improve the performance of the sliding systems, the surface of some silicon carbide cams have been textured with a laser. These micro-structures are expected to keep the fuel in the sliding system on the one hand and to store the particles released by wear on the other hand[15]. The texture used for the investigations has been specially developed and realised at the Institut fuer Werkstoffkunde II at the Karlsruhe Institute of Technology for the present application. This texture is shown in figure 8 and has the following characteristics:

Form: circular micro-dimples

Diameter: 60µm

Depth: 10 µm

Surface ratio: 20%

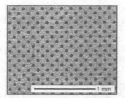

Figure 8. Picture of a silicon carbide cam textured at the Institut fuer Werkstoffkunde II, Karlsruhe Institute of Technology.

INVESTIGATION METHODS

Operating points
    The pump is tested in a sequence of operating points in order to simulate real operation of a high pressure gasoline injection pump driven by the camshaft of a combustion engine. This sequence, shown in figure 9, is repeated as necessary for the desired investigation. One cycle lasts 4 hours for a sliding distance of approximately 6,500 meters in the sliding systems. The pump delivery pressure is varied from 20 MPa to 50 MPa and the rotation speed from 300 rpm (idle camshaft speed) to 2900 rpm. Measurements are performed at each operating point and the mean of the shaft-angle-resolved measurements is calculated over 50 strokes. Unless mentioned, the measurements are performed after the initial running-in phase of the sliding parts.

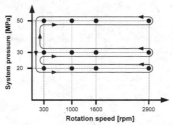

Figure 9. Sequence of operating points used for the tests.

Fuels
    In order to avoid the influence of fluctuations in gasoline composition, isooctane (single component fuel) has been used for all the tests. Complementary tests have been performed with commercial gasoline (RON 95) in order to ensure the transferability of the results. In addition, the performance of self-mated silicon carbide has also been investigated with E100 (99 % ethanol) and two commercial ethanol/gasoline mixtures: E25 and E75 with respectively 25 % and 75 % ethanol.

Table II. Main characteristics of the fuels used for the investigations.

Fuel type	Density [-]	Coefficient of compressibility [10^{-6}/bar]	Dynamic viscosity [10^{-3} Pa.s]
Isooctane	0.692 (at 20°C)	135	0.348 (at 20°C)
Gasoline	0.739 (at 15°C)	110	0.65 (at 20°C)
E100	0.794 (at 15°C)	50	1.2 (at 20°C)
E75	0.779 (at 15°C)	N/A	N/A
E25	0.750 (at 15°C)	N/A	N/A

Calculated values
 In order to compare the performance of the investigated material pairs in the sliding systems, it is necessary to calculate representative values for each measurement. Time-averaging the value of the instant friction coefficient is not satisfactory. Indeed, the instant friction coefficient is especially high in the range where the work of friction (equation 1) is low (between approximately 300 °CA and 360 °CA). For this paper, an integral friction coefficient μ_{Int} has been defined (equation 2). This integral friction coefficient is more representative for the behaviour of the investigated material pairs, as it is closely linked to the work of friction $W_{Friction}$.

$$W_{Friction} = \int \left| F_{Friction} \right| \cdot ds_{Friction}$$

(1)

$$\mu_{Int} = \frac{\int \left| F_{Friction} \right| \cdot ds_{Friction}}{\int \left| F_{Contact} \right| \cdot ds_{Friction}}$$

(2)

 In equation 1 and 2, $F_{Friction}$ is the friction force, $F_{Contact}$ is the contact force and $s_{Friction}$ is the relative displacement of the parts in the sliding system.

RESULTS

Influence of surface roughness
 In order to determine the optimal grinding process for the components in the cam/sliding shoe system, investigations have been performed with 3 self-mated silicon carbide pairs: polished, fine ground, and rough ground.

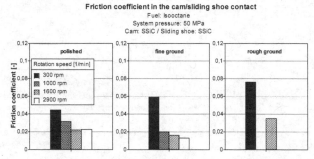

Figure 10. Comparison of silicon carbide material pairs with various surface roughnesses in the cam/sliding shoe system.

 The figure 10 shows the friction coefficient μ_{Int} measured at 50 MPa pump delivery pressure with isooctane. Despite its very smooth surface, the polished material pair (on the left) does not suggest an improved performance in comparison to the fine ground material pair (middle), excepted at 300 rpm. This observation can be explained by the fact that the surfaces are already smoothed at the beginning of the test, so that the fit between the parts cannot be improved by initial wear. The results obtained with the rough ground surfaces (on the right) show higher friction coefficients when compared to the polished and fine ground surfaces. In addition, only two measurements have been performed at 50 MPa pump delivery pressure because of the high and irregular friction forces that were observed during the test. This can be explained by the abrasive effect of the particles released during the running-in which is considerably longer in comparison to the other tests performed. Unless mentioned, all the following results have been obtained with fine ground material pairs, as this surface state shows the best global performance.

Comparison of silicon carbide and SiAlON

The results in figure 11 show the performance of self-mated silicon carbide and self-mated SiAlON. As expected, the friction coefficient measured in the cam-sliding shoe system decreases with increasing speed for both material pair. Both ceramics perform very well especially at 50 MPa delivery pressure and show friction coefficients lower than 0.02 at rotation speeds above 300 rpm. The special feature of the silicon carbide pair is that its friction coefficient decreases with increasing contact force (linked to the pump delivery pressure) in the sliding system, whereas the performance of SiAlON remains almost constant.

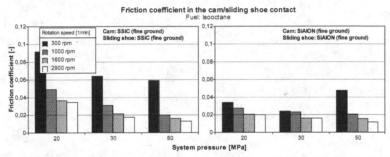

Figure 11. Comparison of self-mated silicon carbide and self-mated SiAlON material pairs in the cam/sliding shoe system.

Combination of silicon carbide and bearing steel

The investigations with the combination of silicon carbide (cam) and hardened bearing steel (sliding shoe) were unsuccessful as the friction forces tended to increase during the test at constant operating conditions. For this reason, no measurements have been performed at high rotation speeds (2900 min^{-1}) or at 50 MPa delivery pressure. The test was stopped after 3 hours of operation because of the too high friction forces in the investigated sliding system. In contrast to the self-mated ceramic pairs, no stabilization of the forces was observed in this time range.

Effect of surface texturing

Figure 12 shows the results obtained with a micro-textured silicon carbide cam self-mated or in combination with hardened bearing steel. As expected, excellent performance can be observed for the self-mated silicon carbide pair (on the left). The friction coefficient measured in the cam/sliding shoe system is approximately 0.01 for all rotation speeds and pump delivery pressures. In comparison to the untextured self-mated silicon carbide pair (figure 11), the improvement is especially high at 300 rpm. Using a textured silicon carbide cam in combination with hardened bearing steel was successful (figure 12, on the right). Measurements at all operating points were possible, as a stabilization of the measured friction forces has been observed after a few hours of operation. Even if the performance is worse than with self-mated silicon carbide or SiAlON, texturing the cam enables the application of this material combination.

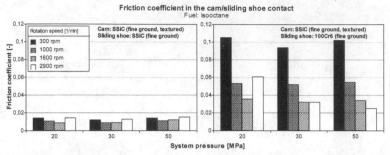

Figure 12. Friction coefficient in the cam/sliding shoe system using a textured cam.

Comparison of various fuels

As modern combustion engines run with gasoline as well as gasoline/ethanol mixtures, complementary measurements were performed with various fuels. The results at 50 MPa pump delivery pressure (figure 13) suggest that the investigations performed with isooctane represent a worst case. The results obtained with gasoline show improved friction coefficients, especially at low rotation speeds. This can be explained by the content of various chemical components which improve the lubricating properties of gasoline. It is also important to notice that the results obtained with pure ethanol and gasoline/ethanol mixtures show similar results. As the commercial gasoline used for the tests also contains approximately 5 % ethanol, it can be supposed that the presence, even in small quantity, of ethanol enhances the performance of silicon carbide components in the sliding systems.

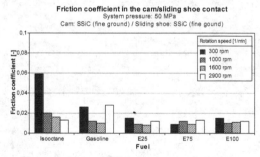

Figure 13. Friction coefficient in the cam/sliding shoe system for isooctane, gasoline, ethanol and gasoline/ethanol mixtures with a self-mated silicon carbide material pair.

A similar effect can be observed with the combination of a textured silicon carbide cam with hardened bearing steel, as shown in figure 14. The performance of this material pair is strongly improved for all operating points when compared to the performance with isooctane as lubricant (figure 12). At 50 MPa, the results are similar to those obtained with self-mated silicone carbide and gasoline.

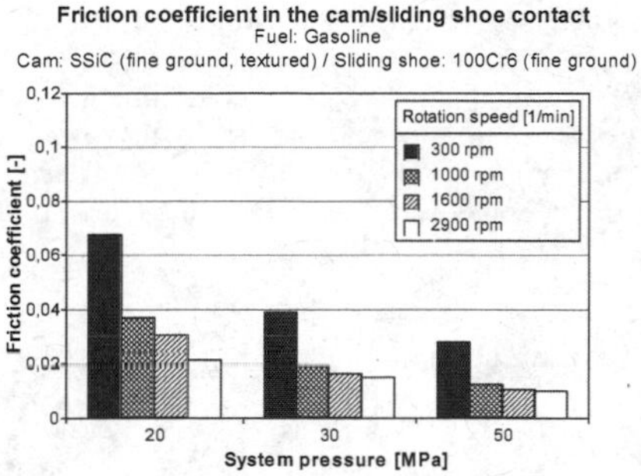

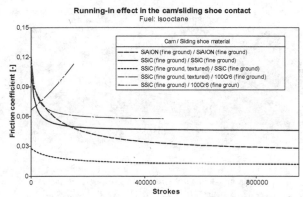

Figure 14. Friction coefficient in the cam/sliding shoe system with silicon carbide (textured) in combination with bearing steel using gasoline as fuel.

Running-in

At the beginning of the tests, the investigated material pairs go through a running-in phase with a high friction coefficient, before reaching stable performance. This observation can be explained by the release of particles in the sliding system during the phase of running-in. These particles have an abrasive action on the surfaces and cause high friction forces and wear[16]. Figure 15 shows the friction coefficient in the cam/sliding shoe system during the running-in phase for various material pairs. The friction coefficients of all material pairs decrease significantly during the first hours of operation except in the case of untextured silicon carbide in combination with bearing steel. As previously explained, the tests with this material pair were unsuccessful because of the increasing friction forces at a single operating point. The SEM analysis, which is presented later, provides an explanation of this observation.

Figure 15. Comparison of running-in effect in the cam/sliding shoe contact for different material pairs (1,000,000 strokes equal to approximately 16 hours of operation).

WEAR ANALYSIS

No measurable wear of the investigated material pairs was observed after the tests, even after 85 hours of operation in the case of self-mated silicon carbide. However, SEM pictures of several parts enable a better understanding of the results presented in this paper.

The evolution of the surface state of a silicon carbide cam (fine ground) is shown in figure 16. A difference can be observed between the 2 first pictures (left: unstressed, middle: after 16 hours): the surface is smoothed during the running-in. There is no visible change in the surface after 85 hours of operation (on the right). The cavities on the surface improve the lubrication of the sliding system as the fuel is stored in these micro-structures.

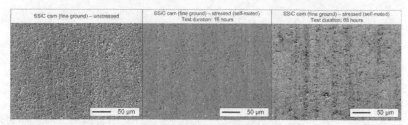

Figure 16. SEM pictures of silicon carbide cams used in combination with silicon carbide sliding shoes after different test durations (left: unstressed; middle: after 16 hours; right: after 85 hours).

Figure 17 shows the evolution of a polished silicon carbide cam before (on the left) and after 12 hours of operation (on the right). There is no visible change in the surface state after 12 hours of operation and the surface looks similar to the stressed fine ground silicon carbide parts. The cavities are smaller than those observed with fine ground surfaces, and keep less fuel in the sliding system. This can explain why the fine ground pair performed better. However, this observation can also be explained by the surface state at the beginning of the test: the surface is already smoothed, so that the fit between the sliding parts cannot be improved by running-in.

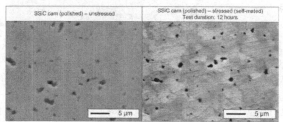

Figure 17. SEM pictures of a silicon carbide cam (polished) before (left) and after 12 hours stress (right).

Figure 18 enables to understand the results observed during the investigations with silicon carbide in combination with bearing steel. In the case of the untextured cam (on the left), a steel layer can be seen on the surface after 3 hours of operation. This layer indicates an adhesion mechanism in the sliding system and explains the increasing friction coefficient shown in figure 15. No steel layer has been observed on the surface of the textured cam (figure 18, on the middle). This confirms that the micro-dimples enhance lubrication and enable to use this material combination for application. On the right, the picture shows the surface of the hardened bearing steel sliding shoe used in combination with the texture silicon carbide cam. Even if the abrasive action of wear particles can be observed, there is no severe wear visible, as these particles are rapidly stored by the micro-dimples.

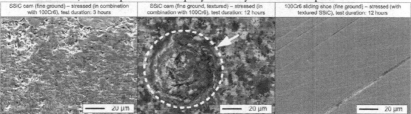

Figure 18. SEM pictures of silicon carbide cams tested in combination with bearing steel (left: untextured; middle: textured), and of the bearing steel sliding shoe tested on textured silicon carbide cam (right). In the white dotted circle (middle): one dimple.

Shown in figure 19 are SEM images of the surface of the SiAlON cam before testing and after 16 hours of testing. On the left, horizontally-oriented grooves left by the grinding process can clearly be seen. The experimentally measured maximum depth of these grooves is approximately 0.5 μm. On the right, a smoothing of the surface in the sliding direction is observed. However, the horizontal grooves left by the grinding process remain in the material, which means that the wear in this sliding system is less than 0.5 μm after 16 hours of operation. This represents an equivalent wear rate of approximately 10^{-8} mm³/Nm, which is well within acceptable limits for such sliding systems.

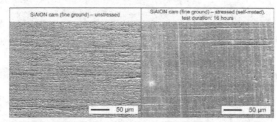

Figure 19. SEM pictures of a SiAlON cam before (left) and after 16 hours stress (right).

CONCLUSION
The results presented in this paper show that the use of ceramic components in high pressure gasoline pumps enables significant increase in the delivery pressure of modern injection systems. Despite the low lubricity of gasoline, silicon carbide as well as SiAlON material pairs perform very well regarding friction coefficient and wear in the cam/sliding shoe system at 50 MPa pump delivery pressure. The best compromise between short running-in, good fitting of the sliding parts and low friction coefficient is offered by fine ground material pairs. Texturing the surface of the cam with circular micro-dimples enhances the performance of self-mated silicon carbide as the texture keeps more fuel in the contact and stores the particles released by wear. The micro-dimples also enable the use of silicon carbide in combination with hardened bearing steel by preventing the problem of adhesion and surface deterioration. Increasing the delivery pressure of gasoline high pressure injection pumps to 50 MPa and more should help engineers to exploit the full advantages of spray-guided GDI-engines.

REFERENCES

[1] Spicher, U., Reissing, J., Kech, J.M., Gindele, J., "Gasoline Direct Injection (GDI) Engines – Development Potentialities", SAE Technical Paper 1999-01-2938, 1999.

[2] Baumgarten, C., Mixture Formation in internal Combustion Engines, Springer, ISBN 3-540-30835, 2005.

[3] Spicher, U., and 21 co-authors, Ottomotor mit Direkteinspritzung, Vieweg, ISBN 978-3-8348-0202-6, 2007.

[4] Nauwerck, A., Pfeil, J., Velji, A., Spicher, U., Richter, B., "A Basic Experimental Study of Gasoline Direct Injection at Significantly High Injection Pressures", SAE Technical Paper 2005-01-0098, 2005.

[5] Schumann, F., Buri, S., Kubach, H., Spicher, U., Hall, M., "Investigation of Particulate Emissions from a DISI Engine with Injection Pressures up to 1000 bar", 19. Aachener Kolloquium Fahrzeug- und Motorentechnik, Aachen, 2010

[6] Czichos, H., Habig, K.-H., Tribologie-Handbuch, Vieweg, ISBN 3-528-16354-2, 2003.

[7] Häntsche, J. P., Krause, G., Velji, A., Spicher, U., "High Pressure Fuel Pump for Gasoline Direct Injection based on Ceramic Components", SAE Technical Paper 2005-01-2103, 2005.

[8] Salmang, H., Scholze, H., Keramik, Springer, ISBN 3-540-63273-5, 2007.

[9] Tietz, H.-D., Technische Keramik, VDI, ISBN 3-18-401204-2, 1994.

[10] Häntsche J. P., "Entwicklung und experimentelle Untersuchungen einer Hochdruckpumpe für Ottokraftstoff basierend auf ingenieurkeramischen Gleitsystemen", Logos Verlag Berlin, ISBN 978-3-8325-2464-7, 2009.

[11] Abo-Naf, S.M., Dulias, U., Schneider, J., Zum Gahr, K.-H., Holzer, S., Hoffmann, M. J., "Mechanical and tribological properties of Nd- and Yb-SiAlON composites sintered by hot isostatic pressing", Journal of Materials Processing Technology 183: 264–272, 2007.

[12] Holzer, S., Huchler, B., Nagel, A., Hoffmann, M. J., "SiAlON ceramics: Processing, microstructure and properties", Ceramic Transactions Vol. 142: 161-175, 2003.

[13] Wöppermann, M., Zum Gahr, K.-H., "Mikrostrukturierung keramischer Funktionsflächen für mediengeschmierte Gleitsysteme", presented at 3. Statuskolloquium des SFB 483, Karlsruhe, GERMANY, October 18, 2007.

[14] Riva, M., "Entwicklung und Charakterisierung von Sialon-Keramiken und Sialon-SiC-Verbunden für den Einsatz in tribologisch hochbeanspruchten Gleitsystemen", KIT Scientific Publishing, ISBN 978-3-86644-587-1.

[15] Wöppermann, M., Zum Gahr, K.-H., Schneider, J., "SiC-Gleitkomponenten mit deterministischen Texturen unter reversierender Beanspruchung in niedrigviskosen Flüssigkeiten", presented at Tribologie-Fachtagung, Gesellschaft für Tribologie, Göttingen, GERMANY, 2008.

[16] Zum Gahr, K.-H., Blattner, R., Hwang, D.-H., Pöhlmann, K., "Micro- and macro-tribological properties of SiC ceramics in sliding contact", Wear 250: 299-310, 2001.

CONTACT INFORMATION

    Dipl.-Ing. Christophe Pfister
    Karlsruhe Institute of Technology (KIT), Institut fuer Kolbenmaschinen
    Kaiserstrasse 12, 76131 Karlsruhe, Germany
    Tel.: +49 721 / 608 48528 Fax.: +49 721 / 608 48578
    e-mail: christophe.pfister@kit.edu

ACKNOWLEDGEMENTS

    This study is funded within the Collaborative Research Centre SFB 483 "High performance sliding and friction systems based on advanced ceramics" by the Deutsche Forschungsgemeinschaft, Federal Ministry of Education and Research, Germany.

DEFINITIONS/ABBREVIATIONS

100Cr6: Bearing steel (equal AISI 52100)
BTC: Bottom Dead Centre
CA: Crank Angle
$F_{Contact}$: Contact force in the sliding system
$F_{Friction}$: Friction force in the sliding system
HP: High Pressure
LP: Low Pressure
RPM: Revolutions Per Minute
$s_{Friction}$: Displacement in the sliding system
SSiC: Sintered Silicon Carbide
TDC: Top Dead Centre
$W_{Friction}$: Friction work in the sliding system

PROPERTY AND MICROSTRUCTURAL CORRELATIONS TO WEAR ON REACTION
BONDED MATERIALS

A.L. Marshall and S. Salamone
M Cubed Technologies, Inc.
1 Tralee Industrial Park
Newark, DE 19711

ABSTRACT
     Composites of silicon carbide (SiC) and silicon (Si); SiC, Si and aluminum (Al); and SiC,
diamond and Si are fabricated by the reactive infiltration of molten alloys into preforms of said
particles and carbon. Depending upon the application, these materials can be used in many situations
due to their favorable properties including high hardness, low thermal expansion, high thermal
conductivity and high stiffness. Particularly, these materials display minimal wear owing to their high
hardness and high stiffness. In this study, grit size, material, and material loading will be studied to
garner an understanding of the effects on wear properties in these systems.

INTRODUCTION
     Reaction Bonded SiC (RBSC) lends itself to many potential applications including: thermal
applications due to the high thermal conductivity, low tailorable thermal expansion, and high specific
stiffness; armor applications due to the high hardness, high Young's modulus, and low density; and
also high-temperature wear applications due to the high hardness and stability at elevated temperatures.
RBSC is a composite material made up of Si and SiC. Some of the properties are easily tailorable to
specific applications depending upon the starting materials used.
     The basic process for making a RBSC part involves four steps. First, one must make a slurry of
SiC, a carbon based binder, and water. Second, a preform is prepared through any number of casting or
molding techniques, including: sedimentation, injection molding, filter casting, slip casting, etc. Third,
the binder in the preform is converted to carbon by heating up the part in a nitrogen rich environment
to cracking temperature of the particular binder. Last, the part is placed in a vacuum furnace in contact
with a Si alloy and heated to a temperature under high vacuum such that the alloy turns molten and
wicks into the preform and reacts with the carbon therein.[1,2] The residual Si fills the interstices,
forming a fully dense composite with minimal dimensional change. A cartoon is provided below in
Figure 1 highlighting this process.

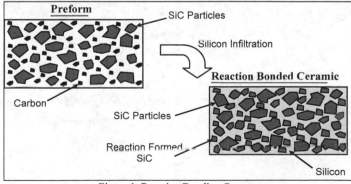

Figure 1. Reaction Bonding Cartoon

EXPERIMENTAL SETUP

This study aims to develop an understanding of how different materials, materials loadings, and material particle size will affect the wear properties of reaction bonded composites. In all cases black SiC was utilized. The experimental test matrix is listed below:

1. 12 μm SiC
2. 12 μm SiC, Carbon Additions
3. 45 μm SiC
4. 150 μm SiC
5. 305 μm SiC
6. 45 μm SiC, Al Additions
7. 150 μm SiC, Al Additions
8. 305 μm SiC, Al Additions
9. 12 μm SiC, Small 22μm Diamond Additions
10. 12 μm SiC, Small 123μm Diamond Additions
11. 150 μm SiC, Small 22μm Diamond Additions
12. 150 μm SiC, Small 123μm Diamond Additions

Composites created in experimental tests 3 through 12 have binary and above particle size distributions. This was necessary to keep the densities similar and to limit the amount of residual silicon in the larger particulate systems, thus allowing grain size comparisons Experiments 1, 3, 4, and 5 were conducted to evaluate grain size effects. Experiment 2 was performed to compare relative loading effects. Experiments 6, 7, and 8 were done to evaluate whether Al additions to the alloy impact wear properties. Experiments 9 through 12 incorporated small amounts of diamond into the composites to evaluate the effect an extremely hard second phase material might provide. The particular wear test used in this study was the dry sand/rubber wheel abrasion wear test described in ASTM G65 – Procedure A.

RESULTS & DISCUSSION

The wear dependence of SiC particulate size in RBSC composites was first examined. The particle size effects are shown in Figure 2 along with some standard wear materials to gain a relative comparison. The results show a generally decreasing volume loss with increasing particle size. However, the 305 μm SiC sample has a slightly higher volume loss than that of the 150 μm sample. This is likely due to the slightly higher loading of the 150 μm sample as shown by the right vertical axis indicating volume percent SiC in the samples. This was calculated with the basic rule of mixture calculation using the densities of Si, SiC, and the composite materials generated. In these samples the volume loss seems to be dependent on the loading of SiC as well as the particle size. As the volume loss decreases from the 12 μm sample to the 150 μm sample, the volume percent of SiC increases. Representative microstructures for the RBSC systems are shown in Figure 3.

In the 12 μm sample when carbon is added, the volume loss also decreases. The extra carbon reacts with Si and consequently forms more SiC during the infiltration process. This effectively increases the loading of the SiC and also slightly increases the effective size of the SiC. This also reduces the amount of residual Si in the system which has a lower hardness than that of SiC. Intuitively this should improve the wear properties as well. Microstructures depicting this are shown in Figure 4.

Figure 2 also shows the grain size effects when examining the results for the 12 μm SiC, Carbon Addition sample with the results for the 45 μm SiC sample. In this case the loading on the carbon addition sample is higher. However, the volume loss of material is higher for this more highly loaded sample as well. When examining this result along with the data for the 105 μm and 305 μm samples, it becomes apparent that particle loading and particle size influence the ability of a reaction bonded SiC material to withstand wear.

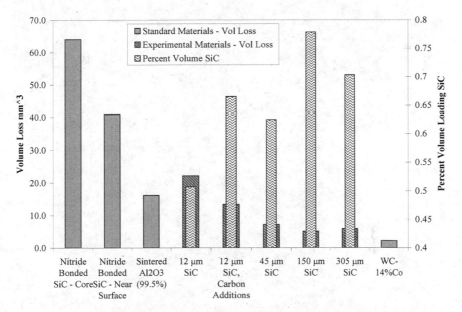

Figure 2. Comparson of Wear Results for RBSC and Traditional Materials

A second set of experiments was performed to determine if aluminum alloying additions would affect the wear properties. This set of experiments was conducted to determine if the Al additions have a deleterious impact on wear, as Al is a much softer material than Si. This is significant as reaction bonded materials can be tailored to meet certain application and specific property criteria; particularly, the thermal conductivity and the coefficient of thermal expansion can be varied to match the properties of substrate materials. The wear results are provided in Figure 5 while two microstructures are provided in Figure 6 for comparison.

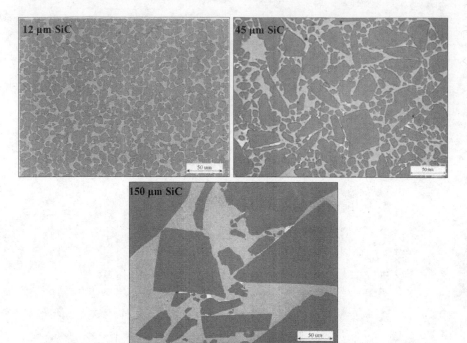

Figure 3. Representative Microstructures for Various RBSC Composites
[Light Shading = Si; Dark Shading = SiC]

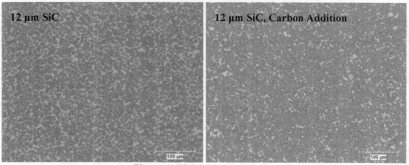

Figure 4. RBSC 12 μm Samples

The samples with Al additions follow a trend of decreasing volume loss with increasing particle size coupled with an increasing particle loading. Upon examining the Si alloy samples, it becomes clear that loading is playing a larger role than particle size over this length scale. The 45 μm SiC has a lower loading and a higher volume loss compared to the 150 μm SiC. However, as previously seen the 305 μm SiC has a lower loading than that of the 150 μm SiC and a higher volume loss. Rahimian et al.

show similar results in sliding wear tests performed on very low loaded Al-Al$_2$O$_3$ samples. Therein, they add 5, 10, and 15 wt.% Al$_2$O$_3$ particles to an Al matrix and show wear loss decreasing with increasing loadings. They also demonstrate sliding wear with Al$_2$O$_3$ particle sizes ranging from 3 μm to 48 μm and demonstrate the reduced affect increasing particle size has at similar particle loadings.

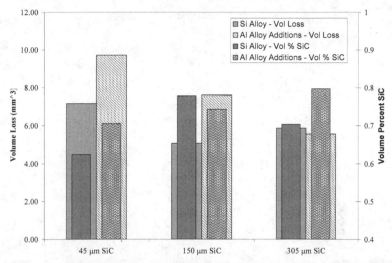

Figure 5. A Comparison of Wear Results for RBSC Materials with Si and Si with Al Additions

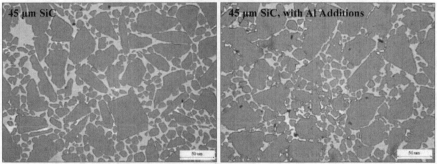

Figure 6. RBSC 45 μm SiC Samples (50 μm scale bar)
[Lightest Shading – Al; Intermediate Shading = Si; Dark Shading = SiC]

The last set of experiments conducted involved small additions of diamond to the 12 μm and the 150 μm SiC samples to evaluate the efficacy an even harder second phase will have on reducing wear. The diamond also reacts with the Si to some degree under the processing conditions reducing the residual Si and thereby increasing properties such as density and Young's modulus. As shown in Figure 7, the diamond has a positive effect on wear in the 12 μm SiC system. The volume loss decreases from 22 mm^3 to approximately 2 mm^3 with either the 22 μm diamond or the 123 μm diamond additions. This is approximately a 90% improvement. The 123 μm diamond has a slightly higher volume loss in this case. This is likely attributed to the volume loading of diamond in the composite. Volume loadings in these diamond containing systems are difficult to determine as some of the diamond reacts with the Si and forms SiC. Difficulties also arise in that the smaller diamond size should react more than the larger size due to the increased surface area, which ends up forming more SiC.[4] Therefore in this system three separate variables are contributing to the reduction in volume loss: diamond loading, SiC loading, and residual Si reduction.

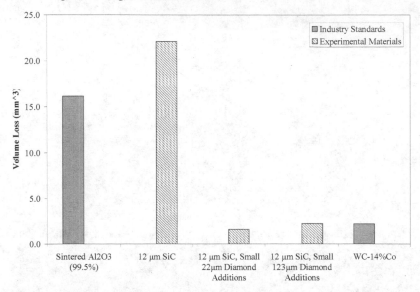

Figure 7. Comparison of Wear Results for the 12 μm RBSC with Diamond Additions and Traditional Materials

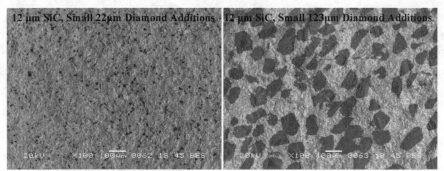

Figure 8. Microstructures of 12 μm SiC with Diamond Additions [Dark Phase = Diamond]

Example microstructures of the two 12 μm SiC diamond systems are provided in Figure 8, above. From these images it is apparent that the diamond is still present and homogeneously dispersed across the sample. The scanning electron microscope (SEM) images were taken in the Back-Scattered mode to differentiate the phases present (the compositional differences) – e.g., Si metal (lightest phase), SiC (intermediate/gray phase), and Diamond (dark phase). All images represent static fracture surfaces due to the difficult of polishing diamond containing specimens.

Figure 9 provides data on diamond additions to the 150 μm family of materials. Small additions of diamond have a positive result on reducing abrasive wear in this case as well, although not as significant of an impact as that of the 12 μm SiC system. It appears as if the 123 μm diamond provides a lower volume loss than the 22 μm diamond; however, as previously seen volume loadings play an important role. The 123 μm diamond version could simply have a higher packing fraction. The 22 μm diamond provided an approximately 30% improvement in the materials resistance to abrasive wear while the 123 μm diamond provided a 66% improvement.

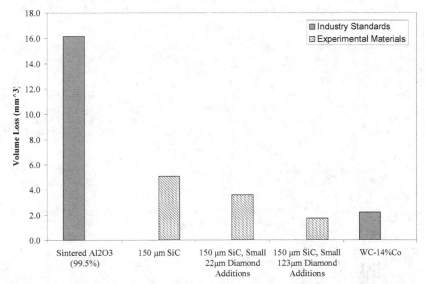

Figure 9. Comparison of Wear Results for the 150 μm RBSC with Diamond Additions and Traditional Materials

Microstructures of the 150 μm SiC systems with diamond additions are provided in Figure 10. In both cases of the 150 μm SiC systems, the diamond seems less homogeneously dispersed than that of the 12 μm SiC systems, which is attributed to the 150 μm SiC presence. Again it is apparent that the diamond survived the processing conditions.

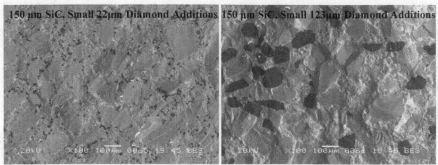

Figure 10. Microstructures of 150 μm SiC with Diamond Additions [Dark Phase = Diamond]

CONCLUSIONS

Reaction bonded SiC samples were tested in the dry sand/rubber wheel abrasion test as described in ASTM G65 – Procedure A. Particle size, particle loading, and material composition were all shown to affect the abrasion wear properties. In the case of the SiC system, particulate loading and particle size inversely affected the volume loss of material, with the particle loading providing a more pronounced outcome. Al additions to the alloy did not change the trend in the wear results; they merely reinforced the findings that the SiC loading dominates the volume loss as compared to particle size effects. Additions of diamond to the RBSC composites drastically reduced the amount of wear in these materials. The diamond containing materials compared well with the WC-14%Co standard.

REFERENCES

1. M. K. Aghajanian, B. N. Morgan, J. R. Singh, J. Mears, R. A. Wolffe, A New Family of Reaction Bonded Ceramics for Armor Applications, *Ceramic Transactions*, **134**, J. W. McCauley et al. editors, 527-40 (2002).

2. P. G. Karandikar, M. K. Aghajanian and B. N. Morgan, Complex, Net-Shape Ceramic Composite Components for Structural, Lithography, Mirror and Armor Applications, *Ceram. Eng. Sci. Proc.*, **24** [4] 561-6 (2003).

3. M. Rahimiana, N. Parvinb, and N. Ensanic, The effect of production parameters on microstructure and wear resistance of powder metallurgy Al-Al₂O₃ composite, *Materials & Design* **32** [2] 1031-1038 (2011).

4. S. Salamone, R. Neill, M. Aghajanian, Si/SiC and Diamond Composites: Microstructure-Mechanical Properties Correlation, *CESP* **31** [2] 97-106 (2010).

Author Index